Advances in Experimental Medicine and Biology

Volume 1466

Advances in Experimental Medicine and Biology provides a platform for scientific contributions in the main disciplines of the biomedicine and the life sciences. This series publishes thematic volumes on contemporary research in the areas of microbiology, immunology, neurosciences, biochemistry, biomedical engineering, genetics, physiology, and cancer research. Covering emerging topics and techniques in basic and clinical science, it brings together clinicians and researchers from various fields.

Advances in Experimental Medicine and Biology has been publishing exceptional works in the field for over 40 years, and is indexed in SCOPUS, Medline (PubMed), EMBASE, Reaxys, EMBiology, the Chemical Abstracts Service (CAS), and Pathway Studio.

2023 CiteScore: 5.9

Hongbo Hu · Xianghui Fu
Editors

Ubiquitination in Immune System and Immune Response

Springer

Editors
Hongbo Hu
Center for Immunology and Hematology, State Key Laboratory of Biotherapy and Cancer Center, West China Hospital, Sichuan University
Chengdu, China

Xianghui Fu
State Key Laboratory of Biotherapy
West China Hospital of Sichuan University
Chengdu, Sichuan, China

ISSN 0065-2598 ISSN 2214-8019 (electronic)
Advances in Experimental Medicine and Biology
ISBN 978-981-97-7287-2 ISBN 978-981-97-7288-9 (eBook)
https://doi.org/10.1007/978-981-97-7288-9

This Springer imprint is published by the registered company Springer Nature Singapore Pte Ltd.
The registered company address is: 152 Beach Road, #21-01/04 Gateway East, Singapore 189721, Singapore

Preface

Ubiquitination is one of the most dynamic and intricate post-translational modifications (PTMs) governing numerous biological processes. In this book, *Ubiquitination in Immune System and Immune Response*, our primary objective is to unveil the profound roles of ubiquitination in shaping the immune system. Across ten enlightening chapters, we embark on a multifaceted exploration of ubiquitination, offering diverse perspectives on how this PTM influences immune cell development, metabolism, homeostasis, and the pathogenesis of various conditions, from autoimmune diseases to immunotherapy.

The mass spectrometry-based methods are crucial to unravel the ubiquitin code. Gong et al. lay the foundation for understanding how these cutting-edge techniques facilitate the identification of proteome-wide ubiquitination sites, the structural intricacies of ubiquitin chains, and the intricate interplay between ubiquitination and other PTMs. As for the function of ubiquitination in innate response, Liu et al. focus on the inflammasomes, multiprotein complexes designed to respond to microbial threats while ensuring immune homeostasis. Here, we explore the roles of ubiquitination in orchestrating inflammasome assembly, providing invaluable insights into sustaining immune balance. Meanwhile, Jie summarizes the studies on how ubiquitination finely tunes the maturation and function of dendritic cells, impacting immune tolerance and immune responses, and the potential target for dendritic cell-based immunotherapy was also discussed. The metabolic reprogramming parallels T cell activation, differentiation, and the pathogenesis of autoimmune diseases regulated by ubiquitination, as reviewed by Du et al., Fan et al., Ma et al., *and* Jin. Peng et al. explore the thymic T cell development and the role of ubiquitination in this vital process. Zhu et al. and Wang reviewed the functions of practical ubiquitin E3 ligases, such as TIF1 family (TRIM24, TRIM28, and TRIM33), and Cullin-RING ligase in various immune cell types and diverse physiological and pathological conditions.

This book offers a profound understanding of how ubiquitination influences immune regulation, autoimmune diseases, and potential therapeutic interventions. The chapters collectively shed light on the dynamic and interconnected world of ubiquitination, from technology-driven discoveries to metabolic insights, innate and adaptive immunity systems, and their applications in tumor immunology.

Chengdu, Sichuan, China — Hongbo Hu
Chengdu, Sichuan, China — Xianghui Fu

Contents

About the Editors

Hongbo Hu is the director of Center for Immunology and Hematology and a full professor at the State Key Laboratory of Biotherapy, West China Hospital, Sichuan University. He received his Ph.D. degree in Immunology from the Chinese Academy of Medical Science/Peking Union Medical College and then worked as a postdoc and instructor at the Baylor College of Medicine and UT MD Anderson Cancer Center. He established his independent laboratory in Sichuan University in 2015. Dr. Hu's research focuses on the regulation of immune cell development and functions, especially the function of protein post-translational modification (ubiquitination) and signaling pathways involved in the development of immune system and immune response, autoimmunity, and cancer immunotherapy.

Xianghui Fu is a full professor at the State Key Laboratory of Biotherapy, West China Hospital, Sichuan University. He received his Ph.D. degree in Biochemistry and Molecular Biology from the Peking Union Medical College and then worked as a postdoc and staff scientist at the University of Bergen and City of Hope National Medical Center, respectively. His laboratory aims to understand the role of metabolism in human life-threatening diseases, especially cancer and metabolic diseases such as diabetes and NAFLD, and to explore their implications in translational and clinical medicine.

Decoding Ubiquitin Modifications by Mass Spectrometry

1

Yanqiu Gong and Lunzhi Dai

Abstract

Protein ubiquitination is a critical and widely distributed post-translational modification (PTM) involved in the regulation of almost every cellular process and pathway in cells, such as proteostasis, DNA repair, trafficking, and immunity. Mass spectrometry (MS)-based proteomics is a robust tool to decode the complexity of ubiquitin networks by disclosing the proteome-wide ubiquitination sites, the length, linkage and topology of ubiquitin chains, the chemical modification of ubiquitin chains, and the crosstalk between ubiquitination and other PTMs. In this chapter, we discuss the application of MS in the interpretation of the ubiquitin code.

Keywords

Tandem mass spectrometry · Post-translational modification · Lysine ubiquitinome

Y. Gong · L. Dai (✉)
National Clinical Research Center for Geriatrics and Department of General Practice, State Key Laboratory of Biotherapy, West China Hospital, Sichuan University, Chengdu, China
e-mail: lunzhi.dai@scu.edu.cn

1.1 Introduction

Canonical protein ubiquitination (also termed ubiquitylation) is a widely distributed post-translational modification (PTM) with ubiquitin covalently attached to the lysine side chain via an isopeptide bond or to the protein's N-terminus via a peptide bond (Rape 2018; Ciechanover 2015; Swatek and Komander 2016; Damgaard 2021). In addition to canonical ubiquitination taking place on amine groups of proteins, ubiquitination also occurs on serine, threonine, and cysteine residues (Wang et al. 2007; De Cesare et al. 2021). Recently, nonprotein ubiquitination occurring on LPS has been reported (Gong et al. 2021a; Otten et al. 2021). The polypeptide ubiquitin, first discovered in 1975 (Goldstein et al. 1975), contains 76 amino acids and is highly conserved among eukaryotes, and 4 genes, UBB, UBC, UBA52, and RPS27A, are known to encode ubiquitin in the human genome (Bianchi et al. 2015). Proteins can be modified with just one ubiquitin (called monoubiquitination) (Hicke 2001; Ronai 2016), multiple single ubiquitins (called multi-monoubiquitination) (Haglund et al. 2003), and polyubiquitin chains (termed polyubiquitination) (Li and Ye 2008). The polyubiquitin chains are usually divided into two categories, a linear polyubiquitin chain (also called M1-linked polyubiquitination (Dittmar and Winklhofer 2020), which means the second ubiquitin molecule conjugates to the

H. Hu, X. Fu (eds.), *Ubiquitination in Immune System and Immune Response*, Advances in Experimental Medicine and Biology 1466, https://doi.org/10.1007/978-981-97-7288-9_1

N-terminal methionine of the previous ubiquitin molecule) and seven nonlinear polyubiquitin chains (K6-, K11-, K27-, K29-, K33-, K48-, and K63-linked polyubiquitination) (French et al. 2021). The number refers to the residue position in the ubiquitin molecule as in K48, K63, or M1. Moreover, the polyubiquitin chains can be linked through the same ubiquitin linkage (called homotypic chains) or a combination of different lysine linkages (called heterotypic chains) (French et al. 2021; Grice et al. 2015; Ohtake and Tsuchiya 2016).

The ubiquitination process is an enzymatic cascade catalyzed by ubiquitin-activating enzymes (E1s), ubiquitin-conjugating enzymes (E2s) and ubiquitin ligases (E3s) (Cruz Walma et al. 2022; Zheng and Shabek 2017). The substrate specificity of ubiquitination is mainly determined by E3 ubiquitin ligases (Yang et al. 2021). To date, more than 600 E3s have been discovered in humans and divided into 3 main categories based on domains: HECT, RING, and RBR ligases (Morreale and Walden 2016; Watanabe et al. 2020). The transfer of ubiquitin chains to proteins is still debatable with two main models: the sequential model and en bloc assembly model (Deol et al. 2019). Protein ubiquitination is a reversible process. The ubiquitin chains can be removed from protein substrates by deubiquitination enzymes (DUBs) (Harhaj and Dixit 2011; Harrigan et al. 2018). Nearly 100 potential DUB genes exist in the human genome and can be classified into 2 main classes: cysteine proteases and metalloproteases (Snyder and Silva 2021). Of the described DUBs, ubiquitin-specific proteases (USPs/UBPs), ubiquitin COOH-terminal hydrolases (UCHs), ovarian tumor proteases (OTUs), Machado-Josephin domain proteases (MJDs), and motifs interacting with the Ub-containing novel DUB family (MINDYs) belong to cysteine proteases (Mevissen and Komander 2017), while the JAB1/MPN/MOV34 family (JAMMs) are Zn^{2+} metalloproteases (Ambroggio et al. 2003). DUBs usually contain a catalytic domain and some accessory domains, such as the ubiquitin-specific protease (DUSP) domain, ubiquitin-like (UBL) domain, meprin and TRAF homology (MATH) domain, and zinc-finger ubiquitin-specific protease (ZnF-UBP) domain (Nijman et al. 2005; Komander et al. 2009).

Protein ubiquitination has diverse functions involved in numerous physiological and pathological processes. Structural studies have revealed that different ubiquitin chain linkages adopt distinct conformations (Bremm et al. 2010; Michel et al. 2017; Kniss et al. 2018). Hence, the functions of ubiquitination and substrate fate are determined by the linkage of ubiquitin chains (French et al. 2021). For example, monoubiquitination has broad functions in regulating trafficking, endocytosis, gene expression, DNA repair, and protein interaction (Deng et al. 2020). K48 polyubiquitination usually targets proteins for proteasomal degradation (Xu et al. 2009). K33 polyubiquitin chains regulate protein interactions and trafficking (Yuan et al. 2014). K63 polyubiquitination is associated with signal transduction, kinase activation, endocytosis, and DNA repair (Harikumar et al. 2014; Wang et al. 2001). M1 ubiquitination is critical for inflammatory pathways (Aalto et al. 2019; Yu et al. 2020).

Many excellent reviews have summarized ubiquitin signaling (Rape 2018; Ciechanover 2015; Swatek and Komander 2016; Damgaard 2021; Ohtake and Tsuchiya 2016; Cruz Walma et al. 2022; Mevissen and Komander 2017; Deng et al. 2020; Yu et al. 2020; Kliza and Husnjak 2020; Hu and Sun 2016). In this chapter, we focus on the mass spectrometry (MS)-based characterization of the ubiquitin code.

1.2 Identification of Lysine Ubiquitination Sites by Tandem Mass Spectrometry

Tandem mass spectrometry (MS/MS) is a robust method for peptide sequencing and PTM characterization (Larsen et al. 2006; Tan et al. 2011; Kim et al. 2006; Yakubu et al. 2019; Gong et al. 2021b). The acquisition of the MS1 spectrum is to measure the molecular weight of the precursor peptides. The selected precursors are then fragmented, which produces a series of daughter ions in the MS/MS spectrum (Steen and Mann 2004). By setting the putative modifications and their theoretical mass shifts in software such as Mascot, Maxquant, and Proteome Discover, the PTMs and the modified positions on peptides can be revealed by comparing the experimentally detected MS/MS spectrum with the theoretical MS/MS spectrum (Larsen et al. 2006) (Fig. 1.1). Most PTMs, such as acylations and methylation, are very stable during sample processing and analysis. Taking the characterization of stable lysine acetylation as an example, lysine acetylation leads to a mass increment of 42 Da of the intact peptide (Kim et al. 2006). In the MS/MS spectrum of a lysine-acetylated peptide, all the daughter ions that contain the acetylated lysine residue exhibit a +42 Da mass increment relative to the corresponding daughter ions of unmodified peptide. For less stable PTMs, such as phosphorylation on serine and threonine residues (Grønborg et al. 2002), the MS/MS fragmentation process often results in a neutral loss of a phosphoric acid and produces one strong signal in the MS/MS spectrum with a 98 Da mass deficit compared with the intact phosphopeptide (Lehmann et al. 2007). This information is very useful for the annotation of PTMs in MS/MS spectra.

Protein ubiquitination is a class of very complicated PTMs causing very large mass shifts, and the analysis of ubiquitination is extremely challenging. The identification of ubiquitination sites is based on the diGly tag with a mass increment of 114.043 Da on ubiquitinated lysine residues after tryptic digestion (Fulzele and Bennett 2018). These diGly remnants on the lysine residues are very stale during the MS/MS fragmentation of the peptide backbone, leading to cleaner MS/MS spectra. However, linear ubiquitinated peptides are unable to be recognized (Kliza and Husnjak 2020). Other ubiquitin-like modifiers, such as NEDD8 and ISG15, can be added to lysine residues of proteins by ligases, and the same diGly remnant is also generated after tryptic digestion. These peptides are indistinguishable from ubiquitinated peptides (Udeshi et al. 2013). Indeed, a systematic assessment of the ubiquitin-modified proteome shows that more than 94% of peptides containing diGly remnants result from ubiquitination (Kim et al. 2011). Other proteases, such as Lys-C or Glu-C, are also used to digest the ubiquitinated proteins, and longer remnants with sequences of ESTLHLVLRLRGG or STLHLVLRLRGG are retained (Xu and Jaffrey 2013), which can unambiguously determine that the modification is from ubiquitin, although the interpretation of the MS/MS becomes more difficult. It should be

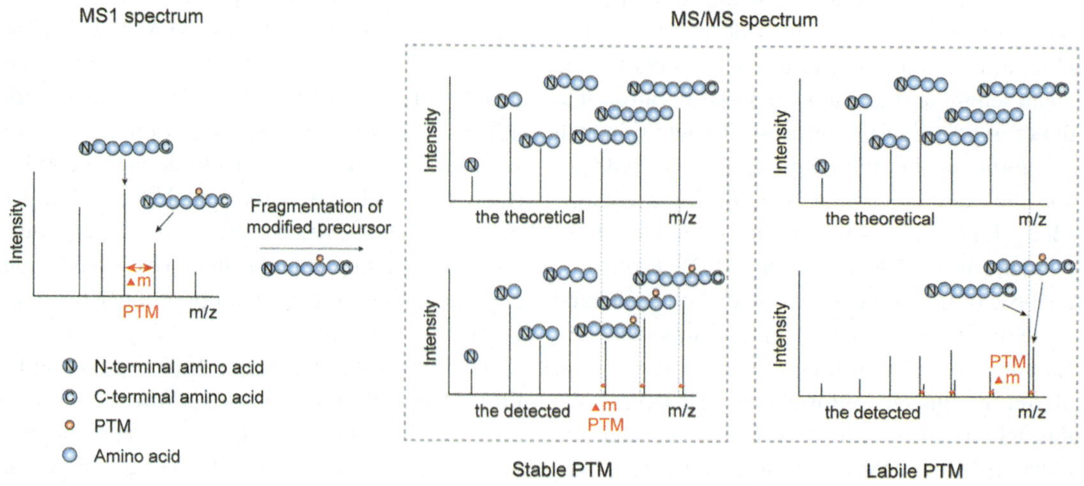

Fig. 1.1 The principle to determine a PTM on a peptide by MS/MS. The PTMs can be revealed by comparing the measured with the theoretical MS/MS spectra. MS/MS of a stable PTM leads to the detection of fragment ions that reveal the identity and position of the modified amino acid residue, while MS/MS of a labile PTM typically leads to the observation of one strong signal in the MS/MS spectrum because of a neutral loss

noted that alkylating chloroacetamide but not iodoacetamide (IAA) is preferred to block the Cys residues in sample digestion procedures to avoid side reactions between IAA and unmodified lysine residues, which produces a chemical modification with the same mass as the diGly remnant and increases the difficulty of interpreting the MS data (Nielsen et al. 2008). To overcome the influence of contaminant diGly remnant peptides derived from NEDD8 and ISG15, a StUbEx PLUS technique has been developed (Akimov et al. 2018). This approach utilizes 6xHIS-tagged ubiquitin to replace endogenous ubiquitin. The ubiquitinated protein can be enriched with anti-His antibodies and tryptically digested, and diGly remnant peptides can be obtained. In the test experiment, more than 41,000 unique diGly remnant peptides in nearly 7800 proteins were identified by StUbEx PLUS in U2OS cells upon proteasome inhibition. However, the requirement of replacing endogenous ubiquitin with 6xHIS-tagged ubiquitin limits the application of StUbEx PLUS.

1.3 Mass Spectrometry Analysis of Ubiquitin Linkages

Compared with the identification of ubiquitination sites, the elucidation of ubiquitin linkages is much more difficult (French et al. 2021). The ubiquitin linkage information, especially the chain length and topology, is missing during the direct tryptic digestion process. To determine the ubiquitin chain topology, ubiquitin linkage-specific antibodies that specifically recognize Met1, K11, K27, K48, and K63 ubiquitin linkages are used (Newton et al. 2012). Proteomic analysis of the immunoprecipitated ubiquitinated proteins with linkage-specific antibodies can reveal both ubiquitin-modified positions and ubiquitin linkage information. However, the throughput of profiling is moderate, and the homotypic and heterotypic ubiquitin chains cannot be clearly distinguished by linkage-specific antibodies (Valkevich et al. 2014). In addition to antibodies, some recombinant proteins with strong binding affinity toward ubiquitin chains have also been developed. The ubiquitin-binding domains (UBDs) show low binding affinity toward the ubiquitin chain. To increase the binding affinity, synthetic multiple repeats of UBDs (tandem ubiquitin-binding entities, TUBEs) and hybrid UBDs (ThUBDs) are developed. Among them, some UBD-based probes, including Lys29/Lys33 linkage-specific (TRABID NZF-based) (Scott et al. 2016), Lys63 linkage-specific (TAB2 NZF- and VPS27 UIM-based) (Johnson and Vert 2016), and Met1-linkage-specific (UBAN-based) TUBEs (Fiil et al. 2013), show good linkage-specific affinity.

Bottom-up proteomics, which analyzes digested peptides, is the more widely used mass spectrometric technique for proteome analysis (Bludau et al. 2021; Zhang et al. 2013). The analysis of the diGly remnant described above is based on bottom-up proteomics. In the pipeline of bottom-up proteomics, the proteins are first digested into small peptides, which are then separated by liquid chromatography, ionized and analyzed by MS/MS. The identification of proteins and peptides is based on the acquired precursors and MS/MS spectra. Limited by the workflows of bottom-up proteomics, the ubiquitin chain length and topology cannot be evaluated due to trypsin digestion. In contrast, middle-down proteomics (Sidoli et al. 2017; Cristobal et al. 2017; Wu et al. 2012), unlike bottom-up proteomics, in which thoroughly digested peptides (MW: 0.7–3 kDa) are analyzed, and top-down proteomics, in which intact proteins (MW: 10–30 kDa) are analyzed, utilizes minimal protease digestion of protein samples and yields relatively longer peptides (more than 3 kDa) (Cristobal et al. 2017). Longer peptides increase the chance of identifying multiple co-occurring neighboring PTMs on a single peptide and provide a good opportunity to study PTM crosstalk (Leutert et al. 2021). Under the optimized conditions, native folded polyubiquitin is tryptically digested only at the Arg74 residue, and the ubiquitin chains are kept largely intact (Xu and Peng 2008). The partially digested ubiquitin chains are enriched by linkage-specific antibodies or recombinant proteins and further analyzed by middle-down proteomics to reveal the branched ubiquitin conjugates (Crowe et al.

2017). To protect the ubiquitin chain from tryptic digestion, an elegant strategy termed Ub-ProT (ubiquitin chain protection from trypsinization) has been developed (Tsuchiya et al. 2018). The trypsin-resistant (TR)-TUBE, consisting of biotin and 6xHIS tags and six tandem repeats of the PLIC1 UBA domain with all the arginine residues replaced by alanine, can bind with the ubiquitin chains and protect them from trypsin digestion. After MS analysis, the length and composition of ubiquitin chains are determined. However, some ubiquitin linkages, such as K6, K27, K29, and K33, are less efficiently protected.

1.4 Mass Spectrometry Analysis of the Lysine Ubiquitinome

Data-dependent acquisition (DDA) and data-independent acquisition (DIA) are two main methods currently used to acquire data on a mass spectrometer (Guo and Huan 2020). In the most widely used DDA acquisition mode (Davies et al. 2021), the mass spectrometer selects the top intense peptide ions, and then the selected precursors are fragmented and analyzed in a second stage of MS/MS. The selection of precursors using a narrow isolation window gives rise to high-quality and higher purity MS/MS spectra. Nevertheless, this selection favors the collection of the most abundant peptide ions, and many low-abundance but biologically important peptides that do not match the defined selection criteria may be missing. Different from the intensity-based precursor selection of DDA, DIA fragments all co-eluting peptides within a selected mass-to-charge (*m*/*z*) range and acquires them simultaneously during the second stage of tandem MS. The resulting fragment ion spectra are highly multiplexed, but this offers more precision, higher identification, and better reproducibility with fewer missing values across samples (Li et al. 2020; Krasny and Huang 2021).

The abundance of most protein ubiquitination events is low in cells. Therefore, to profile the ubiquitinome, ubiquitinated proteins or peptides need to be enriched before MS analysis. His-, Flag-, or Strep-labeled ubiquitin is overexpressed in cultured cells. The ubiquitinated proteins are then purified from the cell lysate with anti-tag antibodies, and the digested peptides were further analyzed by MS (Xu and Jaffrey 2013). The tag-labeled ubiquitin labeling method has its shortcomings. It is just suitable for cultured cells but not for other samples, such as tissues. Additionally, it is an artificial process. To overcome these defects, an FK2 antibody that recognizes both mono- and poly-ubiquitinated proteins has been used for the enrichment of ubiquitinated proteins (Vasilescu et al. 2005). Besides the anti-ubiquitin antibodies used to enrich ubiquitinated proteins, UBDs have also been employed. S5a, a subunit of the 26S proteasome, has 2 UBDs and can bind to polyubiquitin chains containing 4 or more ubiquitins. Using immobilized GST-S5a fusion protein, 17 ubiquitinated proteins were identified in Chang liver cells (Tan et al. 2008). Further developed recombinant proteins, such as ThUBDs, UBA, ThUDQ2, and ThUDA20, show higher binding affinity to ubiquitin chains than naturally occurring UBDs. Remarkably, 1125 ubiquitinated proteins were identified after enrichment with ThUBDs (Gao et al. 2016). Although obviously increased throughputs have been achieved, there are still two inherent issues that restrict the profiling of the ubiquitinome at the protein level. First, because of the large size of the ubiquitin chain, the presence of polyubiquitinated forms, and the low stoichiometry of ubiquitination, the enrichment efficiency is still low. Second, the digested peptides from the putative ubiquitinated proteins also contain many unmodified peptides, which may compromise the identification efficiency of ubiquitinated peptides by MS.

In contrast, the enrichment of K-ε-GG remnant-containing peptides followed by LC–MS/MS analysis is a better method for the profiling of the ubiquitinome (Fig. 1.2). Although the throughput of the initial study was moderate, with the identification of 374 diglycine-modified peptides from 236 proteins (Xu et al. 2010), the throughput of profiling with this strategy was greatly improved in subsequent studies. For example, more than 20,000 unique ubiquitination sites have been identified in murine tissues

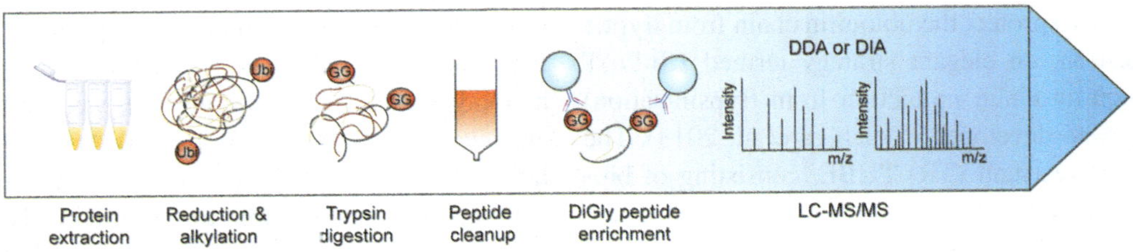

Fig. 1.2 The workflow for profiling the ubiquitinome. Experimental workflow for ubiquitinome profiling. Protein digestion, reduction, alkylation and trypsin digestion are followed by peptide cleanup. The enriched peptides can be analyzed by DDA or DIA mode

(Wagner et al. 2012), and more than 10,000 ubiquitinated sites can be detected at a time (Rose et al. 2016). Although anti-K-ε-GG antibodies have improved the ability to profile the ubiquitinome by MS, the measurement of ubiquitination with sub-milligram sample inputs remains a challenge. Notably, a highly sensitive, rapid, and multiplexed protocol termed UbiFast enables the quantification of ~10,000 ubiquitination sites from as little as 500 μg peptide per sample (Udeshi et al. 2020). The on-antibody tandem mass tag (TMT) labeling shows greater sensitivity than the conventional in-solution TMT labeling of enriched K-ε-GG peptides, probably due to a significant reduction of TMT contaminant side-products. In addition, the UbiFast method enables the analysis of more samples with less protein input and less time but increases the yield of relative K-ε-GG peptides compared to conditional methods.

DIA has been established as an ultrasensitive and reproducible technique for MS measurements of PTMs (Bekker-Jensen et al. 2020). A workflow combining diGly antibody-based enrichment with a DIA method is designed for profiling the ubiquitinome (Hansen et al. 2021). Spectral libraries with more than 90,000 diGly peptides have been built, representing the deepest diGly proteome to date. Under the optimized DIA-based workflow, more than 33,000 distinct diGly sites in single measurements of MG132-treated HEK293 samples were identified. The DIA workflow does not require labeling and offline fractionation, but the depth of coverage and the quantitative accuracy are dramatically increased compared to the same measurements by the DDA method. Because of the antibody-based enrichment step, the DIA workflow currently requires 0.5–1 mg of sample. If the enrichment becomes more efficient and the subsequent peptide purification is eliminated, the sample input can be further reduced. The addition of SDC and chloroacetamide (CAA) during the lysis process and the use of a deep neural network-based data processing method can further increase ubiquitin site coverage (Steger et al. 2021). Notably, CAA barely causes nonspecific lysine di-marbamidomethylation, even at high temperature.

1.5 Mass Spectrometry Analysis of PTMs on Ubiquitin Chains

The ubiquitin chains on proteins can themselves be modified by other PTMs, further increasing the complexity of ubiquitin signaling. Phosphorylation is the most abundant PTM on the ubiquitin chain. Notably, it has been found that phosphorylation of ubiquitin S65 in human cells mediated by PINK1 kinase contributes to the activation of the Parkin E3 ligase, indicating that phosphorylated ubiquitin is a parkin activator (Koyano et al. 2014; Fiesel and Springer 2015). Subtle fluctuation near physiological pH can affect the protonation status of the phosphate group on ubiquitin S65 and alter the ubiquitin linkage (Dong et al. 2017). The ubiquitin system is essential for the regulation of the DNA damage response (DDR), and ubiquitination of H2AK15 can trigger downstream signaling events. Phosphorylation of ubiquitin at Thr12 (pUbT12) controls the DDR by inhibiting the function of 53BP1 (Walser et al. 2020). Mutation of Thr12 prevents the removal of ubiquitin from H2AK15 by USP51, resulting in accumulated ubiquitinated chromatin. Due to the functions of most ubiquitin phosphorylation sites, the identification of ubiquitin phosphorylation sites is important and can be achieved by MS-based phosphoproteomics (Bekker-Jensen et al. 2020; Humphrey et al. 2018). Similar to profiling the ubiquitinome, the enrichment of phosphopeptides is required before the MS measurements. The phosphate group is negatively charged and has a very strong binding affinity toward metal ions. The enrichment of phosphopeptides can be achieved using immobilized-metal affinity chromatography (IMAC (Thingholm and Larsen 2016a; Villén and Gygi 2008)) or metal-oxide affinity chromatography (MOAC (Thingholm et al. 2006; Thingholm and Larsen 2016b), typically using TiO_2). Peptide fractionation with strong cation exchange (SCX) chromatography or high-pH reversed-phase chromatography before phosphopeptide enrichment can significantly increase the identification coverage and throughput (Mohammed and Heck 2011; Batth et al. 2014). Based on large-scale MS-based phosphoproteomics, phosphorylation on nearly all serine, threonine, and tyrosine residues (such as T6, T12 (Walser et al. 2020; Lee et al. 2009), T14 (Ohtake et al. 2015), Y59 (Rikova et al. 2007), and S65 on human ubiquitin (Olsen et al. 2006) and T12, T14, S19, T22, T55, S57, S65, and T66 on yeast ubiquitin) has been revealed (Swaney et al. 2015).

Acetylation is another important PTM on the ubiquitin chain. Notably, six out of the seven lysine residues of ubiquitin, except K29, have been found to be acetylated in response to DNA damage (Elia et al. 2015). Knockout or knockdown of lysine deacetylase results in increased acetylation of ubiquitin (Schölz et al. 2015; Chen et al. 2012; Choudhary et al. 2009; Hansen et al. 2019). Acetylation of ubiquitin at K11, K27, K33, K48, or K63 influences the rate of transthiolation between Uba1 and UBE2D1 (Lacoursiere and Shaw 2021). Moreover, ubiquitin acetylation at K6 or K48 can inhibit K11-, K48-, and K63-linked chain elongation, resulting in accumulated monoubiquitinated substrates (Ohtake et al. 2015). Similar to the identification of ubiquitin phosphorylation, the characterization of ubiquitin acetylation can be achieved by large-scale MS-based acetylomics (Elia et al. 2015). To identify the acetylation sites by MS, the acetylated peptides should be first enriched by pan anti-acetylated antibodies. Peptide fractionation with SCX or high-pH reversed-phase chromatography before enrichment is required to obtain high coverage and throughput.

ADP-ribosylation is a class of widespread PTMs that involves the addition of ADP-ribose groups from nicotinamide adenine dinucleotide (NAD^+) to proteins (Perina et al. 2014) and other macromolecules, such as DNA (Nakano et al. 2015). ADP-ribosylation is involved in many cellular processes and diseases (Palazzo et al. 2017, 2019). Modification of the ubiquitin chain by ADP ribosylation also affects subsequent ubiquitin signaling. CteC from *Chromobacterium violaceum* is an ADP-ribosyltransferase that selectively modifies the ubiquitin chain via threonine ADP ribosylation on residue T66, which prevents the transfer of ubiquitin from E1 to E2, thus inhibiting polyubiquitin synthesis in host cells and the subsequent ubiquitination cascade (Yan et al. 2020). The high-resolution ETD spectra obtained by mass spectrometry confirmed the ADP ribosylation of ubiquitin at T66 by CteC. A mass shift of 541 Da compared with that of unmodified ubiquitin indicates that only one ADP-ribosyl group is added specifically onto the T66 residue of ubiquitin. Parp9 is a mono-ADP-ribosyltransferase that heterodimerizes with the histone E3 ligase Dtx3L and is involved in DNA damage repair. The Dtx3L/Parp9 heterodimer mediates NAD^+-dependent mono-ADP-ribosylation at the carboxyl terminus of ubiquitin Gly76 used for ubiquitin conjugation to substrates and restrains the E3 function of Dtx3L. Mutation of the NAD^+ binding site in Parp9 increases the DNA repair activity of the heterodimer (Yang et al. 2017; Chatrin et al. 2020).

SUMOs belong to the family of ubiquitin-like proteins (UBLs). Protein SUMOylation is known to be involved in protein translocation, DNA damage response, and cell cycle progression (Geoffroy and Hay 2009). The reversible

SUMOylation formed by conjugating SUMO to protein substrates is tightly regulated by SUMO-activating enzyme E1, SUMO E2 conjugation enzyme UBC9, and SUMO E3 ligases (Geiss-Friedlander and Melchior 2007; Celen and Sahin 2020; Han et al. 2018). High-throughput MS analysis identifies more than 4300 SUMOylation sites in thousands of proteins and shows that SUMOylation can occur on multiple lysine residues (K6, K11, K27, K48, and K63) of ubiquitin (Hendriks et al. 2014). Notably, endogenous SUMO can also be ubiquitinated at the K11 position, lying within a SUMOylation consensus site (Danielsen et al. 2011).

Recently, an increasing number of short-chain lysine acylations, including lysine propionylation (Chen et al. 2007), butyrylation (Chen et al. 2007), 2-hydroxyisobutyrylation (Dai et al. 2014), β-hydroxybutyrylation (Xie et al. 2016), succinylation (Xie et al. 2012), malonylation (Xie et al. 2012), glutarylation (Tan et al. 2014), crotonylation (Tan et al. 2011), benzoylation (Huang et al. 2018), lactylation (Zhang et al. 2019), aminoacylation (He et al. 2018), and isonicotinylation (Jiang et al. 2021), have been well characterized. Increasing evidence shows that the concentrations of acyl-CoAs are highly dynamic in response to physiological changes, which is closely associated with the changes in corresponding lysine acylation (Xie et al. 2016; Sabari et al. 2015). The question of whether the functions of ubiquitin are regulated by these novel lysine acylations under certain conditions requires further investigation.

1.6 Crosstalk Between Ubiquitination and Other PTMs

To date, more than 300 classes of protein modifications have been described. These modifications do not exist in isolation. PTMs on the same or on different proteins may take combinatorial action in diverse biological processes and pathways, which is PTM crosstalk. PTM crosstalk can be defined into intra- and inter-protein crosstalk (Leutert et al. 2021) based on PTMs occurring on the same protein and separate proteins, respectively. PTM crosstalk dramatically increases the complexity of functional protein regulation. The crosstalk between PTMs can be either positive or negative in nature (Hunter 2007). In positive crosstalk, one PTM serves as a trigger for the addition or removal of a second PTM or for recruiting a binding protein to add a second modification. In negative crosstalk, PTMs directly compete for the same residue in a protein, or one modification masks the recognition site and prevents its addition and/or removal of a second PTM. Regarding ubiquitination, numerous crosstalk signaling pathways with other PTMs, such as phosphorylation, acetylation, methylation, glycosylation, ADP-ribosylation, and SUMOylation, have been revealed (Fig. 1.3).

Ubiquitination and acetylation crosstalk is very frequent. In most cases, acetylation contributes to increasing protein stability by reducing ubiquitination on the same lysine residue to avoid degradation by the ubiquitin–proteasome system (Caron et al. 2005; Shimizu et al. 2021). In fact, their real interplay goes beyond simple competition, and many proteins of these signaling pathways take part in the process. Interestingly, acetylation can also accelerate protein degradation by modulating protein–protein interactions. For example, pVHL mediates the ubiquitination of Hif1α and promotes its degradation (Lee et al. 2004). The acetylation of Hif1α at specific sites increases the interaction with Hif1α and pVHL (Jeong et al. 2002), thereby inducing subsequent ubiquitination and degradation. In addition, their crosstalk also affects many non-proteasomal functions of ubiquitin. For example, CIITA ubiquitination, which can enhance its transcriptional function, is tightly regulated by HATs and HDACs (Greer et al. 2003).

Ubiquitination and phosphorylation crosstalk at several levels. Phosphorylation can promote or inhibit ubiquitination, which in turn regulates proteasomal degradation or processing or intracellular trafficking of membrane proteins. Phosphorylation can regulate protein ubiquitination in three main ways (Hunter 2007). First, phosphorylation controls the activity of E3 ligase. Second, phosphorylation can create a phospho-

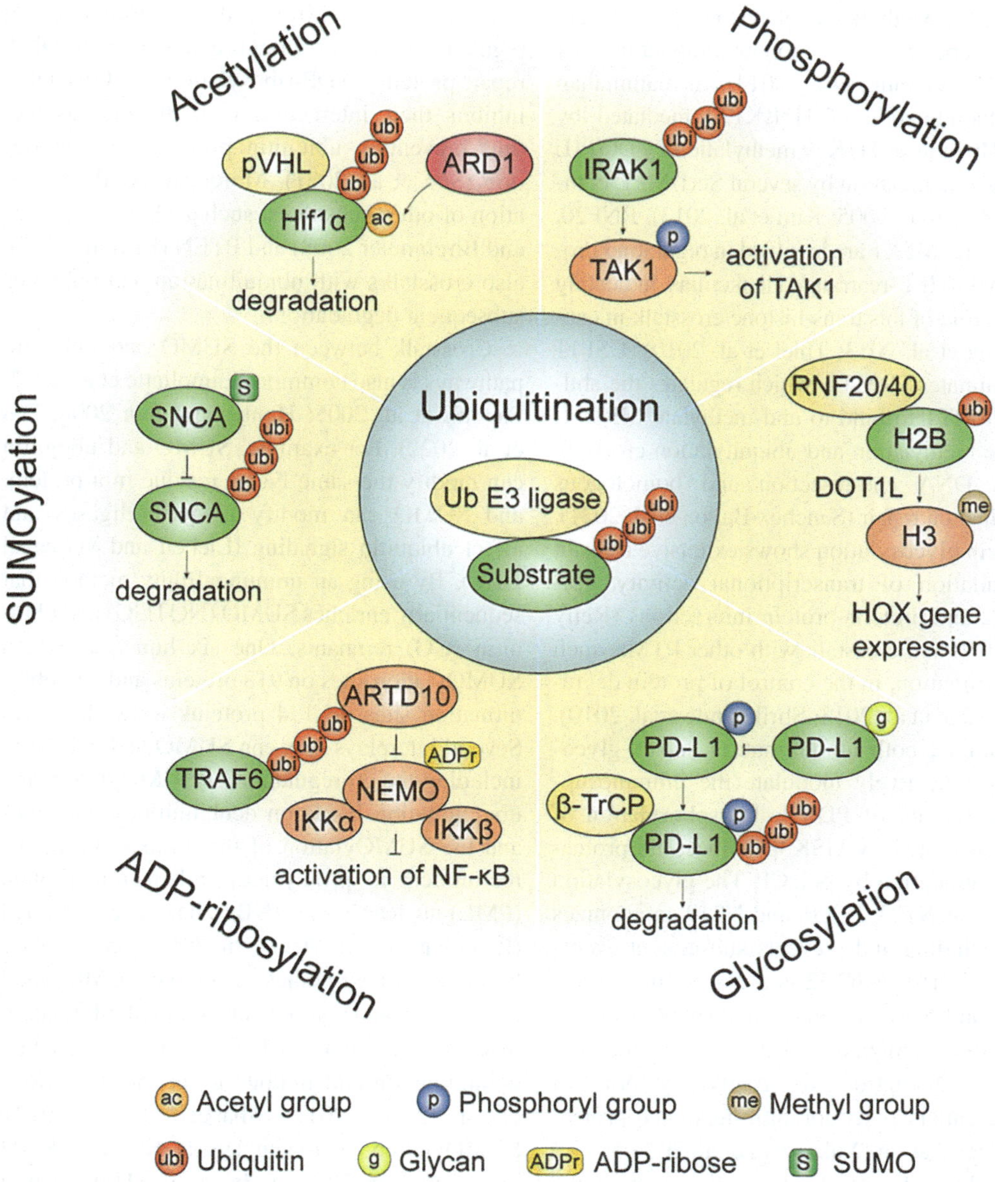

Fig. 1.3 The crosstalk between ubiquitination and other PTMs. Representative crosstalk between ubiquitination and six other major PTMs is presented

degron to increase recognition by E3 ligase. Third, phosphorylation affects ubiquitination by regulating substrate/ligase interactions. Conversely, phosphorylation is also positively or negatively regulated by ubiquitination. For example, the K63-branched ubiquitin chains of TRAF2/6 or IRAK1 can recruit the binding protein TAB2/3 subunits of TAK1 and promote the transphosphorylation and activation of TAK1 via a dimer proximity mechanism (Courtois and Fauvarque 2018). Polyubiquitination of NUAK1 and MARK4 via K29- and K33-branched chains may reduce their kinase activity and the phosphorylation of LKB1 (Al-Hakim et al. 2008).

The most well-known ubiquitination and methylation crosstalk is that between crosstalk histone H2B ubiquitination and histone H3 methylation, which is conserved from yeast to humans (Worden et al. 2019). Methylation of histone H3K79 and H3K4 depends on monoubiquitination of H2BK120 (Vlaming et al. 2014). In mammalian cells, ubiquitination of H2BK120 mediated by RNF20/40 induces H3K79 methylation by DOT1L and H3K4 methylation by several Set1/MLL complexes (Zhu et al. 2005; Kim et al. 2013). RNF20, DOT1L, and MLL1 are involved in regulating proliferation in MLL-rearranged leukemias, indicating a critical role of this trans-histone crosstalk in cancer (Wang et al. 2013; Thiel et al. 2010). USP11 deubiquitinates PRMT1, which regulates the ability of PRMT1 to bind to and methylate MRE11. Arginine methylation and ubiquitylation crosstalk controls DNA end-resection and homologous recombination repair (Sanchez-Bailon et al. 2021).

Protein glycosylation shows extensive roles in the regulation of transcriptional activity, epigenetics, and protein–protein interactions (Reily et al. 2019) and crosstalk with other PTMs, such as ubiquitination, in the control of protein degradation (Ruan et al. 2013; Shrikhande et al. 2010). For example, both ubiquitination and N-glycosylation collectively modulate the immunosuppressive activity of PD-L1. Phosphorylation of PD-L1 mediated by GSK3β induces its proteasomal degradation by β-TrCP. The glycosylation of PD-L1 at N192, N200, and N219 antagonizes GSK3β binding and prevents subsequent events (Li et al. 2016). ABCG2 is a plasma membrane protein, and N-glycosylation at Asn596 is essential for the stabilization of de novo-synthesized ABCG2. Disruption of N-glycosylation of ABCG2 enhances its ubiquitin-mediated proteasomal degradation (Nakagawa et al. 2009).

In addition to the direct modification of the ubiquitin chain by ADP-ribosylation (PARylation) as described above, ADP-ribosylation is identified as a signal for triggering protein degradation through the ubiquitin–proteasome system (Pellegrino and Altmeyer 2016; Vivelo et al. 2019). For example, NEMO in the NF-κB pathway has a role in assembling the IKK complex. ARTD10 binds to the K63-polyubiquitin chain of TRAF6 and interferes with signal propagation by ADP-ribosylating NEMO and preventing NEMO K63-polyubiquitination, thus inhibiting the activation of NF-κB and downstream target genes (Verheugd et al. 2013). ADP-ribosylation can regulate chromatin structure and recruit DNA repair proteins. ADP-ribosylation of TOP1-DPC inhibits their interaction with the proteasome, thus preventing ubiquitin–proteasome degradation (Sun et al. 2021). Moreover, ADP-ribosylation of other substrates, such as Axin1/2 (Klaus and Birchmeier 2008) and PTEN (Li et al. 2015), also crosstalks with ubiquitination and regulates subsequent degradation.

Crosstalk between the SUMO and ubiquitin pathways is also common (Lamoliatte et al. 2017; Papouli et al. 2005; Hunter and Sun 2008; Dai et al. 2022). For example, SUMO and ubiquitin can modify the same lysine residue in a protein, and SUMO can modify ubiquitin ligases and affect ubiquitin signaling (Liebelt and Vertegaal 2016). By using an immunoaffinity method that sequentially enriches SUMO (NQTGG) and ubiquitin (GG) remnants, One six hundred sixteen SUMOylation sites on 918 proteins and 349 ubiquitination sites on 194 proteins were identified. Several interplays between SUMO and ubiquitin, including the co-regulation of SUMOylation and ubiquitination levels on deubiquitinase enzymes and the SUMOylation of the proteasome for its recruitment to promyelocytic leukemia protein (PML) nuclear bodies (NBs), have been revealed (Lamoliatte et al. 2017). In Parkinson's disease brains and Lewy bodies, increased SUMOylated α-synuclein and components of the SUMOylation machinery are shown, which prevents α-synuclein ubiquitination and inhibits its degradation (Rott et al. 2011, 2017; Shabek et al. 2012). SUMOylation of α-synuclein at K96 and K102 only occupies 30% of the total SUMOylation caused by PIAS2, indicating that SUMOylation may occur on lysine residues lacking a consensus sequence for SUMO conjugation.

1.7 Mass Spectrometry Analysis of PTM Crosstalk

Because there is so much PTM crosstalk in the proteome (Minguez et al. 2014), it is a huge task to identify PTM crosstalk one-by-one. MS serves as a robust tool to explore PTM crosstalk at a proteomic scale. Although most proteomics analyses of PTMs are originally not designed to identify PTM crosstalk, efficient enrichment for one modification can also yield peptides harboring other PTMs. For example, the large-scale human SUMO proteome reveals co-modification with phosphorylation, ubiquitination, acetylation, and methylation (Hendriks et al. 2017). Despite tremendous progress, these methods are constrained to examining the co-occurrence of PTMs on the same peptide. To detect distant co-occurring PTMs, a sequential PTM enrichment method consisting of enriching one PTM at the protein level and enriching a second PTM at the peptide level is developed. By using this strategy to investigate ubiquitylation-phosphorylation crosstalk in the yeast *Saccharomyces cerevisiae* (Swaney et al. 2013), 466 proteins with 2100 phosphorylation sites co-occurring with 2189 ubiquitylation sites were identified.

Middle-down proteomics and top-down proteomics are also effective strategies to investigate PTM crosstalk. The interpretation of middle-down and top-down MS data is most challenging because of the co-existing of combinatorial PTMs in a single peptide or protein. In middle-down proteomics, the MS analysis of longer peptides increases the probability of detecting co-occurring PTMs in one peptide and clarifying medium-range intra-protein PTM crosstalk. Middle-down proteomics has widely been used to characterize interdependent and mutually exclusive PTMs in heavily modified proteins such as histones (Sidoli and Garcia 2017). More importantly, the performance of bottom-up and middle-down proteomics strategies is comparable in defining histone PTM relative abundance and stoichiometry (Sidoli et al. 2015). Top-down proteomics, which profiles the "intact" proteome, can characterize proteoforms, identify the combinations of PTMs that are lost in other analysis strategies, and disclose intra-protein PTM crosstalk regardless of their sequence proximity (Leutert et al. 2021; Donnelly et al. 2019). In the gram-negative bacterial pathogen *Salmonella Typhimurium*, top-down proteomics analysis can identify 1665 proteoforms out of 563 unique proteins generated by diverse PTMs and revealed a condition-specific S-glutathionylation-to-S-cysteinylation switch (Ansong et al. 2013). The protein S-thiolation is differentially employed to form S-glutathionylation and S-cysteinylation in response to infection-like conditions versus basal conditions. Comprehensive analysis of histone H3 modifications by top-down and middle down proteomics reveals a series of co-existing PTMs and clarifies the negative crosstalk between H3K36 methylation and H3K23 acetylation in clipped H3 (Tvardovskiy et al. 2015). Native top-down proteomics goes a step further and at least provides three layers of information, including the intact mass of a protein or complex, the masses of its subunits and non-covalent cofactors, and the primary sequence of proteins and their combinations of PTMs, thereby enabling the study of inter-protein PTM crosstalk (Leutert et al. 2021; Schachner et al. 2021; Li et al. 2018).

1.8 Outlooks

Although MS-based techniques have led to significant advances in understanding ubiquitin codes, some limitations still exist. First, there is still a lack of effective approaches to map the non-lysine ubiquitinome. Non-lysine ubiquitination can occur in different organisms and regulate numerous cellular processes and pathways. Moreover, certain E2/E3 enzymes possess esterification activity and preferentially mediate non-lysine ubiquitination (Kelsall et al. 2019; Pao et al. 2018), indicating that these ubiquitin events may not be redundant modifications of lysine ubiquitination (McClellan et al. 2019). The development of methods for the non-lysine ubiquitinome can promote its functional investigation.

Second, the current strategies for the lysine ubiquitinome still require a large amount of sample inputs. The most extensively used procedure for profiling the ubiquitinome requires the enrichment of K-GG remnants from large amounts of extracted proteins (usually milligrams) followed by MS analysis. The current on-antibody labeling and DIA strategy have reduced the sample inputs to approximately 500 μg (Udeshi et al. 2020;

Hansen et al. 2021), but the amounts of many other rare samples, such as stem cells, cannot meet this requirement. Thus, new technologies using fewer sample inputs should be developed.

Third, large-scale ubiquitinomics of clinical samples such as cancer tissues is still lacking. Recently, acetylomics and phosphoproteomics have been extensively used to characterize the molecular features of tumors that may direct the precise diagnosis and therapeutics of cancers (Cao et al. 2021; Satpathy et al. 2021; Wang et al. 2021; Huang et al. 2021; Petralia et al. 2020; Krug et al. 2020; Gillette et al. 2020; Clark et al. 2019; Boja et al. 2018). Ubiquitination is closely associated with human health and diseases (Rape 2018). In cancers, ubiquitination is known to be involved in the mTORC1, AMPK, PTEN-AKT, Wnt, Hippo-YAP, and TLR-, RLR- and STING-dependent signaling pathways (Deng et al. 2020). The profiling of the ubiquitinome of cancers may provide an additional layer of information associated with cancer progression. In addition, ubiquitination shows extensive crosstalk with phosphorylation and acetylation. Resolving the complexity of PTM crosstalk will help us better understand the pathogenesis of cancer and provide novel targets for precision medicine.

Acknowledgments This work was supported by the National Key R&D Program of China (No. 2018YFC2000305), National Natural Science Foundation of China (No. 31870826 and 82073221), and West China Hospital 1.3.5 project for disciplines of excellence (No. ZYYC23013).

Competing Financial Interests There are no competing financial interests.

References

Aalto AL et al (2019) M1-linked ubiquitination by LUBEL is required for inflammatory responses to oral infection in drosophila. Cell Death Differ 26:860–876. https://doi.org/10.1038/s41418-018-0164-x

Akimov V et al (2018) StUbEx PLUS-A modified stable tagged ubiquitin exchange system for peptide level purification and in-depth mapping of ubiquitination sites. J Proteome Res 17:296–304. https://doi.org/10.1021/acs.jproteome.7b00566

Al-Hakim AK et al (2008) Control of AMPK-related kinases by USP9X and atypical Lys(29)/Lys(33)-linked polyubiquitin chains. Biochem J 411:249–260. https://doi.org/10.1042/bj20080067

Ambroggio XI, Rees DC, Deshaies RJ (2003) JAMM: a metalloprotease-like zinc site in the proteasome and signalosome. PLoS Biol 2:e2. https://doi.org/10.1371/journal.pbio.0020002

Ansong C et al (2013) Top-down proteomics reveals a unique protein S-thiolation switch in *Salmonella* typhimurium in response to infection-like conditions. Proc Natl Acad Sci 110:10153–10158. https://doi.org/10.1073/pnas.1221210110

Batth TS, Francavilla C, Olsen JV (2014) Off-line high-pH reversed-phase fractionation for in-depth phosphoproteomics. J Proteome Res 13:6176–6186. https://doi.org/10.1021/pr500893m

Bekker-Jensen DB et al (2020) Rapid and site-specific deep phosphoproteome profiling by data-independent acquisition without the need for spectral libraries. Nat Commun 11:787. https://doi.org/10.1038/s41467-020-14609-1

Bianchi M et al (2015) Dynamic transcription of ubiquitin genes under basal and stressful conditions and new insights into the multiple UBC transcript variants. Gene 573:100–109. https://doi.org/10.1016/j.gene.2015.07.030

Bludau I et al (2021) Systematic detection of functional proteoform groups from bottom-up proteomic datasets. Nat Commun 12:3810. https://doi.org/10.1038/s41467-021-24030-x

Boja E et al (2018) Right data for right patient—a precisionFDA NCI-CPTAC multi-omics mislabeling challenge. Nat Med 24:1301–1302. https://doi.org/10.1038/s41591-018-0180-x

Bremm A, Freund SMV, Komander D (2010) Lys11-linked ubiquitin chains adopt compact conformations and are preferentially hydrolyzed by the deubiquitinase Cezanne. Nat Struct Mol Biol 17:939–947. https://doi.org/10.1038/nsmb.1873

Cao L et al (2021) Proteogenomic characterization of pancreatic ductal adenocarcinoma. Cell 184:5031–5052.e5026. https://doi.org/10.1016/j.cell.2021.08.023

Caron C, Boyault C, Khochbin S (2005) Regulatory cross-talk between lysine acetylation and ubiquitination: role in the control of protein stability. Bioessays 27:408–415. https://doi.org/10.1002/bies.20210

Celen AB, Sahin U (2020) Sumoylation on its 25th anniversary: mechanisms, pathology, and emerging concepts. FEBS J 287:3110–3140. https://doi.org/10.1111/febs.15319

Chatrin C et al (2020) Structural insights into ADP-ribosylation of ubiquitin by Deltex family E3 ubiquitin ligases. Sci Adv 6:eabc0418. https://doi.org/10.1126/sciadv.abc0418

Chen Y et al (2007) Lysine propionylation and butyrylation are novel post-translational modifications in histones. Mol Cell Proteomics 6:812–819. https://doi.org/10.1074/mcp.M700021-MCP200

Chen Y et al (2012) Quantitative acetylome analysis reveals the roles of SIRT1 in regulating diverse substrates and cellular pathways. Mol Cell Proteomics 11:1048–1062. https://doi.org/10.1074/mcp.M112.019547

Choudhary C et al (2009) Lysine acetylation targets protein complexes and co-regulates major cellular functions. Science 325:834–840. https://doi.org/10.1126/science.1175371

Ciechanover A (2015) The unravelling of the ubiquitin system. Nat Rev Mol Cell Biol 16:322–324. https://doi.org/10.1038/nrm3982

Clark DJ et al (2019) Integrated proteogenomic characterization of clear cell renal cell carcinoma. Cell 179:964–983.e931. https://doi.org/10.1016/j.cell.2019.10.007

Courtois G, Fauvarque M-O (2018) The many roles of ubiquitin in NF-κB signaling. Biomedicines 6:43. https://doi.org/10.3390/biomedicines6020043

Cristobal A et al (2017) Toward an optimized workflow for middle-down proteomics. Anal Chem 89:3318–3325. https://doi.org/10.1021/acs.analchem.6b03756

Crowe SO, Rana ASJB, Deol KK, Ge Y, Strieter ER (2017) Ubiquitin chain enrichment middle-down mass spectrometry enables characterization of branched ubiquitin chains in cellulo. Anal Chem 89:4428–4434. https://doi.org/10.1021/acs.analchem.6b03675

Cruz Walma DA, Chen Z, Bullock AN, Yamada KM (2022) Ubiquitin ligases: guardians of mammalian development. Nat Rev Mol Cell Biol 23:350. https://doi.org/10.1038/s41580-021-00448-5

Dai L et al (2014) Lysine 2-hydroxyisobutyrylation is a widely distributed active histone mark. Nat Chem Biol 10:365–370. https://doi.org/10.1038/nchembio.1497

Dai X, Zhang T, Hua D (2022) Ubiquitination and SUMOylation: protein homeostasis control over cancer. Epigenomics 14:43–58. https://doi.org/10.2217/epi-2021-0371

Damgaard RB (2021) The ubiquitin system: from cell signalling to disease biology and new therapeutic opportunities. Cell Death Differ 28:423–426. https://doi.org/10.1038/s41418-020-00703-w

Danielsen JM et al (2011) Mass spectrometric analysis of lysine ubiquitylation reveals promiscuity at site level. Mol Cell Proteomics 10:M110.003590. https://doi.org/10.1074/mcp.M110.003590

Davies V et al (2021) Rapid development of improved data-dependent acquisition strategies. Anal Chem 93:5676–5683. https://doi.org/10.1021/acs.analchem.0c03895

De Cesare V et al (2021) Deubiquitinating enzyme amino acid profiling reveals a class of ubiquitin esterases. Proc Natl Acad Sci 118:e2006947118. https://doi.org/10.1073/pnas.2006947118

Deng L, Meng T, Chen L, Wei W, Wang P (2020) The role of ubiquitination in tumorigenesis and targeted drug discovery. Signal Transduct Target Ther 5:11. https://doi.org/10.1038/s41392-020-0107-0

Deol KK, Lorenz S, Strieter ER (2019) Enzymatic logic of ubiquitin chain assembly. Front Physiol 10:835–835. https://doi.org/10.3389/fphys.2019.00835

Dittmar G, Winklhofer KF (2020) Linear ubiquitin chains: cellular functions and strategies for detection and quantification. Front Chem 7:915. https://doi.org/10.3389/fchem.2019.00915

Dong X et al (2017) Ubiquitin S65 phosphorylation engenders a pH-sensitive conformational switch. Proc Natl Acad Sci 114:6770–6775. https://doi.org/10.1073/pnas.1705718114

Donnelly DP et al (2019) Best practices and benchmarks for intact protein analysis for top-down mass spectrometry. Nat Methods 16:587–594. https://doi.org/10.1038/s41592-019-0457-0

Elia AE et al (2015) Quantitative proteomic atlas of ubiquitination and acetylation in the DNA damage response. Mol Cell 59:867–881. https://doi.org/10.1016/j.molcel.2015.05.006

Fiesel FC, Springer W (2015) Disease relevance of phosphorylated ubiquitin (p-S65-Ub). Autophagy 11:2125–2126. https://doi.org/10.1080/15548627.2015.1091912

Fiil BK et al (2013) OTULIN restricts Met1-linked ubiquitination to control innate immune signaling. Mol Cell 50:818–830. https://doi.org/10.1016/j.molcel.2013.06.004

French ME, Koehler CF, Hunter T (2021) Emerging functions of branched ubiquitin chains. Cell Discov 7:6. https://doi.org/10.1038/s41421-020-00237-y

Fulzele A, Bennett EJ (2018) Ubiquitin diGLY proteomics as an approach to identify and quantify the ubiquitin-modified proteome. Methods Mol Biol 1844:363–384. https://doi.org/10.1007/978-1-4939-8706-1_23

Gao Y et al (2016) Enhanced purification of ubiquitinated proteins by engineered tandem hybrid ubiquitin-binding domains (ThUBDs). Mol Cell Proteomics 15:1381–1396. https://doi.org/10.1074/mcp.O115.051839

Geiss-Friedlander R, Melchior F (2007) Concepts in sumoylation: a decade on. Nat Rev Mol Cell Biol 8:947–956. https://doi.org/10.1038/nrm2293

Geoffroy MC, Hay RT (2009) An additional role for SUMO in ubiquitin-mediated proteolysis. Nat Rev Mol Cell Biol 10:564–568. https://doi.org/10.1038/nrm2707

Gillette MA et al (2020) Proteogenomic characterization reveals therapeutic vulnerabilities in lung adenocarcinoma. Cell 182:200–225.e235. https://doi.org/10.1016/j.cell.2020.06.013

Goldstein G et al (1975) Isolation of a polypeptide that has lymphocyte-differentiating properties and is probably represented universally in living cells. Proc Natl Acad Sci 72:11–15. https://doi.org/10.1073/pnas.72.1.11

Gong Y, Nie L, Dai L (2021a) Beyond proteins: ubiquitylation of lipopolysaccharide to fight bacteria. MedComm 2:855–857. https://doi.org/10.1002/mco2.86

Gong Y, Qin S, Dai L, Tian Z (2021b) The glycosylation in SARS-CoV-2 and its receptor ACE2. Signal Transduct Target Ther 6:396. https://doi.org/10.1038/s41392-021-00809-8
Greer SF, Zika E, Conti B, Zhu X-S, Ting JPY (2003) Enhancement of CIITA transcriptional function by ubiquitin. Nat Immunol 4:1074–1082. https://doi.org/10.1038/ni985
Grice GL et al (2015) The proteasome distinguishes between heterotypic and homotypic lysine-11-linked polyubiquitin chains. Cell Rep 12:545–553. https://doi.org/10.1016/j.celrep.2015.06.061
Grønborg M et al (2002) A mass spectrometry-based proteomic approach for identification of serine/threonine-phosphorylated proteins by enrichment with phospho-specific antibodies: identification of a novel protein, Frigg, as a protein kinase A substrate. Mol Cell Proteomics 1:517–527. https://doi.org/10.1074/mcp.m200010-mcp200
Guo J, Huan T (2020) Comparison of full-scan, data-dependent, and data-independent acquisition modes in liquid chromatography-mass spectrometry based untargeted metabolomics. Anal Chem 92:8072–8080. https://doi.org/10.1021/acs.analchem.9b05135
Haglund K et al (2003) Multiple monoubiquitination of RTKs is sufficient for their endocytosis and degradation. Nat Cell Biol 5:461–466. https://doi.org/10.1038/ncb983
Han ZJ, Feng YH, Gu BH, Li YM, Chen H (2018) The post-translational modification, SUMOylation, and cancer (review). Int J Oncol 52:1081–1094. https://doi.org/10.3892/ijo.2018.4280
Hansen BK et al (2019) Analysis of human acetylation stoichiometry defines mechanistic constraints on protein regulation. Nat Commun 10:1055. https://doi.org/10.1038/s41467-019-09024-0
Hansen FM et al (2021) Data-independent acquisition method for ubiquitinome analysis reveals regulation of circadian biology. Nat Commun 12:254. https://doi.org/10.1038/s41467-020-20509-1
Harhaj EW, Dixit VM (2011) Deubiquitinases in the regulation of NF-kappaB signaling. Cell Res 21:22–39. https://doi.org/10.1038/cr.2010.166
Harikumar KB et al (2014) K63-linked polyubiquitination of transcription factor IRF1 is essential for IL-1-induced production of chemokines CXCL10 and CCL5. Nat Immunol 15:231–238. https://doi.org/10.1038/ni.2810
Harrigan JA, Jacq X, Martin NM, Jackson SP (2018) Deubiquitylating enzymes and drug discovery: emerging opportunities. Nat Rev Drug Discov 17:57–78. https://doi.org/10.1038/nrd.2017.152
He X-D et al (2018) Sensing and transmitting intracellular amino acid signals through reversible lysine aminoacylations. Cell Metab 27:151–166.e156. https://doi.org/10.1016/j.cmet.2017.10.015
Hendriks IA et al (2014) Uncovering global SUMOylation signaling networks in a site-specific manner. Nat Struct Mol Biol 21:927–936. https://doi.org/10.1038/nsmb.2890
Hendriks IA et al (2017) Site-specific mapping of the human SUMO proteome reveals co-modification with phosphorylation. Nat Struct Mol Biol 24:325–336. https://doi.org/10.1038/nsmb.3366
Hicke L (2001) Protein regulation by monoubiquitin. Nat Rev Mol Cell Biol 2:195–201. https://doi.org/10.1038/35056583
Hu H, Sun S-C (2016) Ubiquitin signaling in immune responses. Cell Res 26:457. https://doi.org/10.1038/cr.2016.40
Huang H et al (2018) Lysine benzoylation is a histone mark regulated by SIRT2. Nat Commun 9:3374. https://doi.org/10.1038/s41467-018-05567-w
Huang C et al (2021) Proteogenomic insights into the biology and treatment of HPV-negative head and neck squamous cell carcinoma. Cancer Cell 39:361–379.e316. https://doi.org/10.1016/j.ccell.2020.12.007
Humphrey SJ, Karayel O, James DE, Mann M (2018) High-throughput and high-sensitivity phosphoproteomics with the EasyPhos platform. Nat Protoc 13:1897–1916. https://doi.org/10.1038/s41596-018-0014-9
Hunter T (2007) The age of crosstalk: phosphorylation, ubiquitination, and beyond. Mol Cell 28:730–738. https://doi.org/10.1016/j.molcel.2007.11.019
Hunter T, Sun H (2008) Crosstalk between the SUMO and ubiquitin pathways. Ernst Schering Found Symp Proc (1):1–16. https://doi.org/10.1007/2789_2008_098
Jeong JW et al (2002) Regulation and destabilization of HIF-1alpha by ARD1-mediated acetylation. Cell 111:709–720. https://doi.org/10.1016/s0092-8674(02)01085-1
Jiang Y et al (2021) Isonicotinylation is a histone mark induced by the anti-tuberculosis first-line drug isoniazid. Nat Commun 12:5548. https://doi.org/10.1038/s41467-021-25867-y
Johnson A, Vert G (2016) Unraveling K63 polyubiquitination networks by sensor-based proteomics. Plant Physiol 171:1808–1820. https://doi.org/10.1104/pp.16.00619
Kelsall IR, Zhang J, Knebel A, Arthur JSC, Cohen P (2019) The E3 ligase HOIL-1 catalyses ester bond formation between ubiquitin and components of the Myddosome in mammalian cells. Proc Natl Acad Sci 116:13293–13298. https://doi.org/10.1073/pnas.1905873116
Kim SC et al (2006) Substrate and functional diversity of lysine acetylation revealed by a proteomics survey. Mol Cell 23:607–618. https://doi.org/10.1016/j.molcel.2006.06.026
Kim W et al (2011) Systematic and quantitative assessment of the ubiquitin-modified proteome. Mol Cell 44:325–340. https://doi.org/10.1016/j.molcel.2011.08.025
Kim J et al (2013) The n-SET domain of Set1 regulates H2B ubiquitylation-dependent H3K4 methylation. Mol Cell 49:1121–1133. https://doi.org/10.1016/j.molcel.2013.01.034
Klaus A, Birchmeier W (2008) Wnt signalling and its impact on development and cancer. Nat Rev Cancer 8:387–398. https://doi.org/10.1038/nrc2389

Kliza K, Husnjak K (2020) Resolving the complexity of ubiquitin networks. Front Mol Biosci 7:21. https://doi.org/10.3389/fmolb.2020.00021

Kniss A et al (2018) Chain assembly and disassembly processes differently affect the conformational space of ubiquitin chains. Structure 26:249–258.e244. https://doi.org/10.1016/j.str.2017.12.011

Komander D, Clague MJ, Urbé S (2009) Breaking the chains: structure and function of the deubiquitinases. Nat Rev Mol Cell Biol 10:550–563. https://doi.org/10.1038/nrm2731

Koyano F et al (2014) Ubiquitin is phosphorylated by PINK1 to activate parkin. Nature 510:162–166. https://doi.org/10.1038/nature13392

Krasny L, Huang PH (2021) Data-independent acquisition mass spectrometry (DIA-MS) for proteomic applications in oncology. Mol Omics 17:29–42. https://doi.org/10.1039/d0mo00072h

Krug K et al (2020) Proteogenomic landscape of breast cancer tumorigenesis and targeted therapy. Cell 183:1436–1456.e1431. https://doi.org/10.1016/j.cell.2020.10.036

Lacoursiere RE, Shaw GS (2021) Acetylated ubiquitin modulates the catalytic activity of the E1 enzyme Uba1. Biochemistry 60:1276–1285. https://doi.org/10.1021/acs.biochem.1c00145

Lamoliatte F, McManus FP, Maarifi G, Chelbi-Alix MK, Thibault P (2017) Uncovering the SUMOylation and ubiquitylation crosstalk in human cells using sequential peptide immunopurification. Nat Commun 8:14109. https://doi.org/10.1038/ncomms14109

Larsen MR, Trelle MB, Thingholm TE, Jensen ON (2006) Analysis of posttranslational modifications of proteins by tandem mass spectrometry. BioTechniques 40:790–798. https://doi.org/10.2144/000112201

Lee JW, Bae SH, Jeong JW, Kim SH, Kim KW (2004) Hypoxia-inducible factor (HIF-1)alpha: its protein stability and biological functions. Exp Mol Med 36:1–12. https://doi.org/10.1038/emm.2004.1

Lee HJ et al (2009) Quantitative analysis of phosphopeptides in search of the disease biomarker from the hepatocellular carcinoma specimen. Proteomics 9:3395–3408. https://doi.org/10.1002/pmic.200800943

Lehmann WD et al (2007) Neutral loss-based phosphopeptide recognition: a collection of caveats. J Proteome Res 6:2866–2873. https://doi.org/10.1021/pr060573w

Leutert M, Entwisle SW, Villén J (2021) Decoding post-translational modification crosstalk with proteomics. Mol Cell Proteomics 20:100129. https://doi.org/10.1016/j.mcpro.2021.100129

Li W, Ye Y (2008) Polyubiquitin chains: functions, structures, and mechanisms. Cell Mol Life Sci 65:2397–2406. https://doi.org/10.1007/s00018-008-8090-6

Li N et al (2015) Poly-ADP ribosylation of PTEN by tankyrases promotes PTEN degradation and tumor growth. Genes Dev 29:157–170. https://doi.org/10.1101/gad.251785.114

Li C-W et al (2016) Glycosylation and stabilization of programmed death ligand-1 suppresses T-cell activity. Nat Commun 7:12632. https://doi.org/10.1038/ncomms12632

Li H, Nguyen HH, Ogorzalek Loo RR, Campuzano IDG, Loo JA (2018) An integrated native mass spectrometry and top-down proteomics method that connects sequence to structure and function of macromolecular complexes. Nat Chem 10:139–148. https://doi.org/10.1038/nchem.2908

Li KW, Gonzalez-Lozano MA, Koopmans F, Smit AB (2020) Recent developments in data independent acquisition (DIA) mass spectrometry: application of quantitative analysis of the brain proteome. Front Mol Neurosci 13:564446. https://doi.org/10.3389/fnmol.2020.564446

Liebelt F, Vertegaal ACO (2016) Ubiquitin-dependent and independent roles of SUMO in proteostasis. Am J Phys Cell Phys 311:C284–C296. https://doi.org/10.1152/ajpcell.00091.2016

McClellan AJ, Laugesen SH, Ellgaard L (2019) Cellular functions and molecular mechanisms of non-lysine ubiquitination. Open Biol 9:190147. https://doi.org/10.1098/rsob.190147

Mevissen TET, Komander D (2017) Mechanisms of deubiquitinase specificity and regulation. Annu Rev Biochem 86:159–192. https://doi.org/10.1146/annurev-biochem-061516-044916

Michel MA, Swatek KN, Hospenthal MK, Komander D (2017) Ubiquitin linkage-specific affimers reveal insights into K6-linked ubiquitin signaling. Mol Cell 68:233–246.e235. https://doi.org/10.1016/j.molcel.2017.08.020

Minguez P et al (2014) PTMcode v2: a resource for functional associations of post-translational modifications within and between proteins. Nucleic Acids Res 43:D494–D502. https://doi.org/10.1093/nar/gku1081

Mohammed S, Heck A Jr (2011) Strong cation exchange (SCX) based analytical methods for the targeted analysis of protein post-translational modifications. Curr Opin Biotechnol 22:9–16. https://doi.org/10.1016/j.copbio.2010.09.005

Morreale FE, Walden H (2016) Types of ubiquitin ligases. Cell 165:248–248.e241. https://doi.org/10.1016/j.cell.2016.03.003

Nakagawa H et al (2009) Disruption of N-linked glycosylation enhances ubiquitin-mediated proteasomal degradation of the human ATP-binding cassette transporter ABCG2. FEBS J 276:7237–7252. https://doi.org/10.1111/j.1742-4658.2009.07423.x

Nakano T, Takahashi-Nakaguchi A, Yamamoto M, Watanabe M (2015) Pierisins and CARP-1: ADP-ribosylation of DNA by ARTCs in butterflies and shellfish. Curr Top Microbiol Immunol 384:127–149. https://doi.org/10.1007/82_2014_416

Newton K et al (2012) Using linkage-specific monoclonal antibodies to analyze cellular ubiquitylation. Methods Mol Biol 832:185–196. https://doi.org/10.1007/978-1-61779-474-2_13

Nielsen ML et al (2008) Iodoacetamide-induced artifact mimics ubiquitination in mass spectrometry. Nat Methods 5:459–460. https://doi.org/10.1038/nmeth0608-459

Nijman SM et al (2005) A genomic and functional inventory of deubiquitinating enzymes. Cell 123:773–786. https://doi.org/10.1016/j.cell.2005.11.007

Ohtake F, Tsuchiya H (2016) The emerging complexity of ubiquitin architecture. J Biochem 161:125–133. https://doi.org/10.1093/jb/mvw088

Ohtake F et al (2015) Ubiquitin acetylation inhibits polyubiquitin chain elongation. EMBO Rep 16:192–201. https://doi.org/10.15252/embr.201439152

Olsen JV et al (2006) Global, in vivo, and site-specific phosphorylation dynamics in signaling networks. Cell 127:635–648. https://doi.org/10.1016/j.cell.2006.09.026

Otten EG et al (2021) Ubiquitylation of lipopolysaccharide by RNF213 during bacterial infection. Nature 594:111–116. https://doi.org/10.1038/s41586-021-03566-4

Palazzo L, Mikoč A, Ahel I (2017) ADP-ribosylation: new facets of an ancient modification. FEBS J 284:2932–2946. https://doi.org/10.1111/febs.14078

Palazzo L, Mikolčević P, Mikoč A, Ahel I (2019) ADP-ribosylation signalling and human disease. Open Biol 9:190041. https://doi.org/10.1098/rsob.190041

Pao KC et al (2018) Activity-based E3 ligase profiling uncovers an E3 ligase with esterification activity. Nature 556:381–385. https://doi.org/10.1038/s41586-018-0026-1

Papouli E et al (2005) Crosstalk between SUMO and ubiquitin on PCNA is mediated by recruitment of the helicase Srs2p. Mol Cell 19:123–133. https://doi.org/10.1016/j.molcel.2005.06.001

Pellegrino S, Altmeyer M (2016) Interplay between ubiquitin, SUMO, and poly(ADP-ribose) in the cellular response to genotoxic stress. Front Genet 7:63. https://doi.org/10.3389/fgene.2016.00063

Perina D et al (2014) Distribution of protein poly(ADP-ribosyl)ation systems across all domains of life. DNA Repair 23:4–16. https://doi.org/10.1016/j.dnarep.2014.05.003

Petralia F et al (2020) Integrated proteogenomic characterization across major histological types of pediatric brain cancer. Cell 183:1962–1985.e1931. https://doi.org/10.1016/j.cell.2020.10.044

Rape M (2018) Ubiquitylation at the crossroads of development and disease. Nat Rev Mol Cell Biol 19:59–70. https://doi.org/10.1038/nrm.2017.83

Reily C, Stewart TJ, Renfrow MB, Novak J (2019) Glycosylation in health and disease. Nat Rev Nephrol 15:346–366. https://doi.org/10.1038/s41581-019-0129-4

Rikova K et al (2007) Global survey of phosphotyrosine signaling identifies oncogenic kinases in lung cancer. Cell 131:1190–1203. https://doi.org/10.1016/j.cell.2007.11.025

Ronai ZA (2016) Monoubiquitination in proteasomal degradation. Proc Natl Acad Sci 113:8894–8896. https://doi.org/10.1073/pnas.1610186113

Rose CM et al (2016) Highly multiplexed quantitative mass spectrometry analysis of ubiquitylomes. Cell Syst 3:395–403.e394. https://doi.org/10.1016/j.cels.2016.08.009

Rott R et al (2011) α-Synuclein fate is determined by USP9X-regulated monoubiquitination. Proc Natl Acad Sci 108:18666–18671. https://doi.org/10.1073/pnas.1105725108

Rott R et al (2017) SUMOylation and ubiquitination reciprocally regulate α-synuclein degradation and pathological aggregation. Proc Natl Acad Sci 114:13176–13181. https://doi.org/10.1073/pnas.1704351114

Ruan H-B, Nie Y, Yang X (2013) Regulation of protein degradation by O-GlcNAcylation: crosstalk with ubiquitination. Mol Cell Proteomics 12:3489–3497. https://doi.org/10.1074/mcp.R113.029751

Sabari BR et al (2015) Intracellular crotonyl-CoA stimulates transcription through p300-catalyzed histone crotonylation. Mol Cell 58:203–215. https://doi.org/10.1016/j.molcel.2015.02.029

Sanchez-Bailon MP et al (2021) Arginine methylation and ubiquitylation crosstalk controls DNA end-resection and homologous recombination repair. Nat Commun 12:6313. https://doi.org/10.1038/s41467-021-26413-6

Satpathy S et al (2021) A proteogenomic portrait of lung squamous cell carcinoma. Cell 184:4348–4371.e4340. https://doi.org/10.1016/j.cell.2021.07.016

Schachner LF et al (2021) Reassembling protein complexes after controlled disassembly by top-down mass spectrometry in native mode. Int J Mass Spectrom 465:116591. https://doi.org/10.1016/j.ijms.2021.116591

Schölz C et al (2015) Acetylation site specificities of lysine deacetylase inhibitors in human cells. Nat Biotechnol 33:415–423. https://doi.org/10.1038/nbt.3130

Scott D et al (2016) Mass spectrometry insights into a tandem ubiquitin-binding domain hybrid engineered for the selective recognition of unanchored polyubiquitin. Proteomics 16:1961–1969. https://doi.org/10.1002/pmic.201600067

Shabek N et al (2012) The size of the proteasomal substrate determines whether its degradation will be mediated by mono- or polyubiquitylation. Mol Cell 48:87–97. https://doi.org/10.1016/j.molcel.2012.07.011

Shimizu K et al (2021) Interplay between protein acetylation and ubiquitination controls MCL1 protein stability. Cell Rep 37:109988. https://doi.org/10.1016/j.celrep.2021.109988

Shrikhande GV et al (2010) O-glycosylation regulates ubiquitination and degradation of the anti-inflammatory protein A20 to accelerate atherosclerosis in diabetic ApoE-null mice. PLoS One 5:e14240. https://doi.org/10.1371/journal.pone.0014240

Sidoli S, Garcia BA (2017) Middle-down proteomics: a still unexploited resource for chromatin biology.

Expert Rev Proteomics 14:617–626. https://doi.org/10.1080/14789450.2017.1345632
Sidoli S, Lin S, Karch KR, Garcia BA (2015) Bottom-up and middle-down proteomics have comparable accuracies in defining histone post-translational modification relative abundance and stoichiometry. Anal Chem 87:3129–3133. https://doi.org/10.1021/acs.analchem.5b00072
Sidoli S et al (2017) Metabolic labeling in middle-down proteomics allows for investigation of the dynamics of the histone code. Epigenetics Chromatin 10:34. https://doi.org/10.1186/s13072-017-0139-z
Snyder NA, Silva GM (2021) Deubiquitinating enzymes (DUBs): regulation, homeostasis, and oxidative stress response. J Biol Chem 297:101077. https://doi.org/10.1016/j.jbc.2021.101077
Steen H, Mann M (2004) The ABC's (and XYZ's) of peptide sequencing. Nat Rev Mol Cell Biol 5:699–711. https://doi.org/10.1038/nrm1468
Steger M et al (2021) Time-resolved in vivo ubiquitinome profiling by DIA-MS reveals USP7 targets on a proteome-wide scale. Nat Commun 12:5399. https://doi.org/10.1038/s41467-021-25454-1
Sun Y et al (2021) PARylation prevents the proteasomal degradation of topoisomerase I DNA-protein crosslinks and induces their deubiquitylation. Nat Commun 12:5010. https://doi.org/10.1038/s41467-021-25252-9
Swaney DL et al (2013) Global analysis of phosphorylation and ubiquitylation cross-talk in protein degradation. Nat Methods 10:676–682. https://doi.org/10.1038/nmeth.2519
Swaney DL, Rodríguez-Mias RA, Villén J (2015) Phosphorylation of ubiquitin at Ser65 affects its polymerization, targets, and proteome-wide turnover. EMBO Rep 16:1131–1144. https://doi.org/10.15252/embr.201540298
Swatek KN, Komander D (2016) Ubiquitin modifications. Cell Res 26:399–422. https://doi.org/10.1038/cr.2016.39
Tan F et al (2008) Proteomic analysis of ubiquitinated proteins in normal hepatocyte cell line Chang liver cells. Proteomics 8:2885–2896. https://doi.org/10.1002/pmic.200700887
Tan M et al (2011) Identification of 67 histone marks and histone lysine crotonylation as a new type of histone modification. Cell 146:1016–1028. https://doi.org/10.1016/j.cell.2011.08.008
Tan M et al (2014) Lysine glutarylation is a protein posttranslational modification regulated by SIRT5. Cell Metab 19:605–617. https://doi.org/10.1016/j.cmet.2014.03.014
Thiel AT et al (2010) MLL-AF9-induced leukemogenesis requires coexpression of the wild-type Mll allele. Cancer Cell 17:148–159. https://doi.org/10.1016/j.ccr.2009.12.034
Thingholm TE, Larsen MR (2016a) Phosphopeptide enrichment by immobilized metal affinity chromatography. Methods Mol Biol 1355:123–133. https://doi.org/10.1007/978-1-4939-3049-4_8
Thingholm TE, Larsen MR (2016b) The use of titanium dioxide for selective enrichment of phosphorylated peptides. Methods Mol Biol 1355:135–146. https://doi.org/10.1007/978-1-4939-3049-4_9
Thingholm TE, Jørgensen TJ, Jensen ON, Larsen MR (2006) Highly selective enrichment of phosphorylated peptides using titanium dioxide. Nat Protoc 1:1929–1935. https://doi.org/10.1038/nprot.2006.185
Tsuchiya H et al (2018) Ub-ProT reveals global length and composition of protein ubiquitylation in cells. Nat Commun 9:524. https://doi.org/10.1038/s41467-018-02869-x
Tvardovskiy A et al (2015) Top-down and middle-down protein analysis reveals that intact and clipped human histones differ in post-translational modification patterns. Mol Cell Proteomics 14:3142–3153. https://doi.org/10.1074/mcp.M115.048975
Udeshi ND, Mertins P, Svinkina T, Carr SA (2013) Large-scale identification of ubiquitination sites by mass spectrometry. Nat Protoc 8:1950–1960. https://doi.org/10.1038/nprot.2013.120
Udeshi ND et al (2020) Rapid and deep-scale ubiquitylation profiling for biology and translational research. Nat Commun 11:359. https://doi.org/10.1038/s41467-019-14175-1
Valkevich EM, Sanchez NA, Ge Y, Strieter ER (2014) Middle-down mass spectrometry enables characterization of branched ubiquitin chains. Biochemistry 53:4979–4989. https://doi.org/10.1021/bi5006305
Vasilescu J, Smith JC, Ethier M, Figeys D (2005) Proteomic analysis of ubiquitinated proteins from human MCF-7 breast cancer cells by immunoaffinity purification and mass spectrometry. J Proteome Res 4:2192–2200. https://doi.org/10.1021/pr050265i
Verheugd P et al (2013) Regulation of NF-κB signalling by the mono-ADP-ribosyltransferase ARTD10. Nat Commun 4:1683. https://doi.org/10.1038/ncomms2672
Villén J, Gygi SP (2008) The SCX/IMAC enrichment approach for global phosphorylation analysis by mass spectrometry. Nat Protoc 3:1630–1638. https://doi.org/10.1038/nprot.2008.150
Vivelo CA, Ayyappan V, Leung AKL (2019) Poly(ADP-ribose)-dependent ubiquitination and its clinical implications. Biochem Pharmacol 167:3–12. https://doi.org/10.1016/j.bcp.2019.05.006
Vlaming H et al (2014) Flexibility in crosstalk between H2B ubiquitination and H3 methylation in vivo. EMBO Rep 15:1077–1084. https://doi.org/10.15252/embr.201438793
Wagner SA et al (2012) Proteomic analyses reveal divergent ubiquitylation site patterns in murine tissues. Mol Cell Proteomics 11:1578–1585. https://doi.org/10.1074/mcp.M112.017905
Walser F et al (2020) Ubiquitin phosphorylation at Thr12 modulates the DNA damage response. Mol Cell 80:423–436.e429. https://doi.org/10.1016/j.molcel.2020.09.017

Wang C et al (2001) TAK1 is a ubiquitin-dependent kinase of MKK and IKK. Nature 412:346–351. https://doi.org/10.1038/35085597
Wang X et al (2007) Ubiquitination of serine, threonine, or lysine residues on the cytoplasmic tail can induce ERAD of MHC-I by viral E3 ligase mK3. J Cell Biol 177:613–624. https://doi.org/10.1083/jcb.200611063
Wang E et al (2013) Histone H2B ubiquitin ligase RNF20 is required for MLL-rearranged leukemia. Proc Natl Acad Sci U S A 110:3901–3906. https://doi.org/10.1073/pnas.1301045110
Wang LB et al (2021) Proteogenomic and metabolomic characterization of human glioblastoma. Cancer Cell 39:509–528.e520. https://doi.org/10.1016/j.ccell.2021.01.006
Watanabe M et al (2020) A substrate-trapping strategy to find E3 ubiquitin ligase substrates identifies Parkin and TRIM28 targets. Commun Biol 3:592. https://doi.org/10.1038/s42003-020-01328-y
Worden EJ, Hoffmann NA, Hicks CW, Wolberger C (2019) Mechanism of cross-talk between H2B ubiquitination and H3 methylation by Dot1L. Cell 176:1490–1501.e1412. https://doi.org/10.1016/j.cell.2019.02.002
Wu C et al (2012) A protease for 'middle-down' proteomics. Nat Methods 9:822–824. https://doi.org/10.1038/nmeth.2074
Xie Z et al (2012) Lysine succinylation and lysine malonylation in histones. Mol Cell Proteomics 11:100–107. https://doi.org/10.1074/mcp.M111.015875
Xie Z et al (2016) Metabolic regulation of gene expression by histone lysine β-hydroxybutyrylation. Mol Cell 62:194–206. https://doi.org/10.1016/j.molcel.2016.03.036
Xu G, Jaffrey SR (2013) Proteomic identification of protein ubiquitination events. Biotechnol Genet Eng Rev 29:73–109. https://doi.org/10.1080/02648725.2013.801232
Xu P, Peng J (2008) Characterization of polyubiquitin chain structure by middle-down mass spectrometry. Anal Chem 80:3438–3444. https://doi.org/10.1021/ac800016w
Xu P et al (2009) Quantitative proteomics reveals the function of unconventional ubiquitin chains in proteasomal degradation. Cell 137:133–145. https://doi.org/10.1016/j.cell.2009.01.041
Xu G, Paige JS, Jaffrey SR (2010) Global analysis of lysine ubiquitination by ubiquitin remnant immunoaffinity profiling. Nat Biotechnol 28:868–873. https://doi.org/10.1038/nbt.1654
Yakubu RR, Nieves E, Weiss LM (2019) The methods employed in mass spectrometric analysis of posttranslational modifications (PTMs) and protein-protein interactions (PPIs). Adv Exp Med Biol 1140:169–198. https://doi.org/10.1007/978-3-030-15950-4_10
Yan F et al (2020) Threonine ADP-ribosylation of ubiquitin by a bacterial effector family blocks host ubiquitination. Mol Cell 78:641–652.e649. https://doi.org/10.1016/j.molcel.2020.03.016
Yang C-S et al (2017) Ubiquitin modification by the E3 ligase/ADP-ribosyltransferase Dtx3L/Parp9. Mol Cell 66:503–516.e505. https://doi.org/10.1016/j.molcel.2017.04.028
Yang Q, Zhao J, Chen D, Wang Y (2021) E3 ubiquitin ligases: styles, structures and functions. Mol Biomed 2:23. https://doi.org/10.1186/s43556-021-00043-2
Yu H, Lin L, Zhang Z, Zhang H, Hu H (2020) Targeting NF-κB pathway for the therapy of diseases: mechanism and clinical study. Signal Transduct Target Ther 5:209. https://doi.org/10.1038/s41392-020-00312-6
Yuan WC et al (2014) K33-linked polyubiquitination of coronin 7 by Cul3-KLHL20 ubiquitin E3 ligase regulates protein trafficking. Mol Cell 54:586–600. https://doi.org/10.1016/j.molcel.2014.03.035
Zhang Y, Fonslow BR, Shan B, Baek M-C, Yates JR (2013) Protein analysis by shotgun/bottom-up proteomics. Chem Rev 113:2343–2394. https://doi.org/10.1021/cr3003533
Zhang D et al (2019) Metabolic regulation of gene expression by histone lactylation. Nature 574:575–580. https://doi.org/10.1038/s41586-019-1678-1
Zheng N, Shabek N (2017) Ubiquitin ligases: structure, function, and regulation. Annu Rev Biochem 86:129–157. https://doi.org/10.1146/annurev-biochem-060815-014922
Zhu B et al (2005) Monoubiquitination of human histone H2B: the factors involved and their roles in HOX gene regulation. Mol Cell 20:601–611. https://doi.org/10.1016/j.molcel.2005.09.025

2 Ubiquitination in the T Cell Metabolism-Based Immunotherapy in Diseases

Ke-qi Fan, Yi-yuan Li, and Jin Jin

Abstract

Metabolism refers to the exchange of matter and energy between the organism and the environment and the self-renewal process of matter and energy in the organism. Metabolic activities in cells provide them with energy and various substrates required for development. Naive T cells differentiate into effector T cells and memory T cells after activation, and this process is accompanied by reprogramming of metabolism-related gene expression. These metabolic changes reflect physiological changes in different stages of T cell activation and differentiation. An increasing number of studies have shown that many autoimmune diseases and organ transplantation are accompanied by disorders and imbalances in T cell metabolism. To treat these diseases, related drugs can be used to regulate T cell activation, differentiation, and function. Therefore, T cell metabolism can serve as a new potential target for regulating immune responses.

Keywords

T cells · Ubiquitination · Metabolism Immunotherapy

2.1 Introduction

Biological organisms are always in a dynamic equilibrium with the outside world. The immune system recognizes and excludes invading pathogens or antigenic foreign bodies generated in the body to maintain the stability of the animal's internal environment. T cells are an important part of the immune system. After completing the recognition of the complex structure of the antigen peptide and MHC, the CD3 protein transmits the information of the antigen recognized by the T cell into the cell and initiates the activation process in the cell to further differentiate naive T cells into effector T cells. Helper T cells produce a variety of cytokines to exert cell-mediated cytotoxicity. At the same time, regulatory T cells can inhibit the activation and proliferation of helper T cells. These two pro-

K.-q. Fan
Sir Run Run Shaw Hospital, College of Medicine, Zhejiang University, Hangzhou, China

Y.-y. Li (✉)
Key Laboratory for Developmental Genes and Human Disease, Ministry of Education, Institute of Life Sciences, Jiangsu Province High-Tech Key Laboratory for Bio-Medical Research, Southeast University, Nanjing, China
e-mail: 103200067@seu.edu.cn

J. Jin (✉)
Sir Run Run Shaw Hospital, College of Medicine, Zhejiang University, Hangzhou, China

Center for Neuroimmunology and Health Longevity, The Third Affiliated Hospital of Sun Yat-sen University, Guangzhou, China
e-mail: jjin4@zju.edu.cn

H. Hu, X. Fu (eds.), *Ubiquitination in Immune System and Immune Response*, Advances in Experimental Medicine and Biology 1466, https://doi.org/10.1007/978-981-97-7288-9_2

cesses work together to regulate the immune system and maintain the stability of the internal environment. The function of different T cell subsets is related to their own unique cell metabolism. During the activation and differentiation of naive T cells, changes have been made in the metabolism of glucose, amino acids, fatty acids, cholesterol, and other substances. For example, helper T cells mainly depend on aerobic glycolysis metabolism to produce energy, while regulatory T cells rely primarily on fatty acid oxidation. These differences in metabolism are closely related to the process of immune cell activation and differentiation.

Altered energy sources and metabolic pathways during immune cell activation suggest the possibility of treating immune-related diseases by targeting metabolism. This therapeutic strategy targets only a subpopulation of cells that need to be affected and controls the metabolic regulation of immune cell function to treat autoimmune diseases without affecting basic immune function, which is a frontier and hot area of current immunology research. A deeper understanding of the regulatory mechanisms of T cell metabolism will provide effective molecular targets and potential new clinical therapeutic strategies for the prevention and treatment of related diseases. In the following sections, we will review the changes in metabolism before and after naive T cell activation and recent immunotherapies targeting T cell metabolism.

2.2 Basic T Cell Metabolism

2.2.1 Glucose

Cells take up various substances from the microenvironment that are necessary for life activities. Among them, glucose is the most important substrate for T cell metabolism, entering T cells primarily through glucose transporter 1 (GLUT1), which is then metabolized to produce energy via glycolysis, the tricarboxylic acid cycle, or the pentose phosphate pathway. In T cells, one unit of glucose is degraded in the cytoplasm through glycolysis to generate two units of pyruvate and two units of ATP. In mitochondria, pyruvate is catalyzed by pyruvate dehydrogenase to produce acetyl-CoA. Under aerobic conditions, acetyl-CoA can enter the mitochondrial matrix, and NADH and flavin adenine dinucleotide are generated through the tricarboxylic acid cycle. Flavin adenine dinucleotide provides electrons for the electron transfer chain on the inner membrane of the mitochondria. Then, a proton gradient is generated on the inner membrane of the mitochondria, and ADP is converted into ATP at this site, providing a large number of ATPs for the life activities of cells. Glycolysis is considered to be a rapid but relatively inefficient energy production process, and activated T cells have a significantly increased capacity to metabolize glucose. This metabolic transition is regulated by multiple signaling pathways, such as the AKT and mTOR pathways (Weber et al. 2008). This metabolic reprogramming occurs simultaneously with naive T cell activation, accompanied by increased expression of glucose transporters and key glycolytic enzymes, reflecting the coordinated nature of metabolic reprogramming and T cell activation (Wang et al. 2011). After the initial antigen-induced activation, CD28 protein in helper T cells acts as an adaptor protein to increase the PI3K-Akt signal. The activation of costimulated PI3K results in the upregulation of GLUT1 on the cell surface, thereby promoting the enhancement of glucose inflow. Regulatory T cells that suppress the inflammatory response rarely produce this process (Frauwirth et al. 2002). In vitro experiments showed that GLUT1-deficient T cells had reduced activation-induced proliferation under the same stimulation conditions as the wild-type control group, suggesting that this upregulation of GLUT1 is essential for T cell activation and that glucose metabolism plays an important role in T cell function.

2.2.2 Amino Acids

Cell growth and function are highly dependent on a sufficient source of amino acids. Amino acids are not only an important part of synthetic proteins but also provide an important backbone for de novo synthesis of nucleotides (Altman et al. 2016). Amino acids are mainly transported into T cells through various solute carrier transporters. Changes in amino acid metabolism will affect the activation of immune cells and inflammatory responses. Glutamine is a conditionally essential amino acid and has a high content in T cells. It is necessary for lymphocytes to secrete various cytokines to proliferate and maintain their functions. Glutamine has an important immune promoting effect. In the cytoplasm, glutamine is mainly hydrolyzed by glutaminase-2 to obtain glutamate and ammonium. Glutamate then enters the mitochondria, where glutamate dehydrogenase converts glutamate to α-ketoglutarate (α-KG), which is the substrate of the TCA cycle. Metabolites from glutamine can be used for de novo nucleotide synthesis and for the production of aspartic acid, which can be further synthesized as pyrimidine. T lymphocyte activation is accompanied by increased glutamine uptake. As a precursor and main energy source for nucleic acid biosynthesis, glutamine can promote the differentiation and proliferation of T lymphocytes and increase the production of cytokines such as tumor necrosis factor-α and interleukin-1. Different immune cell subpopulations have different utilization rates of glutamine. In vitro, adding exogenous glutamine to cultured T lymphocytes from mice significantly increased the total number of T lymphocytes and the ratio of CD4/$CD8^+$ T cells. In clinical treatment (Hsiung et al. 2014), patients can be supplemented with glutamine to help improve their immunity (Straaten et al. 2001). Downregulating the uptake of glutamine and leucine can inhibit the differentiation and proliferation of Th1 and Th17 cells, but at the same time, the differentiation and proliferation of Tregs are not affected (Nakaya et al. 2014). In addition, amino acids such as aspartic acid, cystine, serine, and threonine have also been shown to be closely related to the maturation and differentiation of T lymphocytes (Krishnamoorthy et al. 2016).

2.2.3 Fatty Acids

Fatty acids are a type of cell substrate with high energy density. They are the main components of intracellular neutral fats, phospholipids, and glycolipids. Fatty acids can enter cells through various solute carrier transporter proteins and low-density lipoprotein receptor (LDLr) or scavenger receptor/fatty acid transporter (CD36/FAT) gene expression for oxidative decomposition. Short-chain, medium-chain, and long-chain fatty acids are activated in the cytoplasm by acylation of acyl-CoA synthetase to facilitate their transport through the mitochondrial membrane, and then fatty acids undergo β-oxidation in mitochondria. In addition to glucose and glutamine influx, T cell activation also rapidly upregulates fatty acid and cholesterol biosynthesis programs. The metabolic demand of memory T cells is usually low, and their energy requirements are mainly dependent on the metabolism of fatty acids. A sufficient amount of fatty acids could help memory T cells activate and proliferate rapidly after reinduction. Fatty acid metabolism can also affect the abundance and function of helper and regulatory T cells. Adding exogenous fatty acids to mouse T lymphocytes cultured in vitro can reduce the production of cytokines in Th1, Th2, and Th17 cells, increase the expression of FoxP3 in regulatory T cells at the same time, and regulate the function of the immune system in different ways (Melnik and Plewig 1992).

2.2.4 Cholesterol

Cholesterol is a cell membrane component necessary for cell growth and proliferation. It can participate in the formation of lipid rafts to regulate the fluidity of cell membranes and play an important role in the activation of immune cells (Amersfoort and Kuiper 2017; Lafaille et al. 1994). T lymphocytes can synthesize neopentyl cholesterol through the mevalonate pathway. In this process, 3-hydroxy-3-methylglutaryl coenzyme A reductase (HMGCR) and squalene epoxidase (SLE) are rate-limited enzymes. After TCR stimulation, T cells reduce cholesterol efflux through the ATP binding cassette (ABC) transporter and increase cholesterol synthesis by upregulating the expression of HMGCR and SQLE to ensure that sufficient cholesterol is available for membrane production. During the development and proliferation of T cells, cholesterol is also essential for the biological production of the endoplasmic reticulum (Heusch et al. 2004) (Figs. 2.1 and 2.2).

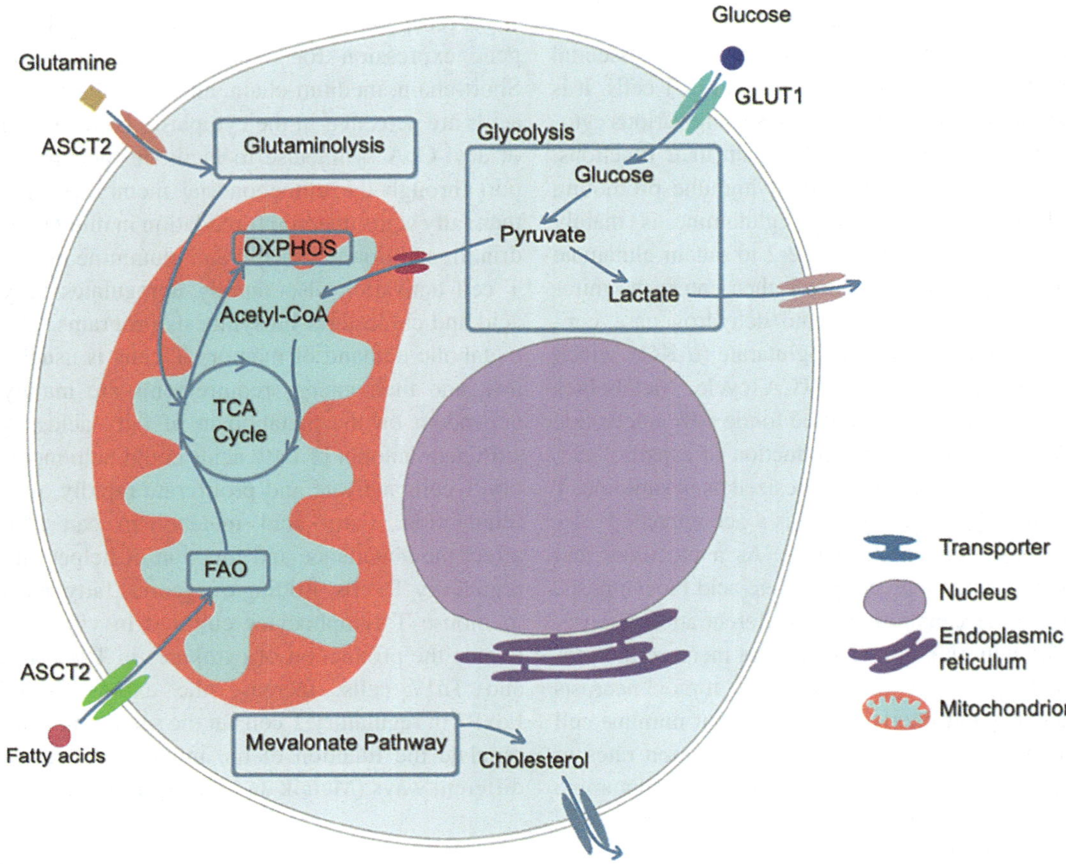

Fig. 2.1 T cell substrate uptake and metabolic pathway. After uptake of different metabolic substrates, cells undergo metabolic anabolism and catabolism of glucose, amino acids, fatty acids, etc. in the cytoplasm and mitochondria to obtain substances and energy necessary for life activities. Several major metabolic processes are illustrated in the figure

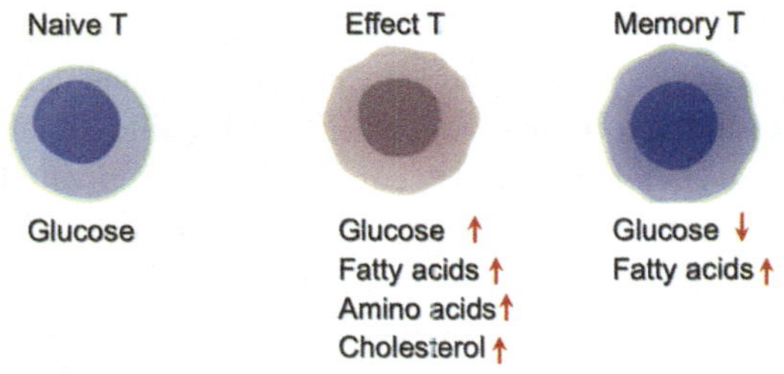

Fig. 2.2 T cell metabolism changes after differentiation. The process of activation and differentiation of T cells is accompanied by metabolic reprogramming, with altered uptake of different substrates in the microenvironment

2.3 Immunotherapy Based on T Cell Metabolism

As mentioned above, naive T cells are mainly in a low metabolic state that depends on a low degree of glycolysis, fatty acid oxidation, and oxidative phosphorylation. After activation, naive T cells switch to a high metabolic state, which is characterized by an increased glycolytic rate, upregulated glutamine metabolism, and a high degree of oxidative phosphorylation. These metabolic processes provide energy for cell activities and metabolic intermediates to produce large molecules. In addition, effector T cells eventually differentiate into memory T cells after the immune response is complete. Through metabolic changes, naive T cells eventually differentiate into effector T cells and differentiate into memory T cells after the immune response. In general, the activation, differentiation, proliferation, and function of T cells have obvious metabolic characteristics. Disturbances and imbalances of T cell metabolism can cause a variety of diseases. Many studies have tried to regulate the immune effect of T cells by regulating the

T cell metabolic process. When a certain metabolic condition is changed, the cell subgroup that is most dependent on this metabolic pathway will be affected the most. Understanding this process will help us target T cells to design treatments to regulate T cell metabolism and inhibit the development of metabolic-related diseases. Many studies have tried to regulate the immune effect of T cells by using different drugs to change the metabolism of T cells. When a certain metabolic process is changed, the cell subgroup that is most dependent on this metabolic pathway will be affected most. Understanding this process will help us design immunotherapies that target T cell metabolism and inhibit the development of related diseases.

2.3.1 Targeting Glycolysis

The energy characteristic of activated T cells often manifests as an upregulation of glycolysis. 2-Deoxy-D-glucose (2-DG) is a competitive inhibitor of glucose that can inhibit glucose from entering the metabolic pathway. Activated helper T cells promote growth and proliferation by enhancing glucose uptake and upregulating glycolysis, while regulatory T cells obtain energy mainly through fatty acid oxidation and oxidative phosphorylation metabolism pathways. Therefore, the injection of 2-DG should have an inhibitory effect on helper T cells without affecting regulatory T cells, and it has a negative regulatory effect on immune responses. Some studies have confirmed this inference. In a mouse EAE model, the MOG antigen was injected into the mouse body and caused autoimmune encephalomyelitis. The injection of MOG antigen into mice leads to autoimmune encephalomyelitis, which manifests as immune cells infiltrating and attacking the central nervous system to gradually paralyze the animal. Injection of 2-DG can inhibit the differentiation and proliferation of Th17 cells and simultaneously enhance the generation of Tregs, which significantly reduces the phenotype of the mouse's inability to move (Shi et al. 2011). Systemic lupus erythematosus (SLE) is the most common type of connective tissue disease. Damage often affects the central nervous system, and 20–30% of patients exhibit psychiatric symptoms. Dysregulation of glucose metabolism is also crucial in the pathogenesis of this disease. In patients with normal brain MRI findings, decreases in glucose metabolism coupled with decreases in regional cerebral blood flow are associated with serious neuropsychiatric SLE (NPSLE) presentations (Kao et al. 2010). Use drugs targeting glucose metabolism may be helpful in treating NPSLE by regulating the metabolism of the immune system. Adoptive therapy of T cells is an effective new method for treating cancer found by the National Cancer Institute, that is, selecting T cells with the strongest ability to kill cancer cells in patients, culturing and proliferating T cells in vitro, improving their recognition and resistance ability of the tumor cells, and then transferring these T cells into the patient to play a therapeutic role. The physiological activity of helper T cells is more dependent on glycolysis. Therefore, adding 2-DG when expanding tumor-specific T cells in vitro can inhibit regulatory T cells and enrich memory T cells, which is beneficial to the persistence of immune cells after adoptive transfer (Sukumar et al. 2013). In $CD4^+$ T cells, glucose metabolism may also occur not through the TCA cycle but through the pentose phosphate pathway to synthesize purines from scratch, with the resulting xanthines being excreted extracellularly and reaching the brain via the blood circulation to activate the amygdala, ultimately leading to an anxious phenotype in animals. This process is associated with mitochondrial fragmentation and can be treated with purine synthesis inhibitors such as BCX1777 (Fan et al. 2019).

The product of glycolysis, pyruvate, is an important substrate for the TCA cycle. Another potential method for controlling autoimmune diseases related to glucose metabolism is to regulate pyruvate metabolism. Using dichloroacetate to inhibit pyruvate dehydrogenase kinase PDHK1 can inhibit glycolysis and significantly reduce the paralysis phenotype of mice in the EAE model. 3-PO is a small molecule inhibitor of a key regulator of glycolysis, 6-phosphofructo-2-kinase (PFKFB3). Using 3-PO can inhibit PFKFB3 and thereby inhibit glycolysis, significantly attenuating the proliferation of several malignant hematopoietic cell lines and adenocarcinoma cell lines (Yalcin et al. 2007). Studies have also shown that treatment of mice with 3-PO after bone marrow transplantation can suppress the proliferation of helper T cells, significantly reduce rejection after transplantation, and improve the survival expectancy of mice (Herijgers et al. 1997). In the human blood microenvironment, the blood glucose level is maintained at 80–120 mg. Some studies have shown that intermittent fasting can keep the blood glucose level at a low level to reduce metabolism and promote anticancer effects, which is helpful in clinical treatment (Roky et al. 2004; Anson et al. 2003). T1DM is a type of hyperglycemia caused by T cell-mediated islet autoimmunity. Islet autoantibodies are markers of immune destruction of islet β cells and are the key indicators for the diagnosis of T1DM, including islet cell antibody (ICA) and glutamic acid decarboxylase antibody (GAD-Ab). The result of immune destruction of pancreatic islet β cells is insufficient insulin, which causes T cells in T1DM patients to be in a microenvironment with high blood sugar for a long time. Under hyperglycemic conditions, T cell subtypes that highly express GLUT1 (such as Th1 cells) are more activated than regulatory T cells, which ultimately leads to an inflammatory response, leading to an increased risk of atherosclerosis and cardiovascular disease in T1DM patients. In this situation, therapeutic intervention can be achieved by reducing hyperglycemia, which causes inflammation (Middlebrook et al. 2007). AMPK (adenosine 5′-monophosphate (AMP)-activated protein kinase) is a key enzyme in the regulation of biological metabolism. Genetic and pharmacological studies have shown that AMPK can regulate glucose metabolism in various ways by increasing glucose uptake and glycolysis, inhibiting glycogen synthesis and inhibiting gluconeogenesis and is necessary for animals to maintain glucose balance. Studies have found that the AMPK agonist metformin regulates T cell metabolism and inhibits the proliferation of regulatory T cells; therefore, metformin has been routinely used as an antidiabetic drug. Metformin can also increase insulin-mediated glucose utilization and increase glucose uptake by noninsulin-dependent tissues. Research on mouse models shows that pharmacological effects have therapeutic potential for tumors. Metformin can play an antitumor role in a low-sugar microenvironment but has no effect in a high-sugar microenvironment (Elgendy et al. 2019). Another AMPK agonist, 5-aminoimidazole-4-carboxamide ribonucleoside (AICAR), can reduce T cell proliferation and inhibit the secretion of proinflammatory cytokines, shows good inflammatory inhibitory properties, and can be used for the treatment of autoimmune diseases (Jung et al. 2004).

Many autoimmune diseases, such as Crohn's disease, multiple sclerosis, systemic lupus erythematosus, and rheumatoid arthritis, are caused by overactivation of the immune system. This process of immune cell activation is accompanied by altered energy sources and metabolic pathways, with T cells typically consuming more glucose and glutamine through glycolysis. Inhibiting T cell glycolysis may be a promising approach to treating autoimmune diseases. Recent work has identified dimethyl fumarate, a derivative of the Krebs cycle metabolite fumaric acid, for the treatment of human autoimmune diseases such as psoriasis and multiple sclerosis (MS), producing anti-inflammatory effects by inhibiting aerobic glycolysis (Kornberg et al. 2018). In summary, targeting glycolysis to regulate T cell metabolism has important therapeutic potential in the therapeutic intervention of immune diseases such as autoimmune diseases and transplant rejection reactions.

T cell metabolism also plays an important role in fighting viral infections. Through metabolomic characterization, investigators confirmed a strong correlation between aerobic glycolytic pathway reprogramming and reduced mitochondrial function, oxidative stress, and HIV-related immune function parameters in T cells, suggesting that metabolic differences could be used as biomarkers and therapeutic targets for rapid screening of HIV infection (Tarancón-Diez et al. 2019). In a study of clinical treatment options for SARS-CoV-2, some investigators analyzed the differences in protein levels in infected cells at different time points and found that the regulatory protein changes were mainly associated with RNA shearing as well as metabolic processes. Further studies confirmed that pladienolide B (an inhibitor of splicing factor SF3B1) and 2-DG (an inhibitor of hexokinase) effectively inhibited SARS-CoV-2 replication in Caco-2 cells, revealing that glycolysis may also be a potential therapeutic target for SARS-CoV-2 infection (Bojkova et al. 2020).

2.3.2 Targeting Amino Acids

After the activation of T cells, the expression of amino acid transporters and the uptake of various amino acids increase, which is necessary for biosynthesis during the proliferation of T cells (Nakaya et al. 2014; Sinclair et al. 2013). The increased expression of amino acid transporters is mediated by the transcription factor nuclear factor of activated T cells (NFAT) and the metabolic regulator Myc (Wang et al. 2011). Myc is a transcription factor that regulates various genes involved in proliferation, metabolism, apoptosis, senescence, and differentiation and is essential for the expansion of T cells (Dang 1999). The expression of Myc increases during T cell activation, and Myc promotes glucose transporter 1, hexokinase 2, pyruvate kinase, and lactate dehydrogenase expression to induce glycolysis and glutamine metabolism. Using the small molecule, Myc inhibitor 10058-F4 can inhibit the proliferation of activated T cells, proving that Myc's regulation of glutamine metabolism plays a key role in promoting an effective T cell response, and drugs targeting glutamine metabolism can be used to treat inflammation caused by excessive immune responses (Madapura et al. 2016). 6-Diazo-5-oxo-L-norleucine (DON) is a structural analog of glutamine. This small molecule can very broadly and effectively inhibit the activity of enzymes that use glutamine as a substrate. In mouse models of cerebral malaria and viral encephalitis, studies have shown that DON can effectively inhibit glutamine-dependent metabolism, suppress severe inflammation, and promote survival (Gordon et al. 2015; Manivannan et al. 2016). Acetate plays an important role in maintaining the activity of CD8+ T cells. Studies have found that T cells can also use acetate as a carbon source to supply the TCA cycle after glutamine metabolism is inhibited. Acetate may also become one of the potential targets of immunotherapy (Balmer et al. 2016).

2.3.3 Targeting Fatty Acids and Cholesterol

Studies have shown that in the hyperlipidemia microenvironment, T cell proliferation in the peripheral circulation of obese diabetic Wistar fatty rats is significantly reduced, and their immune response function is also reduced due to the reduction of immune cells. Compared with the wild-type control group, obese diabetic Wistar fatty rats are more likely to be infected by pathogens (Tanaka et al. 2000). The fatty acids contained in the blood of human hyperlipidemia patients are mostly saturated fatty acids, which are also involved in the regulation of cellular energy metabolism and affect T cell metabolism by changing the microenvironment. Excessive blood lipids in the microenvironment can change the content of lipids in T cells. After excessive lipids are taken into cells, lipids may be stored in intracellular lipid droplets or shuttled to the cell membrane to promote the formation of membrane rafts, thereby affecting the immune process.

Hypercholesterolemia is mainly manifested by increased low-density lipoprotein in the blood, which may promote T cell proliferation by providing components for membrane synthesis and promoting T cell-mediated autoimmunity. Immune activation caused by hyperlipidemia and hypercholesterolemia may lead to a wide range of inflammatory autoimmune diseases, including atherosclerosis and systemic lupus erythematosus (Monica et al. 2018). Therefore, inhibiting T cell metabolism in patients with hyperlipidemia and hypercholesterolemia may help to treat T cell-mediated autoimmune diseases. Clinically, mTOR inhibitors such as sirolimus (rapamycin) and everolimus can be used to downregulate mTOR signaling to inhibit fatty acid synthesis in effector T cells and promote the generation of regulatory T cells to stop the development of atherosclerosis. Statins such as simvastatin, simvastatin, lovastatin, and atorvastatin can also be used to inhibit the synthesis of cholesterol to produce a therapeutic effect. In obese patients, visceral adipose tissue (VAT) contains disproportionately low amounts of regulatory T cells that may cause low-grade inflammation. Rapamycin can also be used to suppress obesity-related VAT inflammation and may be beneficial in the preventive treatment of obesity-related diseases (Lam et al. 2011). Studies have shown that effector T cells activated during graft-versus-host disease (GVHD) mainly rely on fatty acids as an energy source to support their activation in vivo. In this process, using Etormus to inhibit fatty acid oxidation can effectively suppress the allogeneic T cell response, and fatty acid metabolism may become the therapeutic target of GVHD (Byersdorfer et al. 2013). As a key component of cell membrane lipids, cholesterol also plays an important role in T cell signaling and function. Researchers found that inhibiting the cholesterol esterase ACAT1 increased cholesterol levels in the plasma membrane of CD8+ T cells and enhanced their antitumor activity. A small molecule inhibitor of ACAT1, avasimibe, has been used to treat cancer in mice and has shown promising antitumor effects (Yang et al. 2016). However, a series of side effects may occur during long-term treatment with these compounds. Long-term rapamycin treatment may cause dyslipidemia, and long-term metformin treatment may cause gastrointestinal diseases. Ongoing research includes in vitro treatment of T cells isolated from peripheral blood with the proposed compounds, followed by adoptive transfer, or using nanoparticles to transport drugs targeting T cells. These studies may lead to more T cell-specific method to avoid potential off-target effects.

2.3.4 Targeting Mitochondria

Mitochondria are ubiquitous in eukaryotic cells and are mainly composed of proteins, lipids, and a small amount of DNA. They synthesize ATP through aerobic respiration to provide energy to the cells. The basic structure of mitochondria can be divided into four parts: mitochondrial matrix, mitochondrial inner membrane, mitochondrial membrane gap, and mitochondrial outer membrane from inside to outside. Changes in the mitochondrial physiological state can cause metabolic disorders in immune cells and may lead to mental illness (Wakabayashi et al. 2015; Filiano et al. 2016). Metabolic processes can occur in the cytoplasm or mitochondria. Mitochondria are an important hub of metabolism, and their key role is to generate large amounts of ATP through the electron transport chain (ETC) on the inner membrane of the mitochondria. When the energy requirements change, mitochondrial morphology also changes. Activated effector T cells possess divided mitochondria, while naive T cells and memory T cells maintain their mitochondria in a confluent state. The mechanism of mitochondrial proliferation is generally considered to have two signaling pathways: the AMPK pathway and the SIRT1 pathway. The former is activated by AMP, and the latter is activated by NAD. Among them, AMPK activation leads to mTOR repression. Studies have confirmed that the structure of mitochondria is essential for regulating the metabolism of immune cells. Mitochondrial dynamics regulate T cell fate through metabolic reprogramming. After activation from the resting state, the expression of various mitochondrial-related genes in T cells changes, and the mitochondria change from the fusion state to the division state. The mitochondrial fusion protein optic atrophy 1 is necessary for memory T cells but not effector T cells. Mitochondria play an important role in the regulation of T cell metabolism through changes in the fusion/division state. Regulating the mitochondrial morphology of immune cells can enhance immune cells' ability to recognize and destroy tumor cells (Buck et al. 2016; O'Sullivan et al. 2014).

Metformin is an inhibitor of ETC complex I, which can inhibit mitochondrial activity, thereby reducing cellular energy and restricting the transport of molecules through the nuclear pore. During this process, mTORC1 is inhibited, resulting in the activation of acyl-CoA dehydrogenase family member-10 (ACAD10). Treating nematodes with metformin can inhibit mitochondrial activity in this animal, slow its growth and prolong its life (Wu et al. 2016). The combination of metformin and 2-DG can effectively inhibit allograft rejection by inhibiting mitochondrial activity and glycolysis, and the inhibition of graft-versus-host disease (GVHD) by etormus is also related to its function of inhibiting mitochondrial metabolism (Byersdorfer et al. 2013). Long-term surviving memory T cells rely on oxidative phosphorylation and fatty acid oxidation to provide energy. This metabolic process relies on normal mitochondrial activity in T cells. Mitochondrial dysfunction will prevent memory T cells from being produced (van der Windt et al. 2013; Pearce et al. 2013), resulting in a weak immune response to tumors and chronic viral infection (Bengsch et al. 2016). Peroxisome proliferator-activated receptor-γ coactivator-1α (PGC1α) is a nuclear transcription coactivator discovered in recent years that can promote mitochondrial biogenesis and maintain mitochondrial function. Increased expression of PGC1α can lead to increased mitochondrial biogenesis, enhance the T cell response, and lead to a stronger antitumor response (Scharping et al. 2016). The morphology of mitochondria regulates T cell metabolism and thus affects the ability of T cells to fight cancerous cells. Mitochondria are important hubs of cellular metabolic activity and are responsible for energy supply. Studies have also found that activated effector T cells possess dividing mitochondria, while memory T cells maintain their mitochondria as fused networks. Forcing mitochondrial fusion of effector T cells by deleting mitochondrial outer membrane proteins Mfn1/2 can cause these treated cells to develop the characteristics of memory T cells, gaining a longer potential lifespan and greater immunity to tumor cells (Buck et al. 2016). Furthermore, a variety of technologies, including CRISPR gene editing and gene sequencing, can also be used to target mitochondria to regulate the biochemical properties and behavior of T cells (Bailis et al. 2019). In summary, the regulation of mitochondria in T cells may become a target for immunotherapy in the future.

2.3.5 Targeting mTOR to Regulate T Cell Metabolism

mTOR can regulate the expression of GLUT1 and enzymes required for glycolysis through hypoxia-inducible factor 1α (HIF1α) and indirectly regulate metabolism by regulating the YY1-PGC1α complex and its downstream target S6 kinase. mTOR is also involved in the effector functions of other metabolic regulators, such as Myc and SREBP. mTOR activation promotes glycolysis, fatty acid synthesis, and mitochondrial biogenesis. The upstream and downstream mTOR signaling pathways provide potential therapeutic targets for metabolic regulation (Cunningham et al. 2007; Pollizzi and Powell 2015). There are at least two functional polyprotein complex forms of mTOR, mTORC1 and mTORC2, which can play different biological roles. The differentiation and proliferation of T cells requires the integration of multiple, sometimes opposite, cytokine signals. The mTOR signaling pathway plays a decisive role in the fate of T cells. mTOR-deficient $CD4^+$ T cells can be activated and produce IL-2 after stimulation, but these cells cannot differentiate into Th1, Th2, or Th17 cells. The reason for the inability to differentiate may be related to the reduction in STAT transcription factors (Delgoffe et al. 2009). Activation of mTORC1 promotes the production of $CD8^+$ effector T cells with high glycolysis. $CD8^+$ T cells lacking mTORC1 cannot differentiate into effector T cells after activation, exhibiting characteristics similar to memory T cells. Inhibition of mTORC2 activity leads to T cell metabolic reprogramming and promotes the differentiation of $CD8^+$ T cells into memory T cells (Pollizzi et al. 2015). Based on these results, immune function can be adjusted by controlling the differentiation of T cells with mTORC1 or mTORC2 as the target.

TH17 cells, a subpopulation of CD4+ T cells, are highly plastic and can differentiate into Treg cells or TH1-like cells in response to stimulation by characteristic cytokines or antigens. In the mouse autoimmune disease model (EAE model), TH17 cells can be divided into CD27+ T cells and CD27− T cells. CD27+ T cells have memory cells properties and low levels of metabolic activity, while CD27− T cells behave as effector cells and have high levels of metabolic activity. Meanwhile, the transcription factor TCF-1 was highly expressed in CD27+ T cells, and T-bet was highly expressed in CD27− T cells. It was demonstrated that knockdown of Raptor in TH17 cells resulted in a blockade of mTORC1 signaling, which blocked both the differentiation of TH17 cells into TH1-like cells and the induction of autoimmune neuroinflammation. Mechanistically, mTORC1 signaling and cellular metabolism affect the differential expression of TCF-1 and T-bet. Thus, mTORC1 signaling plays a central role in regulating TH17 cell fate decisions by integrating metabolic and transcriptional programs, confirming that metabolism can control helper T cell plasticity (Karmus et al. 2019). As a specific inhibitor of mTOR, rapamycin can block the mTOR signaling pathway, regulate the differentiation, development, proliferation, and function of T cells, and have an effective immunosuppressive effect. In addition to the metabolic regulation and anti-inflammatory effects on T cells in various autoimmune disease models, metformin may also induce T cell autophagy by inhibiting mTOR, which may be related to the PP2A-GSK3β-MCL-1 pathway (Elgendy et al. 2019). In summary, the specific regulation of mTOR is a promising method to control T cell metabolism.

2.3.6 Target Protein Ubiquitination

Ubiquitin is a very conserved polypeptide consisting of 76 amino acid residues that can be covalently linked to one or more lysine residues on intracellular target proteins catalyzed by a series of enzymatic reactions, such as E1, E2, and E3 enzymes, causing the substrate protein to be recognized and degraded by the proteasome. The ubiquitin proteins themselves also contain seven lysine residues, so they can also be linked to each other through these sites to form polyubiquitin protein chains (Hershko et al. 1983). A growing body of evidence suggests that ubiquitination and its reverse process, deubiquitination, play a key role in the regulation of innate and adaptive immune responses by modulating the function of different cell types in the immune system, thus affecting many major human diseases, such as autoimmune diseases, infectious diseases, and malignant tumors.

The c-Myc protein is a very important transcription factor involved in the regulation of cell growth, differentiation, and apoptosis. Disturbances in its metabolism are often associated with abnormal cell differentiation and can even lead to cancers, such as T cell lymphoma. Although c-Myc accumulates in T cell lymphoma, its transcriptional levels are not overtly abnormal, suggesting that it is regulated by post-translational modifications of the protein. In T lymphocytes, the $Ca2^{+}$/calmodulin-dependent protein kinase II (CAMKII) signaling pathway plays a very important role, and it has been found that CAMKII kinase inhibits the ubiquitin degradation of the c-Myc protein at serine 62 (S62) by directly phosphorylating it, leading to c-Myc accumulation in patients. If its activity is inhibited by the CAMKII inhibitor berbamine, it can significantly inhibit the proliferation of T cell lymphoma in mice, suggesting its potential clinical application (Gu et al. 2017).

T cell receptor (TCR) signal intensity is a key determinant of the direction of T cell differentiation, and ubiquitination also modulates the onset of allergic diseases by altering T cell receptor signal intensity and the direction of T cell differentiation. Mice that knock down the ubiquitin ligase genes Itch and WWP2 in their T cells differentiate more TH2 cells and spontaneously develop autoimmune diseases and lung inflammation. It has been shown that Itch and WWP2 synergistically regulate the strength of the TCR signal by forming a complex and catalyzing the ubiquitination of the dephosphorylated protease SHP1, thereby attenuating its binding to the tyrosine kinase Lck, which ultimately regulates the TCR proximal signal and influences the direction of T cell differentiation (Daisuke et al. 2018). Upon binding of the TCR receptor to exogenous peptide-MHC molecules, the TCR complex initiates a tyrosine phosphorylation cascade that activates serine-threonine kinase PKCθ. PKCθ kinase induces the formation of a CBM complex containing CARMA1, BCL10, and MALT1. The MALT1 protein recruits TRAF6 protein to the CBM complex and activates its E3 ligase activity by promoting the oligomerization of TRAF6 protein. Several TRAF6 target proteins have been identified, and these target proteins have the potential to activate the TAK1-IKK complex following polyubiquitination pathway modifications in response to TCR signaling stimulation. The activation of IKK and MAPK in mouse thymus and peripheral T cells is severely attenuated after TCR stimulation by knocking out TAK1 or UBC13 (Bhoj and Chen 2009). These studies elucidate the role of protein ubiquitination in the regulation of TCR signal intensity and T cell differentiation direction and provide new ideas and targets for the treatment of allergic and autoimmune diseases.

2.3.7 Target Autophagy

Autophagy is the process of phagocytosis of one's own cytoplasmic proteins or organelles, thereby fulfilling the cell's own metabolic needs and the renewal of certain organelles. The induction of autophagy in immune cells is often associated with anti-inflammatory and dysfunctional autophagy that disrupts Treg function. Once Tregs are activated and start working, the autophagy machinery degrades unneeded molecules. Atg7 or Atg5 is essential for Treg autophagy. Atg7-deficient Tregs show impaired cell survival and stability, with a corresponding increase in apoptotic markers and a decrease in FoxP3 stability. In addition, Atg7-deficient Tregs exhibited increased glycolytic metabolism and produced more IFNγ and IL-17 than WT Tregs, suggesting that autophagy is required for immune homeostasis (Tian et al. 2017). Autophagy may be related to the degradation of intracellular lipid droplets and thus increase Treg survival. Therefore, stimulation of cellular autophagy with metformin or rapamycin may be an effective way to increase and stabilize Tregs to reduce autoimmunity (Quan-Gang et al. 2016; Park et al. 2016). Scientists also found that Atg7- or Atg5-knockout mice display some important features of regulatory T cell dysfunction, including inflammation and autoimmune disease. Because their immune systems were activated, these mice could also clear tumors from their bodies more easily (Wei et al. 2016). Understanding the role of autophagy in regulatory T cells has the potential to lead to anticancer immunotherapy that is synergistic with other strategies.

2.3.8 Special Microenvironment

The tumor microenvironment refers to the cells and tissues surrounding the tumor and is closely related to the development of cancer. Several features of the tumor microenvironment, such as low glucose levels, low pH, hypoxia, and production of inhibitory metabolites, can lead to immunosuppression and limit antitumor immune responses. Although current cancer therapies primarily target tumor cells, there is a growing awareness of the importance of the tumor microenvironment. Abnormal tissues and blood vessels restrict gas exchange within the tumor, creating an oxygen-poor tumor environment. Lack of oxygen enhances the glycolysis ability of tumor cells and increases the growth of lactic acid. In contrast to the mechanism of immune activation, the effect of the microenvironment on T cells is nonantigen specific. The large amount of lactic acid in the tumor microenvironment affects the immune system's recognition of and response to tumor cells; nutritional deprivation makes various cells compete with each other to meet their own needs, while immune cells are less adaptable to nutritional deprivation (Liyanage et al. 2002). Studies have shown that glucose consumption by tumors metabolically restricts T cells, inhibiting their mTOR activity, glycolytic capacity, and IFN-γ production, allowing further tumor development. Thus, clinically used CTLA-4, PD-1, and PD-L1 test point blocking antibodies can alter metabolic homeostasis in tumors, restore glucose levels in the tumor microenvironment, and promote T cell glycolysis and IFN-γ production (Chang et al. 2015).

Unlike the mechanism of immune activation, the effect of the microenvironment on T cells is nonantigen-specific, which provides ideas for T cell-targeted therapeutic regimens. $CD8^+$ T cells are the main effector cells executing antigen-specific killing of tumor cells, mainly relying on glycolytic pathways to maintain their expansion and function. Acylglycerol kinase (AGK) in CD8+ T cells was found to inhibit PTEN activity by mediating PTEN phosphorylation, thereby activating the PI3K/mTOR signaling pathway induced by TCR signaling and enhancing glycolytic metabolism to promote the antitumor activity of $CD8^+$ T cells, which provides new ideas to improve the therapeutic efficacy of monoclonal antibodies in tumor immune monitoring sites (Hu et al. 2019).

Reactive oxygen species (ROS) are a class of molecules produced by the body in response to oxidative stress that are expressed at abnormally high levels at the tumor site and are closely associated with tumor immune tolerance. As a subpopulation of T cells with immunosuppressive functions, regulatory T cells (Tregs) play an important role in tumor immune tolerance. It was found that TCR signal-induced ROS specifically restricts the protein degradation of the deubiquitination enzyme SENP3, which in turn mediates the deubiquitination modification of the transcription factor BACH2 and its nuclear localization, thereby maintaining BACH2 to inhibit the expression of effector cytokines such as IFN-γ and IL-17 and control the immunosuppressive activity of regulatory T cells. Targeting SENP3 can effectively break the regulatory T cell-mediated tumor immune tolerance mechanism and enhance the function of T cells to kill tumor cells (Yu et al. 2018). Recent studies on the metabolic restriction of T cells in the tumor microenvironment have provided important information for antitumor strategies in immunotherapy. Therapeutic approaches targeting metabolic restriction have shown promise as comprehensive cancer treatment options.

The intestine is also a special microenvironment in the body with complex changes in flora structure and metabolic activity. A protein called integrin β7 (integrin β7) is known to direct immune cells into this environment (Neri et al. 2011). A new study found that β7-positive T cells present in the small intestinal epithelium can reduce the levels of GLP-1 protein to inhibit insulin secretion and glucose uptake, ultimately slowing systemic metabolism and prompting the conversion of food into fat storage rather than being rapidly converted to energy. Further studies of such β7-positive small intestinal intraepithelial T cells may be beneficial in the treatment of diet metabolism-related diseases such as obesity, diabetes, hypertension, and atherosclerosis (He et al. 2019).

Common inflammatory bowel diseases include Crohn's disease, which affects the colon tissue, and ulcerative colitis, which affects the entire digestive tract. The development of inflammatory bowel disease may be associated with an abnormal immune response of the body to multiple antigenic stimuli in the intestine. Earlier studies found that antigens in the intestinal colonization flora, particularly flagellins expressed by bacteria of the Lachnospiraceae, are the dominant antigens in the heterogeneous immune response to Crohn's disease. The phenomenon of increased memory $CD4^+$ T cells in peripheral blood against these proteins correlates significantly with the complex disease course of the patient. Initial T cells and memory T cells have low levels of metabolism, but upon encountering antigen activation, the cells rapidly upregulate mTOR activity and initiate aerobic glycolysis-based metabolism to meet the enormous energy demands of cell differentiation and proliferation (Lennard-Jones 2009). A study found that the use of the metabolic inhibitory drug rapamycin inhibited the mTOR complex that regulates the metabolism of these T cells, while the use of metformin activated AMPK kinase upstream of mTOR and thus inhibited mTOR. Rapamycin also promoted the induction of regulatory T cells and ultimately apoptosis and deactivation of these entero-inflammatory $CD4^+$ T cells, achieving the goal of preventing the recurrence of Crohn's disease (Zhao et al. 2020) (Fig. 2.3).

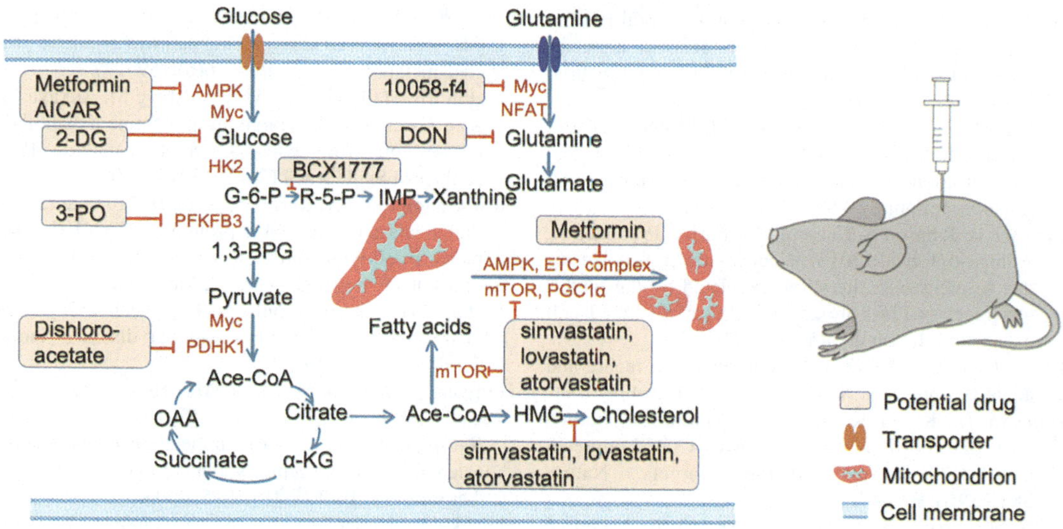

Fig. 2.3 Some potential immunotherapies for metabolism-related diseases. Using some potential drugs to affect the metabolic processes of T cells in different ways can regulate the proportion of T cell subpopulations and immune responsiveness in animal models and clinical patients. This figure illustrates some potential immunotherapies for metabolism-related diseases by targeting key enzymes or signaling pathways involved in T cell metabolism. The potential drug targets are indicated in red font

2.4 Conclusions

T cells are responsible for the body's defense and homeostasis, and their cellular metabolic state plays an important role in the regulation of specific organs and tissues, as well as in the differentiation, development, and function of different immune cells. The importance of cell metabolism in the fate of T cells is unquestionable. Recent studies have shown that metabolic reprogramming plays an important role in T cell fate determination. This article discusses the effects of different metabolic pathways on T cell proliferation, differentiation, and function and summarizes the metabolic changes that occur after T cell activation and the mechanisms that regulate these metabolic changes. Different T cell subpopulations have different metabolic reprogramming pathways, so specific T cell subpopulations can be selectively enhanced or inhibited by using drugs to target different metabolic pathways.

Therefore, exploring the chemical metabolomics of T lymphocytes in the microenvironment of the organism and researching the metabolic pathways that may be regulated may provide new drug treatment targets for the treatment of autoimmune diseases related to common metabolic diseases such as diabetes, atherosclerosis, systemic lupus erythematosus, obesity, and cardiovascular disease. After understanding the metabolic mechanism of T lymphocytes, we can design new drugs based on these mechanisms or improve the utilization of existing drugs and reduce side effects by combining multiple drugs and discovering new uses. Effective immunotherapy is expected to be designed by controlling different metabolic processes in T cells for autoimmune diseases, cancer, and transplant rejection.

References

Altman BJ, Stine ZE, Dang CV (2016) From Krebs to clinic: glutamine metabolism to cancer therapy. Nat Rev Cancer 16:619–634

Amersfoort J, Kuiper J (2017) T cell metabolism in metabolic disease-associated autoimmunity. Immunobiology 222:925

Anson RM, Guo Z, Cabo RD, Iyun T, Mattson MP (2003) Intermittent fasting dissociates beneficial effects of dietary restriction on glucose metabolism and neuronal resistance to injury from calorie intake. Proc Natl Acad Sci U S A 100(10):6216–6220

Bailis W, Shyer JA, Zhao J, Canaveras J, Khazal FA, Qu R et al (2019) Distinct modes of mitochondrial metabolism uncouple T cell differentiation and function. Nature 571(7765):403–407

Balmer M, Ma E, Bantug G, Grählert J, Pfister S, Glatter T et al (2016) Memory $CD8^{+}$ T cells require increased concentrations of acetate induced by stress for optimal function. Immunity 44(6):1312

Bengsch B, Johnson A, Kurachi M, Odorizzi P, Pauken K, Attanasio J et al (2016) Bioenergetic insufficiencies due to metabolic alterations regulated by the inhibitory receptor PD-1 are an early driver of $CD8^{+}$ T cell exhaustion. Immunity 45:358–373

Bhoj VG, Chen ZJ (2009) Ubiquitylation in innate and adaptive immunity. Nature 458(7237):430–437

Bojkova D, Klann K, Koch B, Widera M, Münch C (2020) Proteomics of SARS-COV-2-infected host cells reveals therapy targets. Nature 583(7816):469–472

Buck M, O'Sullivan D, Geltink RIK, Curtis J, Pearce E (2016) Mitochondrial dynamics controls T cell fate through metabolic programming. Cell 166(1):63–76

Byersdorfer CA, Tkachev V, Opipari AW, Goodell S, Swanson J, Sandquist S et al (2013) Effector T cells require fatty acid metabolism during murine graft-versus-host disease. Blood 122(18):3230–3237

Chang CH, Qiu J, O'Sullivan D, Buck M, Noguchi T, Curtis J et al (2015) Metabolic competition in the tumor microenvironment is a driver of cancer progression. Cell 162:1229–1241

Cunningham JT, Rodgers JT, Arlow DH, Vazquez F, Puigserver P (2007) mTOR controls mitochondrial oxidative function through a YY1-PGC-1alpha transcriptional complex. Nature 450(7170):736–740

Daisuke A, Hui L, Wen Z et al (2018) The E3 ligases itch and WWP2 cooperate to limit $T_{H}2$ differentiation by enhancing signaling through the TCR. Nat Immunol 19(7):766–775

Dang CV (1999) c-Myc target genes involved in cell growth, apoptosis, and metabolism. Mol Cell Biol 19(1):1–11

Delgoffe GM, Kole TP, Zheng Y, Zarek PE, Matthews KL, Xiao B et al (2009) The mTOR kinase differentially regulates effector and regulatory T cell lineage commitment. Immunity 30(6):832–844

Elgendy M, Cirò M, Hosseini A, Weiszmann J, Mazzarella L, Ferrari E et al (2019) Combination of hypoglycemia and metformin impairs tumor metabolic plasticity and growth by modulating the PP2A-GSK3β-MCL-1 axis. Cancer Cell 35:798

Fan KQ, Li YY, Wang HL, Mao XT, Jin J (2019) Stress-induced metabolic disorder in peripheral CD4+ T cells leads to anxiety-like behavior. Cell 179(4):864–879.e19

Filiano AJ, Xu Y, Tustison NJ, Marsh RL, Baker W, Smirnov I, Overall CC, Gadani SP, Turner SD, Weng Z et al (2016) Unexpected role of interferon-γ in regulating neuronal connectivity and social behaviour. Nature 535:425–429

Frauwirth KA, Riley JL, Harris MH, Parry RV, Thompson CB (2002) The CD28 signaling pathway regulates glucose metabolism. Immunity 16(6):769–777

Gordon EB, Hart GT, Tran TM, Waisberg M, Akkaya M, Kim AS et al (2015) Targeting glutamine metabolism rescues mice from late-stage cerebral malaria. Proc Natl Acad Sci U S A 112(42):13075–13080

Gu Y, Zhang J, Ma X, Kim BW, Wang H, Li J et al (2017) Stabilization of the c-Myc protein by CAMKIIγ promotes T cell lymphoma. Cancer Cell 32(1):115–128

He S, Kahles F, Rattik S, Nairz M, Mcalpine CS, Anzai A et al (2019) Gut intraepithelial T cells calibrate metabolism and accelerate cardiovascular disease. Nature 566(7742):115–119

Herijgers N, Van Eck M, Groot PHE, Hoogerbrugge PM, van Berkel TJC (1997) The effect of bone marrow transplantation on lipoprotein metabolism and atherosclerosis in LDL receptor knockout mice. Arterioscler Thromb Vasc Biol 17(10):1995–2003

Hershko A, Heller H, Elias S, Ciechanover A (1983) Components of ubiquitin-protein ligase system. Resolution, affinity purification, and role in protein breakdown. J Biol Chem 258:8206–8214

Heusch MD, Oldenhove G, Urbain J, Thielemans K, Maliszewski C, Leo O et al (2004) Depending on their maturation state, splenic dendritic cells induce the differentiation of $CD4^{+}$ T lymphocytes into memory and/or effector cells in vivo. Eur J Immunol 34(7):1861

Hsiung Y-C, Liu J-J, Hou Y-C, Yeh C-L, Yeh S-L (2014) Effects of dietary glutamine on the homeostasis of $CD4^{+}$ T cells in mice with dextran sulfate sodium-induced acute colitis. PLoS One 9(1):e84410

Hu Z, Qu G, Yu X, Jiang H, Teng XL, Ding L et al (2019) Acylglycerol kinase maintains metabolic state and immune responses of CD8+ T cells. Cell Metab 30(2):290–302.e5

Jung JE, Lee JW, Ha JH, Kim SS, Kang IS (2004) 5-Aminoimidazole-4-carboxamide-ribonucleoside enhances oxidative stress-induced apoptosis through activation of nuclear factor-kappaB in mouse Neuro 2a neuroblastoma cells. Neurosci Lett 354(3):197–200

Kao C-H et al (2010) Discrepancy between regional cerebral blood flow and glucose metabolism of the brain in systemic lupus erythematosus patients with normal brain magnetic resonance imaging findings. Arthritis Rheum 42(1):61–68

Karmus PWF, Vogel P, Chi H et al (2019) Metabolic heterogeneity underlies reciprocal fates of TH17 cell sternness and plasticity. Nature 565(7737):101–105

Kornberg MD, Bhargava P, Kim PM, Putluri V, Snowman AM, Putluri N et al (2018) Dimethyl fumarate targets GAPDH and aerobic glycolysis to modulate immunity. Science 360(6387):449–453

Krishnamoorthy RV, Venkataramiah A, Lakshmi GJ, Biesiot P (2016) J Food Sci 1(6):314–315

Lafaille JJ, Nagashima K, Katsuki M, Tonegawa S (1994) High incidence of spontaneous autoimmune encephalomyelitis in immunodeficient anti-myelin basic protein T cell receptor transgenic mice. Cell 78(3):399–408

Lam YY, Janovská A, Mcainch AJ, Belobrajdic DP, Hatzinikolas G, Game P et al (2011) The use of adipose tissue-conditioned media to demonstrate the differential effects of fat depots on insulin-stimulated glucose uptake in a skeletal muscle cell line. Obes Res Clin Pract 5(1):e43–e54

Lennard-Jones JE (2009) Classification of inflammatory bowel disease. Scand J Gastroenterol 24(s170):2–6

Liyanage UK, Moore TT, Joo HG, Tanaka Y, Herrmann V, Doherty G et al (2002) Prevalence of regulatory T cells is increased in peripheral blood and tumor microenvironment of patients with pancreas or breast adenocarcinoma. J Immunol 169(5):2756–2761

Madapura HS, Salamon D, Wiman KG, Lain S, Klein E, Nagy N (2016) cMyc-p53 feedback mechanism regulates the dynamics of T lymphocytes in the immune response. Cell Cycle 15(9):1267–1275

Manivannan S, Baxter VK, Schultz KL, Slusher BS, Griffin DE (2016) Protective effects of glutamine antagonist 6-diazo-5-oxo-l-norleucine in mice with alphavirus encephalomyelitis. J Virol 90:9251–9262

Melnik B, Plewig G (1992) Are disturbances of ω-6-fatty acid metabolism involved in the pathogenesis of atopic dermatitis? Acta dermato-venereologica Supplementum 176:77–85

Middlebrook AJ, Lebsack T, Deluca D (2007) TNF-α mediated modulation of T cell development and exacerbation of in vitro T1DM in fetal thymus organ culture. J Autoimmun 29(2–3):134–145

Monica C, Prokopec KE, Garimella MG, Katrin H, Lisa H, Stark JM et al (2018) Acute loss of apolipoprotein E triggers an autoimmune response that accelerates atherosclerosis. Arterioscl Thromb Vasc Biol 38(8):e145–e158. https://doi.org/10.1161/ATVBAHA.118.310802

Nakaya M, Xiao Y, Zhou X, Chang JH, Chang M, Cheng X et al (2014) Inflammatory T cell responses rely on amino acid transporter ASCT2 facilitation of glutamine uptake and mTORC1 kinase activation. Immunity 40(5):692–705

Neri P, Ren L, Azab AK, Brentnall M, Gratton K, Klimowicz AC et al (2011) Integrin β7-mediated regulation of multiple myeloma cell adhesion, migration, and invasion. Blood 117(23):6202

O'Sullivan D, Van Der Windt GW, Huang SC, Curtis J, Chang CH, Buck M et al (2014) Memory $CD8^+$ T cells use cell-intrinsic lipolysis to support the metabolic programming necessary for development. Immunity 41(1):75–88

Park MJ, Lee SY, Moon SJ, Son HJ, Lee SH, Kim EK et al (2016) Metformin attenuates GVHD via restricting mTOR/STAT3 and promoting AMPK-autophagy for the balance between TH17 and Tregs. Transl Res 173:115–130

Pearce EL, Poffenberger MC, Chang CH, Jones RG (2013) Fueling immunity: insights into metabolism and lymphocyte function. Science 342(6155):210

Pollizzi KN, Powell JD (2015) Regulation of T cells by mTOR: the known knowns and the known unknowns. Trends Immunol 36(1):13–20

Pollizzi KN, Patel CH, Sun IH, Oh MH, Powell JD (2015) mTORC1 and mTORC2 selectively regulate $CD8^+$ T cell differentiation. J Clin Invest 125(5):2090–2108

Quan-Gang C, Ren-Jin C, Hong-Hua Y (2016) Rapamycin regulateing Treg cells differentiation and function via autophagy. Hlongjiang Med J

Roky R, Houti I, Moussamih S, Qotbi S, Aadil N (2004) Physiological and chronobiological changes during ramadan intermittent fasting. Ann Nutr Metab 48(4):296–303

Scharping N, Menk A, Moreci R, Whetstone R, Dadey R, Watkins S et al (2016) The tumor microenvironment represses T cell mitochondrial biogenesis to drive intratumoral T cell metabolic insufficiency and dysfunction. Immunity 45(2):374–388

Shi LZ, Wang R, Huang G, Vogel P, Chi H (2011) HIF1alpha-dependent glycolytic pathway orchestrates a metabolic checkpoint for the differentiation of TH17 and Treg cells. J Exp Med 208(7):1367–1376

Sinclair LV, Rolf J, Emslie E, Shi YB, Taylor PM, Cantrell DA (2013) Corrigendum: control of amino-acid transport by antigen receptors coordinates the metabolic reprogramming essential for T cell differentiation. Nat Immunol 15(1):109

Straaten OV, Bosman RJ, Treskes M, Spoel HJIVD, Zandstra DF (2001) Plasma glutamine depletion and patient outcome in acute ICU admissions. Intensive Care Med 27(1):84–90

Sukumar M, Liu J, Ji Y, Subramanian M, Crompton JG, Yu Z et al (2013) Inhibiting glycolytic metabolism enhances $CD8^+$ T cell memory and antitumor function. J Clin Invest 123(10):4479–4488

Tanaka SI, Isoda F, Kiuchi Y, Ikeda H, Mobbs CV, Yamakawa T (2000) T lymphopenia in genetically obese-diabetic Wistar fatty rats: effects of body weight reduction on T cells. Metabolism 49(10):1261–1266

Tarancón-Diez L, Rodríguez-Gallego E, Rull A, Peraire J, Viladés C, Portilla I et al (2019) Immunometabolism is a key factor for the persistent spontaneous elite control of HIV-1 infection. EBioMedicine 42:86–96

Tian Y, Liang X, Wu Y (2017) The alternation of autophagy/apoptosis in CD4+CD25+Foxp3+ Tregs on the developmental stages of atherosclerosis. Biomed Pharmacother 97:1053–1060

van der Windt GJW, O'Sullivan D, Everts B, Huang CC, Buck MD, Curtis JD et al (2013) CD8 memory T cells have a bioenergetic advantage that underlies their rapid recall ability. Proc Natl Acad Sci 110(35):14336–14341

Wakabayashi C, Numakawa T, Odaka H, Ooshima Y, Kiyama Y, Manabe T, Kunugi H, Iwakura Y (2015) IL-1 receptor-antagonist (IL-1Ra) knockout mice show anxiety-like behavior by aging. Neurosci Lett 599:20–25

Wang R, Dillon C, Shi L, Milasta S, Carter R, Finkelstein D et al (2011) The transcription factor Myc controls metabolic reprogramming upon T lymphocyte activation. Immunity 35(6):871–882

Weber YG, Storch A, Wuttke TV, Brockmann K, Lerche H (2008) Glut1 mutations are a cause of parox-

ysmal exertion-induced dyskinesias and induce hemolytic anemia by a cation leak. J Clin Invest 118(6):2157–2168
Wei J, Long L, Yang K, Guy C, Shrestha S, Chen Z et al (2016) Autophagy enforces functional integrity of regulatory T cells by coupling environmental cues and metabolic homeostasis. Nat Immunol 17:277–295
Wu L, Zhou B, Oshiro-Rapley N, Li M, Paulo JA, Webster CM et al (2016) An ancient, unified mechanism for metformin growth inhibition in C. elegans and cancer. Cell 167(7):1705–1718.e13
Yalcin A, Simmons A, Clem A et al (2007) Nuclear compartmentalization of a key regulator of glycolysis, 6-phosphofructo-2-kinase (PFKFB3). FASEB J 20(s 1–3):156–181
Yang W, Bai Y, Xiong Y, Zhang J, Chen S, Zheng X et al (2016) Potentiating the antitumour response of CD8+ T cells by modulating cholesterol metabolism. Nature 531:651–655
Yu X, Lao Y, Zou Q et al (2018) SENP3 maintains the stability and function of regulatory T cells via BACH2 deSUMOylation. Nat Commun 9(1):3157
Zhao Q, Duck LW, Huang F, Alexander KL, Elson CO (2020) CD4+ T cell activation and concomitant mTOR metabolic inhibition can ablate microbiota-specific memory cells and prevent colitis. Sci Immunol 5(54):eabc6373

3 Ubiquitination of Immune System and Cancer Therapy

Yizhou Du, Huiyuan Zhang, and Hongbo Hu

Abstract

Ubiquitination is a post-translational modification mechanism which regulates a variety of signaling pathways and crucial biological processes. It has long been known that ubiquitination regulates the fundamental cellular processes through the induction of proteasomal degradation of target proteins. Meanwhile, the nondegradative types of polyubiquitination modification have been appreciated as important regulatory machinery by modulating the activity or subcellular localization of key signaling proteins. The function of ubiquitination plays an important role in immune responses, which helps to maintain the stability of the internal environment and to control over protein stability and function and are thus critical for the regulation of both innate and adaptive immunity. Furthermore, ubiquitination also regulates both tumor-suppressing and tumor-promoting pathways in cancer. In this review, we will discuss recent progress regarding how ubiquitination regulates immune responses, focusing on Toll-like receptors signaling in innate immunity, T cell activation, TCR signaling, and tumor immunotherapy.

Keywords

Ubiquitination · Immune response · Immunotherapy · TLR · TCR

3.1 Introduction

Post-translational modifications (PTMs) of proteins extend their function, which allow rapid changes in their behavior in response to stimuli and without the need for protein de novo synthesis (Rittinger and Ikeda 2017). Many post-translational modification mechanisms of proteins have been found and studied, such as phosphorylation, methylation, acetylation, and glycosylation (Yakubu et al. 2019). Besides, another important post-translational modification mechanism for regulating protein function is ubiquitination, which covalent attachment of the ubiquitin to substrate proteins (Liu et al. 2016). Ubiquitin is a highly conserved protein composed of 76 amino acids (8.6 kDa) (Kulathu and

Y. Du · H. Zhang
Center for Immunology and Hematology, Department of Biotherapy, Cancer Center and State Key Laboratory of Biotherapy, West China Hospital, Sichuan University, Chengdu, China

H. Hu (✉)
Center for Immunology and Hematology, State Key Laboratory of Biotherapy and Cancer Center, West China Hospital, Sichuan University, Chengdu, China

Chongqing International Institute for Immunology, Chongqing, China
e-mail: hongbohu@scu.edu.cn

H. Hu, X. Fu (eds.), *Ubiquitination in Immune System and Immune Response*, Advances in Experimental Medicine and Biology 1466, https://doi.org/10.1007/978-981-97-7288-9_3

Komander 2012). It is transferred to the surface of the substrate through the continuous E1, E2, and E3 enzyme cascade reaction, involving ubiquitin-activating enzyme E1, ubiquitin-conjugating enzyme E2, and ubiquitin-ligating E3 (Davis and Gack 2015). The ubiquitinated substrates subsequently participate in different cellular processes, whose fate is mainly determined by the covalent attachment of different ubiquitin chains (Swatek and Komander 2016). The most studied one is the K48-linked polyubiquitination, which usually targets proteins for 26s proteasomal degradation (Grice and Nathan 2016). But ubiquitination does not necessarily target proteins for their degradation in proteasomes, other lysine residues (such as Lys6, Lys11, Lys27, Lys29, Lys33, and Lys63) also can be ubiquitinated (Chen and Sun 2009). These polyubiquitinated proteins are involved in different fates and the regulation of different cellular processes subsequently (Buneeva and Medvedev 2016).

The immune system is a very diverse system of the host that evolved during evolution to cope with various pathogens presented in the vicinity of environmental surroundings. Immune responses are physiological and biochemical processes in which organisms recognize and respond to "non-self" antigens, including innate immune response and adaptive immune responses (Borad and Ward 2010). The innate immune system provides the first line of defense against foreign microbes and particulate materials, while adaptive immunity can provide a repertoire of recognition for both "self" and "non-self" antigens (Bonilla and Oettgen 2010). T and B lymphocytes play a central role in protecting the human body from infectious pathogens, but occasionally they can escape immune tolerance, become activated, and induce autoimmune diseases (Sujashvili 2016). The complexity, versatility, and specificity of an immune response are accomplished not only by the variety of different cell types and receptors, but also by post-translational modifications of proteins involved in immune signaling pathways (Zinngrebe et al. 2014). Cancer immunotherapy has been found pivotal in changing the prognoses of malignant tumors by causing durable immune responses, which aim to design effective treatments to improve the specificity and strength of the immune system against cancer (Cha et al. 2018). Recent studies have shown that dysregulation of molecules involved in the ubiquitination-proteasome system (UPS) can directly initiate and promote the progress of various human diseases, such as cancer (Mansour 2018). The ubiquitination is crucial for a plethora of physiological processes, including tumor immunotherapy.

3.2 Ubiquitination of Innate Immune Responses

3.2.1 Innate Immune Responses and PRRs

The innate immune system provides the first line of defense against foreign microbes and particulate materials. The human innate immune system mainly comprises innate immune cells and its humoral components that are secreted by innate immune cells, which provide critical mechanisms for the rapid sensing and elimination of pathogens (Vijay 2018). Furthermore, innate immunity is also important for the activation of adaptive immunity (Beutler 2004). The innate immune recognition is based on pattern recognition receptors (PRRs), such as Toll-like receptors (TLRs), RIG-I-like receptors (RLRs), C-type lectin receptors (CLRs), and NOD-like receptors (NLRs). These PRRs recognize a wide range of pathogen-associated molecular patterns (PAMPs) of the invading pathogens (Beutler 2004; Takeuchi and Akira 2010). PRRs are responsible for sensing the presence of microorganisms. Upon recognition of their ligand species, PRRs specifically recruit intracellular adaptor proteins to initiate signaling pathways culminating in the activation of nuclear factor-kappa B (NF-κB), mitogen-activated protein (MAP) kinases, and interferon regulatory factors (IRFs) that control the transcription of genes (Zhu et al. 2018). There is evidence showing that PRRs are also responsible for recognizing endogenous molecules released from damaged cells, termed damage-associated molecular patterns (DAMPs) (Takeuchi and Akira 2010). These sensors constitute a network of the innate immune system, and regulate various physiological processes via immune responses (Thaiss et al. 2016).

3.2.2 Toll-Like Receptors (TLRs)

Toll-like receptors, one of the best-characterized PRRs family, are responsible for sensing invading pathogens outside of the cell and in intracellular endosomes and lysosomes. Various organisms express the TLRs family, especially mammals, and 13 types of TLRs have been reported combinedly in mice and humans (Moresco et al. 2011). TLR1 to 9 are conserved in both the mouse and human (Takeda and Akira 2005). However, the distribution of active TLRs is different. TLR1, TLR2, TLR4, TLR5, TLR6, and TLR10 are expressed on the cell surface, whereas TLR3, TLR7, TLR8, TLR9, TLR11, TLR12, and TLR13 are expressed within intracellular vesicles or endosomes (Moresco et al. 2011; Takeda and Akira 2005). TLR signaling plays an essential role in the innate immune response at the early stage of infection, occurring in various immune cells. TLRs are involved in the recognition of pathogens including bacteria, viruses, fungi, and parasites in animals as soon as they come in contact with the host (Vijay 2018). TLRs are type I transmembrane proteins with ectodomains containing leucine-rich repeats that mediate the recognition of PAMPs derived from pathogens, and DAMPs from dead or injured cells (Satoh and Akira 2016). Each TLR has a toll-interleukin-1 receptor (TIR) domain, and forms a signaling platform to follow ligand bind and contain adaptors (Zhou et al. 2017). After occurring the recognition of pathogen-associated molecular patterns, TLRs initiate signaling cascades that lead to activation of transcription factors such as NF-κB and IRFs, which culminate in the transcription of downstream effector genes and induction of innate immune responses and inflammatory responses (Chen and Sun 2009).

3.2.3 Ubiquitination of TLRs Signaling

The signal transduction from the recognition of TLRs to induce the expression of numerous genes which are ultimately required for the immune response is a cascade occurrence, and the process requires the participation of multiple sensors. Upon recognition of PAMPs, TLRs induce downstream signaling pathways. TLRs recruit a specific set of adaptor molecules that harbor the TIR domain, such as MyD88 and TRIF, which are essential adaptor proteins required for innate immune responses mediated by TLRs, thereafter initiate downstream signaling events (Wu et al. 2019). These adapters initiate two signaling pathways for the induction of proinflammatory cytokines and typeI IFNs, namely the myeloid differentiation primary response 88 (MyD88) dependent and TIR domain-containing adapter inducing interferon-β (TRIF) dependent (Takeda and Akira 2005; Kim et al. 2018; Kawai and Akira 2011). The signaling pathway through MyD88 is utilized by all TLRs (except for TLR3) family and transmits signals culminating in NF-kB and MAP kinase activation (Kawai and Akira 2011). TRIF is a TIR domain-containing adaptor, which is essential for TLR3 and TLR4 mediated signaling pathways (Hu et al. 2015). TLR4 mediates both MyD88-dependent and TRIF-dependent signal transduction pathways, leading to the activation of NF-κB and IRF3, then mediates the induction of inflammatory cytokine productions and type I IFNs (Barton and Kagan 2009).

TLR1, TLR2, and TLR4 recruit MyD88 via the TIR domain-containing adaptor protein (TIRAP) as an additional adaptor; meanwhile, the TRIF-related adaptor molecule (TRAM) acts as a bridge between TLR4 and TRIF (Buneeva and Medvedev 2016). Although the signaling downstream of MyD88 and TRIF differs, the outcome of each pathway is conceptually similar, they lead to the induction of cytokine genes together (Barton and Kagan 2009). After binding to the TLR1, TLR2, and TLR4, TIRAP recruits MyD88 and forms a complex of IRAKs and TRAF6 subsequently (Takeuchi and Akira 2010).

TRAF6 acts as an E3 ubiquitin ligase forming an unconjugated free K63-linked polyubiquitin chain and undergoes self-ubiquitination (Kobayashi et al. 2004). TLR3 and TLR4 recruit TRIF by TRAM and activate TRIF (Takeda and Akira 2005). Then TRIF activates the downstream RIP1 via mediating K63-linked polyubiquitin by Peli1, an E3 ligase that is activated via its phosphorylation by TBK1 or IKKε (Chang et al. 2009). TBK1 or IKKε also activates IRF3, then activates IRF3 into the nucleus resulting in the induction of type I IFNs (tenOever et al. 2004). The activated TRAF6 and RIP1 recruit and activate a complex composed of TAK1, TAB1, and TAB2/3, which is a ubiquitin-blinding protein that is ubiquitinated by the unconjugated free K63-linked polyubiquitin chain (Morlon et al. 2005). The activated TAK1, TAB1, and TAB2/3 complex result in the activation of an IKK complex composed of IKKα, IKKβ, and NEMO (Sebban et al. 2006). Mechanistically, the IKKβ is phosphorylated by the TAK1, the ubiquitin-blinding protein NEMO is ubiquitinated by the TAB2/3, and the activated IKKα of the activated IKK complex promotes activation of NF-κB by mediating phosphorylation and K48-linked ubiquitination of IκB, then the ubiquitinated IκB is degraded, and the freed NF-κB translocate into the nucleus where it drives expression of proinflammatory cytokines (Sebban et al. 2006; Walsh et al. 2008). Simultaneously, the activated TAK1 activates MAP kinase (MAPK) cascades leading to the activation of transcription factor AP-1, which is also critical for the induction of proinflammatory cytokines (Walsh et al. 2008). These type I IFNs and proinflammatory cytokines were generated through these signaling pathways to mediate the innate immune responses.

3.3 Ubiquitination of Adaptive Immune Responses

3.3.1 Adaptive Immune Responses and T Cell

Adaptive immune response refers to the whole process in which antigen-specific T/B lymphocytes are activated, proliferate, and differentiate into effector cells and produce a series of biological effects. Another principal feature of adaptive immunity is the generation of immunological memory. In subsequent encounters with the same pathogen, the memory cells are quickly activated to yield a more rapid and robust protective response (Bonilla and Oettgen 2010). The T cells are major components of the adaptive immune system. Naive T cells migrate from the fetal liver and adult bone marrow to the thymus where they undergo differentiation, ultimately emerge as mature T cells and migrate to the periphery and circulate between the bloodstream and lymphoid organs. T cells respond to foreign antigens presented by antigen-presenting cells (APCs) of the innate immune system, and activated T cells then initiate adaptive immune responses (Sujashvili 2016). Activation of T cells occurs upon ligation of T cell receptors (TCRs) by major histocompatibility complex (MHC) molecules expressed on APCs which present peptides from either endogenously encoded self-molecules or encoded pathogen molecules (Brownlie and Zamoyska 2013). The T cell recognizes both peptides associated with MHC molecules on APCs and other molecules, followed by the joint formation of a complex structure called the immunological synapse, and T cell activation requires ligation of both the TCR and other costimulatory molecules, such as CD3 and CD28 (Smith-Garvin et al. 2009). The signaling mechanisms by TCR mediation are very important to our understanding of immunity and harnessing T cells therapeutically, so ubiquitination plays a critical role in TCR signaling.

3.3.2 Ubiquitination of T Cell Activation

The mechanisms of T cell activation are crucial to adaptive immune responses. Here is a brief introduction to the process of T cell activation signaling. The second costimulatory signal provided by TCR and CD28 is necessary to achieve complete activation of T cells. Upon recognizing MHC-peptide complexes presented by APCs, specific TCR signal transduction is triggered by the binding of the complexes (Gaud et al. 2018). In most T cells, TCR acts as a complex with the costimulatory proteins CD3 at the surface (Yokosuka et al. 2002). The TCR-CD3 complex is composed of homologous α chain and β chain of TCR noncovalently associated with the invariant CD3 dimers CD3εγ, CD3εγ, and CD3ζζ (Mariuzza et al. 2020).

When TCR ligand a relevant short peptide bound to MHC class I or class II molecules (called pMHC), TCR signaling is initiated. Mechanistically, the protein tyrosine kinase Lck is recruited to the TCR complex by the colocalization of CD4 or CD8 co-receptors to pMHC molecules where Lck can phosphorylate immunoreceptor tyrosine-based activation motifs (ITAMs) signaling motifs (Courtney et al. 2018). Phosphorylated ITAMs enable recruitment and activation of Syk kinase family through binding to dual SH2 domains on these kinases (Underhill and Goodridge 2007). Lck phosphorylates the ITAMs of the TCR complex, leading to the recruitment of TCRζ-chain-associated protein kinase of 70 kDa (Zap70) and its phosphorylation. Prior to being recruited and phosphorylated by Lck, Zap70 predominately resides within the cytoplasm where it is autoinhibited (Brownlie and Zamoyska 2013). However, the Nrdp1, as a RING-type E3 ligase, mediates the K33-linked polyubiquitination of Zap70 via the negative regulatory phosphatase Sts1 and Sts2 and thereby terminates early TCR signaling (Yang et al. 2015). The Casitas B-cell lymphoma (Cbl), as an E3 ubiquitin ligase, plays an important role as a negative regulator of activated receptor tyrosine kinases by facilitating their ubiquitination, which blind with the SH3 domain of Lck and mediate Lck degradation (Panigada et al. 2002; Rao et al. 2002), and the Cbl-b, an E3 ligase of Cbl family, has been found to be a negative regulatory factor of the tyrosine kinases in Zap70 family (Paolino et al. 2011). The active conformation of Zap70 is further stabilized through phosphorylation and activation by Lck, and its phosphorylation at multiple tyrosine residues, such as Tyr315, Tyr319, and Tyr493 (Hu and Sun 2016). Phosphorylated and activated Zap70 phosphorylate the transmembrane adaptor protein linker for activation of T cells (LAT) subsequently, leading to the recruitment of multiple adaptors cytosolic adapters or linker proteins and the formation of the LAT signalosome (Sommers et al. 2004). LAT contains major Zap70 phosphorylation sites on tyrosine residues: 132, 171, 191, and 226. Phosphorylation of Tyr132 is required for the activation of PLCγ1, Tyr171, and Tyr191 associate Ras signaling pathways, such as cytosolic adapters or linker proteins Grb2 and Gads (Paz et al. 2001). The Src homology 2 domain-containing leukocyte protein of 76 kDa (SLP76) is phosphorylated by Zap70, forming a complex with phosphorylated LAT via Gads, and is an important adaptor molecule for regulating the signaling of LAT in the signaling cascade (Brownlie and Zamoyska 2013). Upon activation of Zap70, the deubiquitinase Otud7b is rapidly recruited to the Zap70, which facilitates Zap70 activation by deubiquitinating and preventing the association of Zap70 with the negative regulatory phosphatases Sts1 and Sts2 (Hu et al. 2016). Ubiquitin specific protease 9X (Usp9X) is able to deubiquitinate ZAP70 through TCR-dependent phosphorylation and enhancement of its catalytic activity and association with the LAT signalosome (Naik and Dixit 2016). Activated LAT recruits PLC-γ1, then activates PLC-γ1 to provide for protein kinase C (PKC), and Ras and calcium pathways activation (Wange 2000). These pathways lead to the activation of transcription factors and cooperate in the induction of genes involved in the activation of T cells.

Mechanistically, activated PLC-γ1 catalyzes the hydrolysis of the lipid phosphatidylinositol-4,5-bisphosphate (PIP2) to the second messenger molecules inositol-1,4,5-trisphosphate (IP3) and diacylglycerol (DAG) (Sommers et al. 2004). IP3 binds to IP3 receptors in the membrane of the endoplasmic reticulum (ER) and induces the release of ER Ca^{2+} stores into the cytoplasm (Oh-hora 2009). In T cell activation, sustained Ca^{2+} entry is necessary for complete and long-lasting activation of nuclear factor of activated T cells (NFAT), leading to their translocation to the nucleus, and forming cooperative transcriptional complexes with a variety of transcription factors

(Hogan 2017). The mouse double minute 2 homolog (MDM2), as an E3 ubiquitin ligase, can mediate the K48-linked ubiquitination degradation of transcription factors NFAT (Zou et al. 2014). DAG pathway results in the activation of two major pathways involving the Ras and PKC (Balagopalan et al. 2010). Ras is a guanine nucleotide-binding protein that leads to the activation of the mitogen-associated protein kinases (MAPKs) and extracellular signal-regulated kinase (ERK) (Roose et al. 2005). The activation of Ras is facilitated by Ras guanine nucleotide exchange factor (GEF), son of sevenless (SOS) and Ras guanyl nucleotide-releasing protein (RasGRP) in T cells (Sommers et al. 2004). DAG recruits and activates RasGRP leading to the activation of Ras. In addition, the recruitment of SOS to LAT via Grb2 also results in the activation of the Ras pathway (Genot and Cantrell 2000), which lead to the activation of ERK, which activates activator protein-1 (AP-1), a transcription factor that is important for the activation of T cell (D'Ambrosio et al. 1994).

CD28 activates the key effector downstream phosphatidylinositol 3-kinase (PI3K) and also leads to the phosphorylation and activation of PKCθ (Nolz et al. 2007). Cbl-b from the Casitas B-cell lymphoma (Cbl) family ubiquitin ligases has been found to be a key negative regulator of NF-κB activation by mediating K48-linked ubiquitination of PI3K and PKCθ (Qiao et al. 2008). The activated PKCθ is able to phosphorylate and activate the intermediate signaling complex, which is composed of CARMA1, BCL10, and MALT1, namely the CARMA1-BCL10-MALT1 (CBM) signalosome (Meininger and Krappmann 2016). USP9X, as a deubiquitinase, interacts with the CBM complex and removes the TCR-induced ubiquitin chain from BCL10, which facilitates the association of CARMA1 with BCL10-MALT1 (Park et al. 2013). CBM then mediates the activation of the transcription factor NF-κB, which is important for T cell proliferation and activation during the adaptive immune response (Zhao et al. 2014). In the quiescent state, NF-κB is sequestered in the cytoplasm with its inhibitor (IκB). Upon activation, IκB is phosphorylated by the IκB kinase (IKK) complex, composed of IKKα, IKKβ, and a regulatory subunit called IKKγ or NEMO (Liu et al. 2012). BCL10-mediated MEMO K63-linked ubiquitination promotes the activation of IKK, BCL10 promotes the activation of transcription factors NF-kB through the ubiquitin-conjugating enzyme Uev1A and Ubc13-dependent ubiquitination of NEMO (Bastow et al. 2004). In addition, TRAF6 associates with Malt1 as an E3 ligase, mediating K63-linked ubiquitination of Malt1, and providing a docking surface for the recruitment of the IKK regulatory subunit NEMO (Oeckinghaus et al. 2007). After phosphorylation, the IKK phosphorylated sites on IκB serve as an essential part of a specific recognition site for SCF-type E3 ubiquitin ligase, and mediate IκB K48-linked ubiquitination and degradation (Kanarek et al. 2010). Moreover, ubiquitination is not only the mediator of the IκB degradation, but also the NF-κB precursors, such as p105 and p100, and processes into mature forms by the proteasome (Bastow et al. 2004). Peli1, as an E3 ubiquitin ligase, mediating K48-linked ubiquitination, negatively regulates c-Rel, a member of the NF-κB family of transcription factors (Chang et al. 2011). Under the strict regulation of these ubiquitinases and deubiquitinases, the activation signaling pathway of T cells can proceed orderly.

3.4 Ubiquitination of Cancer Therapy

Cancer immunotherapy aims to design effective treatments to improve the specificity and strength of the immune system against cancer. Ubiquitination is crucial for a plethora of physiological processes, and recent studies have shown that dysregulation of molecules involved in the UPS can directly initiate and promote the progress of various human diseases, such as cancer (Mansour 2018). Thus, ubiquitination can regulate both tumor-suppressing and tumor-promoting pathways in cancer. E3 ubiquitin ligases and deubiquitinases form a family to regulate the function of diverse immune cells and to be exploited for enhancing antitumor immunity in tumor cells (Zhou and Sun 2021). Here we will discuss how genetic defects in components of the Ub system can mediate the progression or suppression of the tumorigenic pathways in different types of cancer and shed light on the current and future perspectives of cancer therapeutics that depend on either activation or deactivation of the ubiquitination on target proteins.

3.4.1 PD-1 and PD-L1

The programmed cell death protein 1 (PD-1, also known as CD279) as a type I transmembrane protein is broadly expressed on activated T cells, B cells, and myeloid cells (Ishida et al. 1992). PD-1 plays a vital role in tumor immune responses and peripheral self-tolerance through regulating the activity of T cells (Salmaninejad et al. 2018). The programmed cell death ligand 1 (PD-L1) is a coinhibitory factor expressed by APCs and other nonlymphoid tissues, which combines with PD-1 and leads to the inhibition of TCR-mediated signaling and at least suboptimal levels of CD28-mediated costimulation (Freeman et al. 2000). PD-L1 also plays an important role in various malignancies where cancer cells and APCs express PD-L1 to inhibit the activity of effector T cells through PD-1 engagement, which is responsible for cancer immune escape and makes a huge effect on cancer therapy (Han et al. 2020). Hence, many cancers evade immune surveillance by overexpressing PD-L1 (Wang et al. 2020). Mechanistically, ligation of PD-1 during the stimulation of T cells through the TCR-CD3 complex and CD28 inhibited activation of the PI3K-Akt and Ras-MEK-ERK pathways and blocked expression of gene (Patsoukis et al. 2012). Thus, PD-1 and PD-L1 are important targets for tumor immunotherapy. A growing amount of evidence suggests that ubiquitination and deubiquitination of PD-1 and PD-L1 play crucial roles in the regulation of PD-1 and PD-L1 protein stabilization and dynamics (Hu et al. 2021).

PTMs of PD-1 and PD-L1 such as glycosylation, phosphorylation, ubiquitination, SUMOylation and acetylation have emerged as important regulatory mechanisms of PD-1 and PD-L1 protein stability and translocation as well as protein-protein interactions, which modulate immunosuppression in cancer (Hsu et al. 2018). Ubiquitination is the major way of regulating protein degradation. A growing amount of evidence suggests that ubiquitination and deubiquitination of PD-1 and PD-L1 are vital in tumor immunotherapy (Li et al. 2020). Ubiquitination of PD-1 and PD-L1 removes the inhibition of T cell signaling, and the ubiquitination of PD-L1 is affected by many factors. The FBXO38 is a special E3 ligase of PD-1 that mediates K48-linked polyubiquitination and subsequent proteasome degradation, which is essential for maintaining the antitumor activity of T cells (Meng et al. 2018). Whereas the transcription of FBXO38 is suppressed in the tumor microenvironment, it is possible that IL-2 signaling in T cells results in STAT5-mediated transcriptional activation of FBXO38 gene (Fbxo38) expression (Serman and Gack 2019). The E3 ubiquitin ligase c-Cbl in immune cells binds and ubiquitinates PD-1 to target it for proteasomal degradation resulting in the downregulation of cell surface expression of PD-1 and the inhibition of T cells, which leads to the suppression of tumor growth (Lyle et al. 2019).

The expression E3 ubiquitin ligases Cbl-b and c-Cbl is negatively correlated with PD-L1 expression in non-small cell lung cancer (NSCLC), and both inhibit PD-L1 by inactivating STAT, AKT and ERK signaling (Wang et al. 2018). The E3 ligase RBX1 induces ubiquitination of PD-L1 protein in hepatitis B virus (HBV)-related hepatocellular carcinoma (HCC) cells by activating the adenosine 5′-monophosphate-activated protein kinase pathway, and promotes the degradation of PD-L1 (Chen et al. 2020). The glycogen synthase kinase 3β (GSK3β) is a serine-threonine kinase, which mediates phosphorylation and often facilitates ubiquitin E3 ligase recognition (Hart et al. 1999). The beta transducin repeat-containing protein (β-TrCP) is an E3 ligase, and PD-L1 contains a β-TrCP targeting box. β-TrCP polyubiquitinates non-glycosylated and phosphorylated PD-L1 by GSK3b and leads to degradation of PD-L1 in the cytoplasm (Li et al. 2016). The STIP1 homology and U-box containing protein 1 (STUB1) is an E3 ubiquitin ligase that induces ubiquitination and destabilization of PD-L1 (Hu et al. 2021). The cyclin-dependent kinase 4/6 (CDK4/6) also plays important roles in antitumor treatment, which phosphorylates Speckle-type POZ protein (SPOP), an E3 ligase that destabilizes PD-L1 via ubiquitination, and controls cancer immune surveillance (Zhang et al. 2018). The ER-associated degradation (ERAD) E3 ligase HMG-CoA reductase degradation protein 1 (HRD1) bound to PD-L1 recognizes abnormal glycosylation of PD-L1 and promotes its degradation in ER (Cha et al. 2018). These E3 ubiquitin ligases bind and ubiquitinate PD-1 or PD-L1, which results in proteasomal degradation of PD-1 or PD-L1. The degradation of PD-1 or PD-L1 promotes and maintains the antitumor activity of T cells.

Rather, the ubiquitin-specific peptidase 22 (USP22), as a deubiquitinating enzyme, interacts with PD-L1 and deubiquitinates PD-L1 resulting in inhibition of its proteasome degradation and promotion of its stability (Wang et al. 2020). The ubiquitin-specific peptidase 9, X-linked (USP9X) could be combined with PD-L1 to induce its deubiquitination and stabilize its protein expression in oral squamous cell carcinoma (OSCC) (Jingjing et al. 2018). The CKLF (Chemokine-like factor)-like MARVEL transmembrane domain-containing family member 4/6 (CMTM4/6) is present at the cell surface, which associates with PD-L1 protein, reduces its ubiquitination, and increases PD-L1 protein half-life, enhancing T cell inhibitory capacity of PD-L1 expressing tumor cells (Mezzadra et al. 2017). The constitutive photomorphogenesis 9 (COP9) signalosome 5 (CSN5), induced by TNF-α-mediated NF-κB p65, inhibits the ubiquitination and degradation of PD-L1, whereas curcumin could inhibit CSN5 and diminish cancer cell PD-L1 expression (Lim et al. 2016). Moreover, USP22 also stabilizes CSN5, and coordinates with CSN5 to regulate PD-L1 (Wang et al. 2020). The deubiquitinated PD-L1 reduces T cell proliferation and activation as well as cancer cell death mediated by T cells thereafter. Furthermore, the depletion of these deubiquitinases inhibits tumorigenesis and promotes T cell cytotoxicity. Hence, these deubiquitinases could be therapeutic targets, and they may play a crucial role in tumor therapy.

3.4.2 Proteolysis Targeting Chimera (PROTAC)

Besides tumor immunotherapy, small molecule drug therapy for cancer is also involved in the treatment of cancer (Bushweller 2019). Recent years have seen a new class of small molecules that can target proteins for degradation, especially the proteolysis targeting chimera (PROTAC) has been developed to be a useful technology for targeted protein degradation (Cromm and Crews 2017). PROTAC is a small heterozygous molecule with dual functions, it connects the target protein ligand with E3 ubiquitin ligase ligand through a suitable connecting chain, which promotes the formation of a ternary complex between the target protein of interest (POI) and E3 ligase, thus making the target protein close to the ubiquitination system, and then being recognized by the 26S subunit of proteasome, eventually leading to the degradation of the target protein (Li and Song 2020; Raina and Crews 2017). PROTAC also mediates the degradation of critical oncoproteins in tumor cells, which may be a potential therapeutic strategy for cancer (Khan et al. 2020).

Compared with conventional chemotherapeutics, PROTAC has several advantages as a new technology of protein level regulation. Firstly, PROTAC can degrade the whole protein, and clear the enzymatic and nonenzymatic functions, which can overcome drug resistance of tumors. Drug resistance caused by tumor-targeted therapy is a major problem in cancer research (Raina et al. 2016). In addition, degradation of protein is catalytic. After the ubiquitination of the target protein, PROTAC can be dissociated from the complex and enter the next catalytic cycle. Therefore, PROTAC can work at very low doses (Smith et al. 2019). Traditional drugs nonspecifically inhibit cell proliferation including normal cells and almost focus on directly controlling the activation of proteins, which may cause undesired toxicities and side effects (Zhou et al. 2020). The targeted cancer therapeutics suppresses cancer cell proliferation and progression by its protein of interest (POI) but not normal cells, where E3 ligases are heavily dependent on (Kargbo

2019). Although there are more than 600 E3 ubiquitin ligases have been found, only very few with small molecule ligands have been used for designing PROTACs, such as VHL, CRBN, IAPs, and MDM2 (Gao et al. 2020). With the rapid development of PROTAC, there are also some problems to be solved, but it is sure that through the joint efforts of scientific research, these problems can be solved satisfactorily in the near future.

3.5 Prospective

This review highlights some of the major functions of ubiquitination in the regulation of immune responses and tumor therapy. We appreciate the crucial and paradoxical function of ubiquitination in regulating the immune response. Ubiquitination participates in different aspects of immune functions, ranging from PRR signaling in innate immunity and inflammation to T cell activation and differentiation, and ubiquitination can regulate both tumor-suppressing and tumor-promoting pathways in cancer. Ubiquitin signaling pathways play pivotal roles in immune homeostasis and chronic inflammation, especially autoimmune diseases, tumorigenesis, and chronic inflammatory diseases. More and more efforts have been put into determining how ubiquitination functions in autoimmune diseases and anti-tumor processes, which can offer a molecular basis to improve the clinical response rate and efficacy in patients suffering from cancer and other diseases.

References

Balagopalan L et al (2010) The LAT story: a tale of cooperativity, coordination, and choreography. Cold Spring Harb Perspect Biol 2(8):a005512

Barton GM, Kagan JC (2009) A cell biological view of Toll-like receptor function: regulation through compartmentalization. Nat Rev Immunol 9(8):535–542

Bastow R et al (2004) Vernalization requires epigenetic silencing of FLC by histone methylation. Nature 427(6970):164–167

Beutler B (2004) Innate immunity: an overview. Mol Immunol 40(12):845–859

Bonilla FA, Oettgen HC (2010) Adaptive immunity. J Allergy Clin Immunol 125(2 Suppl 2):S33–S40

Borad A, Ward H (2010) Human immune responses in cryptosporidiosis. Future Microbiol 5(3):507–519

Brownlie RJ, Zamoyska R (2013) T cell receptor signalling networks: branched, diversified and bounded. Nat Rev Immunol 13(4):257–269

Buneeva OA, Medvedev AE (2016) Atypical ubiquitination of proteins. Biomed Khim 62(5):496–509

Bushweller JH (2019) Targeting transcription factors in cancer—from undruggable to reality. Nat Rev Cancer 19(11):611–624

Cha JH et al (2018) Metformin promotes antitumor immunity via endoplasmic-reticulum-associated degradation of PD-L1. Mol Cell 71(4):606–620.e7

Chang M, Jin W, Sun SC (2009) Peli1 facilitates TRIF-dependent Toll-like receptor signaling and pro-inflammatory cytokine production. Nat Immunol 10(10):1089–1095

Chang M et al (2011) The ubiquitin ligase Peli1 negatively regulates T cell activation and prevents autoimmunity. Nat Immunol 12(10):1002–1009

Chen ZJ, Sun LJ (2009) Nonproteolytic functions of ubiquitin in cell signaling. Mol Cell 33(3):275–286

Chen Z et al (2020) 2,5-dimethylcelecoxib improves immune microenvironment of hepatocellular carcinoma by promoting ubiquitination of HBx-induced PD-L1. J Immunother Cancer 8(2):e001377

Courtney AH, Lo W-L, Weiss A (2018) TCR signaling: mechanisms of initiation and propagation. Trends Biochem Sci 43(2):108–123

Cromm PM, Crews CM (2017) Targeted protein degradation: from chemical biology to drug discovery. Cell Chem Biol 24(9):1181–1190

D'Ambrosio D, Cantrell DA, Frati L, Santoni A, Testi R (1994) Involvement of p21ras activation in T cell CD69 expression. Eur J Immunol 24(3):616–620

Davis ME, Gack MU (2015) Ubiquitination in the antiviral immune response. Virology 479–480:52–65

Freeman GJ et al (2000) Engagement of the PD-1 immunoinhibitory receptor by a novel B7 family member leads to negative regulation of lymphocyte activation. J Exp Med 192(7):1027–1034

Gao H, Sun X, Rao Y (2020) PROTAC technology: opportunities and challenges. ACS Med Chem Lett 11(3):237–240

Gaud G, Lesourne R, Love PE (2018) Regulatory mechanisms in T cell receptor signalling. Nat Rev Immunol 18(8):485–497

Genot E, Cantrell DA (2000) Ras regulation and function in lymphocytes. Curr Opin Immunol 12(3):289–294

Grice GL, Nathan JA (2016) The recognition of ubiquitinated proteins by the proteasome. Cell Mol Life Sci 73(18):3497–3506

Han Y, Liu D, Li L (2020) PD-1/PD-L1 pathway current researches in cancer. Am J Cancer Res 10(3):727–742

Hart M et al (1999) The F-box protein beta-TrCP associates with phosphorylated beta-catenin and regulates its activity in the cell. Curr Biol 9(4):207–210

Hogan PG (2017) Calcium-NFAT transcriptional signalling in T cell activation and T cell exhaustion. Cell Calcium 63:66–69

Hsu JM et al (2018) Posttranslational modifications of PD-L1 and their applications in cancer therapy. Cancer Res 78(22):6349–6353

Hu H, Sun SC (2016) Ubiquitin signaling in immune responses. Cell Res 26(4):457–483

Hu W et al (2015) Differential outcome of TRIF-mediated signaling in TLR4 and TLR3 induced DC maturation. Proc Natl Acad Sci U S A 112(45):13994–13999

Hu H et al (2016) Otud7b facilitates T cell activation and inflammatory responses by regulating Zap70 ubiquitination. J Exp Med 213(3):399–414

Hu X et al (2021) Emerging role of ubiquitination in the regulation of PD-1/PD-L1 in cancer immunotherapy. Mol Ther 29:908

Ishida Y, Agata Y, Shibahara K, Honjo T (1992) Induced expression of PD-1, a novel member of the immunoglobulin gene superfamily, upon programmed cell death. EMBO J 11(11):3887–3895

Jingjing W et al (2018) Deubiquitination and stabilization of programmed cell death ligand 1 by ubiquitin-specific peptidase 9, X-linked in oral squamous cell carcinoma. Cancer Med 7(8):4004–4011

Kanarek N et al (2010) Ubiquitination and degradation of the inhibitors of NF-kappaB. Cold Spring Harb Perspect Biol 2(2):a000166

Kargbo RB (2019) PROTAC-mediated degradation of estrogen receptor in the treatment of cancer. ACS Med Chem Lett 10(10):1367–1369

Kawai T, Akira S (2011) Toll-like receptors and their crosstalk with other innate receptors in infection and immunity. Immunity 34(5):637–650

Khan S et al (2020) PROteolysis TArgeting Chimeras (PROTACs) as emerging anticancer therapeutics. Oncogene 39(26):4909–4924

Kim AY et al (2018) Differential regulation of MyD88- and TRIF-dependent signaling pathways of toll-like receptors by cardamonin. Int Immunopharmacol 64:1–9

Kobayashi T, Walsh MC, Choi Y (2004) The role of TRAF6 in signal transduction and the immune response. Microbes Infect 6(14):1333–1338

Kulathu Y, Komander D (2012) Atypical ubiquitylation—the unexplored world of polyubiquitin beyond Lys48 and Lys63 linkages. Nat Rev Mol Cell Biol 13(8):508–523

Li X, Song Y (2020) Proteolysis-targeting chimera (PROTAC) for targeted protein degradation and cancer therapy. J Hematol Oncol 13(1):50

Li CW et al (2016) Glycosylation and stabilization of programmed death ligand-1 suppresses T-cell activity. Nat Commun 7:12632

Li SM et al (2020) Recent findings in the posttranslational modifications of PD-L1. J Oncol 2020:5497015

Lim SO et al (2016) Deubiquitination and stabilization of PD-L1 by CSN5. Cancer Cell 30(6):925–939

Liu F et al (2012) IKK biology. Immunol Rev 246(1):239–253

Liu J, Qian C, Cao X (2016) Post-translational modification control of innate immunity. Immunity 45(1):15–30

Lyle C et al (2019) C-Cbl targets PD-1 in immune cells for proteasomal degradation and modulates colorectal tumor growth. Sci Rep 9(1):20257

Mansour MA (2018) Ubiquitination: friend and foe in cancer. Int J Biochem Cell Biol 101:80–93

Mariuzza RA, Agnihotri P, Orban J (2020) The structural basis of T-cell receptor (TCR) activation: an enduring enigma. J Biol Chem 295(4):914–925

Meininger I, Krappmann D (2016) Lymphocyte signaling and activation by the CARMA1-BCL10-MALT1 signalosome. Biol Chem 397(12):1315–1333

Meng X et al (2018) FBXO38 mediates PD-1 ubiquitination and regulates anti-tumour immunity of T cells. Nature 564(7734):130–135

Mezzadra R et al (2017) Identification of CMTM6 and CMTM4 as PD-L1 protein regulators. Nature 549(7670):106–110

Moresco EM, LaVine D, Beutler B (2011) Toll-like receptors. Curr Biol 21(13):R488–R493

Morlon A, Munnich A, Smahi A (2005) TAB2, TRAF6 and TAK1 are involved in NF-kappaB activation induced by the TNF-receptor, Edar and its adaptator Edaradd. Hum Mol Genet 14(23):3751–3757

Naik E, Dixit VM (2016) Usp9X is required for lymphocyte activation and homeostasis through its control of ZAP70 ubiquitination and PKCbeta kinase activity. J Immunol 196(8):3438–3451

Nolz JC, Fernandez-Zapico ME, Billadeau DD (2007) TCR/CD28-stimulated actin dynamics are required for NFAT1-mediated transcription of c-rel leading to CD28 response element activation. J Immunol 179(2):1104–1112

Oeckinghaus A et al (2007) Malt1 ubiquitination triggers NF-kappaB signaling upon T-cell activation. EMBO J 26(22):4634–4645

Oh-hora M (2009) Calcium signaling in the development and function of T-lineage cells. Immunol Rev 231(1):210–224

Panigada M et al (2002) Constitutive endocytosis and degradation of the pre-T cell receptor. J Exp Med 195(12):1585–1597

Paolino M et al (2011) Essential role of E3 ubiquitin ligase activity in Cbl-b-regulated T cell functions. J Immunol 186(4):2138–2147

Park Y, Jin HS, Liu YC (2013) Regulation of T cell function by the ubiquitin-specific protease USP9X via modulating the Carma1-Bcl10-Malt1 complex. Proc Natl Acad Sci U S A 110(23):9433–9438

Patsoukis N et al (2012) Selective effects of PD-1 on Akt and Ras pathways regulate molecular components of the cell cycle and inhibit T cell proliferation. Sci Signal 5(230):ra46

Paz PE, Wang S, Clarke H, Lu X, Stokoe D, Abo A (2001) Mapping the Zap-70 phosphorylation sites on LAT (linker for activation of T cells) required for recruitment and activation of signalling proteins in T cells. Biochem J 356(Pt 2):461–471

Qiao G et al (2008) T-cell receptor-induced NF-kappaB activation is negatively regulated by E3 ubiquitin ligase Cbl-b. Mol Cell Biol 28(7):2470–2480

Raina K, Crews CM (2017) Targeted protein knockdown using small molecule degraders. Curr Opin Chem Biol 39:46–53

Raina K et al (2016) PROTAC-induced BET protein degradation as a therapy for castration-resistant prostate cancer. Proc Natl Acad Sci U S A 113(26):7124–7129

Rao N, Miyake S, Reddi AL, Douillard P, Ghosh AK, Dodge IL, Zhou P, Fernandes ND, Band H (2002) Negative regulation of Lck by Cbl ubiquitin ligase. Proc Natl Acad Sci U S A 99(6):3794–3799

Rittinger K, Ikeda F (2017) Linear ubiquitin chains: enzymes, mechanisms and biology. Open Biol 7(4):7

Roose JP et al (2005) A diacylglycerol-protein kinase C-RasGRP1 pathway directs Ras activation upon antigen receptor stimulation of T cells. Mol Cell Biol 25(11):4426–4441

Salmaninejad A et al (2018) PD-1 and cancer: molecular mechanisms and polymorphisms. Immunogenetics 70(2):73–86

Satoh T, Akira S (2016) Toll-like receptor signaling and its inducible proteins. Microbiol Spectr 4(6)

Sebban H, Yamaoka S, Courtois G (2006) Posttranslational modifications of NEMO and its partners in NF-kappaB signaling. Trends Cell Biol 16(11):569–577

Serman TM, Gack MU (2019) FBXO38 drives PD-1 to destruction. Trends Immunol 40(2):81–83

Smith BE et al (2019) Differential PROTAC substrate specificity dictated by orientation of recruited E3 ligase. Nat Commun 10(1):131

Smith-Garvin JE, Koretzky GA, Jordan MS (2009) T cell activation. Annu Rev Immunol 27:591–619

Sommers CL, Samelson LE, Love PE (2004) LAT: a T lymphocyte adapter protein that couples the antigen receptor to downstream signaling pathways. BioEssays 26(1):61–67

Sujashvili R (2016) Advantages of extracellular ubiquitin in modulation of immune responses. Mediat Inflamm 2016:4190390

Swatek KN, Komander D (2016) Ubiquitin modifications. Cell Res 26(4):399–422

Takeda K, Akira S (2005) Toll-like receptors in innate immunity. Int Immunol 17(1):1–14

Takeuchi O, Akira S (2010) Pattern recognition receptors and inflammation. Cell 140(6):805–820

tenOever BR et al (2004) Activation of TBK1 and IKKvarepsilon kinases by vesicular stomatitis virus infection and the role of viral ribonucleoprotein in the development of interferon antiviral immunity. J Virol 78(19):10636–10649

Thaiss CA et al (2016) The microbiome and innate immunity. Nature 535(7610):65–74

Underhill DM, Goodridge HS (2007) The many faces of ITAMs. Trends Immunol 28(2):66–73

Vijay K (2018) Toll-like receptors in immunity and inflammatory diseases: past, present, and future. Int Immunopharmacol 59:391–412

Walsh MC et al (2008) TRAF6 autoubiquitination-independent activation of the NFkappaB and MAPK pathways in response to IL-1 and RANKL. PLoS One 3(12):e4064

Wang S et al (2018) E3 ubiquitin ligases Cbl-b and c-Cbl downregulate PD-L1 in EGFR wild-type non-small cell lung cancer. FEBS Lett 592(4):621–630

Wang Y et al (2020) The deubiquitinase USP22 regulates PD-L1 degradation in human cancer cells. Cell Commun Signal 18(1):112

Wange RL (2000) LAT, the linker for activation of T cells: a bridge between T cell-specific and general signaling pathways. Sci STKE 2000(63):re1

Wu X et al (2019) Regulation of TRIF-mediated innate immune response by K27-linked polyubiquitination and deubiquitination. Nat Commun 10(1):4115

Yakubu RR, Nieves E, Weiss LM (2019) The methods employed in mass spectrometric analysis of posttranslational modifications (PTMs) and protein-protein interactions (PPIs). Adv Exp Med Biol 1140:169–198

Yang M et al (2015) K33-linked polyubiquitination of Zap70 by Nrdp1 controls CD8(+) T cell activation. Nat Immunol 16(12):1253–1262

Yokosuka T et al (2002) Predominant role of T cell receptor (TCR)-alpha chain in forming preimmune TCR repertoire revealed by clonal TCR reconstitution system. J Exp Med 195(8):991–1001

Zhang J et al (2018) Cyclin D-CDK4 kinase destabilizes PD-L1 via cullin 3-SPOP to control cancer immune surveillance. Nature 553(7686):91–95

Zhao Y et al (2014) TCR-induced, PKC-theta-mediated NF-kappaB activation is regulated by a caspase-8-caspase-9-caspase-3 cascade. Biochem Biophys Res Commun 450(1):526–531

Zhou X, Sun SC (2021) Targeting ubiquitin signaling for cancer immunotherapy. Signal Transduct Target Ther 6(1):16

Zhou Y et al (2017) Post-translational regulation of antiviral innate signaling. Eur J Immunol 47(9):1414–1426

Zhou X et al (2020) PROTAC: a promising technology for cancer treatment. Eur J Med Chem 203:112539

Zhu G et al (2018) Targeting pattern-recognition receptors to discover new small molecule immune modulators. Eur J Med Chem 144:82–92

Zinngrebe J et al (2014) Ubiquitin in the immune system. EMBO Rep 15(1):28–45

Zou Q et al (2014) USP15 stabilizes MDM2 to mediate cancer-cell survival and inhibit antitumor T cell responses. Nat Immunol 15(6):562–570

Ubiquitination and Metabolic Disease

4

Meilin Ma, Rong Cao, Yan Tian, and Xianghui Fu

Abstract

The increasing incidence of metabolic diseases, including obesity, type 2 diabetes mellitus (T2DM), and non-alcoholic fatty liver disease (NAFLD), in the past decade is a serious concern worldwide. Disruption of cellular protein homeostasis has been considered as a crucial contributor to the pathogenesis of metabolic diseases. To maintain protein homeostasis, cells have evolved multiple dynamic and self-regulating quality control processes to adapt new environmental conditions and prevent prolonged damage. Among them, the ubiquitin proteasome system (UPS), the primary proteolytic pathway for degradation of aberrant proteins via ubiquitination, has an essential role in maintaining cellular homeostasis in response to intracellular stress. Correspondingly, accumulating evidences have shown that dysregulation of ubiquitination can aggravate various metabolic derangements in many tissues, including the liver, skeletal muscle, pancreas, and adipose tissue, and is involved in the initiation and progression of diverse metabolic diseases. In this part, we will summarize the role of ubiquitination in the pathogenesis of metabolic diseases, including obesity, T2DM and NAFLD, and discuss its potential as a therapeutic target.

Keywords

Obesity · Ubiquitination · Metabolic disease · Diabetes

Abbreviations

A204T	A functional FBXW7 variant
ACC	Acetyl–coenzyme A carboxylase
AMPK	AMP-activated protein kinase
APC/C	Anaphase promoting complex
ARE	Antioxidant-responsive element
ATGL	Adipose triglyceride lipase
BAT	Brown adipose tissue
BCL-2	B-cell lymphoma 2
Cbl-b	Casitas B-cell lymphoma-b
CEBPα	CCAAT enhancer-binding protein α
CHIP	C-terminus of HSC70-interacting protein
Clec16a	C-type lectin domain family 16, member A
CLU	Clusterin

M. Ma · R. Cao · Y. Tian
Division of Endocrinology and Metabolism, State Key Laboratory of Biotherapy and Cancer Center, West China Hospital, Sichuan University and Collaborative Innovation Center of Biotherapy, Chengdu, Sichuan, China

X. Fu (✉)
State Key Laboratory of Biotherapy, West China Hospital of Sichuan University, Chengdu, Sichuan, China
e-mail: xfu@scu.edu.cn

H. Hu, X. Fu (eds.), *Ubiquitination in Immune System and Immune Response*, Advances in Experimental Medicine and Biology 1466, https://doi.org/10.1007/978-981-97-7288-9_4

COP1	Constitutive photomorphogenic protein 1
CRC	Colorectal cancer
CREB	cAMP-responsive element-binding protein
CRL4B	CUL4B-RING E3 ligase
DM	Diabetes mellitus
DNA-PKcs	DNA-dependent protein kinase, catalytic subunit
DNL	De novo lipogenesis
DUBs	Deubiquitinating enzymes
E1	Ubiquitin activating enzyme
E2	Ubiquitin conjugating enzyme
E3	Ubiquitin ligase
ER	Endoplasmic reticulum
ERAD	ER-associated degradation
Ets-1	E26 oncogene homolog 1
FAs	Fatty acids
FASN	Fatty acid synthase
FBPs	F-box proteins
FBXO4	F-box only protein 4
FBXW2	F-box and WD repeat domain containing 2
FFAs	Free fatty acids
FKHR/Foxo1	Forkhead transcription factor
FTO	Fat mass and obesity-associated
G6Pase	Glucose-6-phosphatase
GLP-1	Glucagon-like peptide-1
Glut4	Glucose transporter 4
GSIS	Glucose-stimulated insulin secretion
HDM2	Homolog of mouse double minute 2
HECT	Homologous to E6-AP carboxyl terminus
HFD	High fat diet
HIF-1α	Hypoxia-inducible factor 1α
HIFs	Hypoxia-inducible factors
HRD1	HMG-CoA reductase degradation 1
HSCs	Hepatic stellate cells
IAPP	Islet amyloid polypeptide
IGF-1	Insulin-like growth factor 1
IKK complex	IκB kinase complex
IKKβ	IκB kinase β
IL-1	Interleukin-1
IR	Insulin receptor
IRS	Insulin receptors substrate
IRX3	Iroquois homeobox gene 3
IκBα	Inhibitor of nuclear factor κB α
JNK	c-Jun N-terminal kinase
KLF	Kruppel-like factor family
KO	Knockout
KSRP	KH-type splicing regulatory protein
LDLR	Low density lipoprotein receptor
LILRB4	Leukocyte immunoglobulin-like receptor B4
lncRNA	Long non-coding RNA
LPS	Lipopolysaccharide
Lys	Lysine
MafA	Musculoaponeurotic fibrosarcoma oncogene homolog A
MAPK	Mitogen-activated protein kinase
MCP1/Ccl2	Monocyte chemotactic protein 1
MDM2	Murine double minute 2
Met1	N-terminal methionine
MKRN1	Makorin Ring Finger Protein 1
MST1	Mammalian sterile 20-like kinase-1
MSTs	Mammalian sterile 20-like kinases
mTORC1	Mammalian target of the rapamycin complex 1
NAD	Nicotinamide adenine dinucleotide
NAFLD	Non-alcoholic fatty liver disease
NASH	Non-alcoholic steatohepatitis
NF-κB	Nuclear factor κB
Nrdp1	Neuregulin receptor degradation protein 1
Nrf2	Nuclear factor erythoid 2-related factor 2
Nrg4	Neuregulin 4
NSCLC	Non-small cell lung carcinoma
PC	Pyruvate carboxylase
PCK	Phosphoenolpyruvate carboxy kinase
PDX1	Pancreatic duodenal homeobox-1
PEDF	Pigment epithelium-derived factor
PGC-1	Peroxisome proliferator-activated receptor-γ coactivator-1
PI3K	Phosphoinositide 3-kinase

Pin1	Prolyl isomerase
PPARγ	Peroxisome proliferator-activated receptor γ
PPDPF	Pancreatic progenitor cell differentiation and proliferation factor
PROTAC	Proteolysis targeting chimera
PRDM16	PRD1-BF1-RIZ1 homologous domain-containing protein 16
p-Smad3	Phosphorylated Smad 3
RNF5	Ring finger protein 5
ROS	Reactive oxygen species
S6K1	S6 kinase 1
SCF	Skp1-Cul1-F-box ubiquitin ligase
SIRT1	Sirtuin-1
SMCs	Smooth muscle cells
SNX8	Sorting nexin 8
SREBP	Sterol regulatory element binding protein
T2DM	Type 2 diabetes mellitus
TAK1	TGF-β activated kinases 1
TPD	Targeted protein degradation
TBK1	TANK-binding kinase 1
TGFβ1	Transforming growth factor beta 1
TIPE2	Tumor necrosis factor α-induced protein 8-like 2
TLRs	Toll-like receptors
TNF-R1	Tumor necrosis factor receptor 1
TNFα	Tumor necrosis factor α
TNIP3	TNFAIP3 interacting protein 3
TPR	Tetratricopeptide-repeat motif
TRAF	Tumor necrosis factor receptor (TNFR)-associated factor
TRAF6	TNF receptor associated factor 6
TRB3	Tribbles 3
TRIM28	Tripartite motif containing 28
TRUSS	Tumor necrosis factor receptor-associated ubiquitous scaffolding and signaling protein
TZDs	Thiazolidinediones
Ub	Ubiquitin
UCHL1	Ubiquitin carboxyl-terminal esterase L1
UCP1	Uncoupling protein 1
UFD	Ubiquitin fusion degradation
UPR	Unfolded protein response
UPS	Ubiquitin-proteasome system
USP21	Ubiquitin-specific peptidase 21
WAT	White adipose tissue
XBP1s	X-box binding protein 1 spliced
βFaar	β cell function and apoptosis regulator

4.1 Introduction

The global incidence of metabolic derangements, including obesity, T2DM and NAFLD, has increased worldwide in the past ~50 years, reaching pandemic levels, imposing an enormous burden on people's health. A long-term energy imbalance between too many calories consumed and too few calories expended causes many physiological changes in body composition, such as increased visceral fat and decreased muscle mass and strength; these changes are strongly associated with metabolic diseases (Bluher 2019). The accumulation of body fat mass, especially fat in visceral adipose tissues and non-adipose tissues, including the liver and skeletal muscle, is an important risk factor for the development of obesity-related metabolic disorders (Bluher 2019). In addition, impaired pancreatic β cell function is crucial for the pathogenesis of T2DM, which is also associated with obesity (Cnop et al. 2005; Oh 2015).

A hallmark of obesity is a state of chronic inflammation in the adipose tissues, which in turn leads to insulin resistance, β cell dysfunction, and ultimately T2DM and NAFLD (Reilly and Saltiel 2017). Inflammation is generated by recruitment and activation of immune cells, especially pro-inflammatory macrophages (Weisberg et al. 2003; Lumeng and Saltiel 2011). Monocyte chemotactic protein 1 (MCP1/Ccl2), produced by obese adipocytes, plays a key role in macrophage infiltration (Kamei et al. 2006; Kanda et al. 2006). Consequently, pro-inflammatory cytokines secreted by immune cells in the adipose tissues disrupt insulin action in adipocytes (Hotamisligil et al. 1995).

Ubiquitination is mediated by the cooperative action of three enzymes: the ubiquitin activating enzyme (E1), the ubiquitin conjugating enzyme (E2), and the ubiquitin ligase (E3) in the presence of ATP to covalently modify substrate protein with ubiquitin (Ub), and plays an important role in protein stability, localization, activity, interaction, and degradation (Rape 2018; Hershko and Ciechanover 1998). Under conditions of metabolic disorders, cellular protein homeostasis is disrupted in pancreatic cells and insulin target tissues, including the adipose tissue, liver, and skeletal muscle. Impaired ubiquitination is closely related to the pathogenesis of several metabolic diseases, such as obesity, T2DM, and NAFLD. Therefore, targeting ubiquitination is a potential strategy for the treatment of obesity and obesity-related metabolic derangements.

In this part, we outline the principles of ubiquitination and its regulation, and elaborate on the molecular mechanisms by which E3 ligase and direct substrates, including transcriptional factors, molecules of signaling pathways and metabolic enzymes, which intertwine to influence metabolic homeostasis. We also discuss the potential of targeting ubiquitination for metabolic dysregulation and diseases.

4.2 Ubiquitination

Ub is an evolutionarily conserved 76 amino acid protein that can be conjugated onto proteins to generate a degradation signal or exert degradation-independent regulatory functions. Ub is covalently attached to a substrate, and it could also be ubiquitinated on any of its seven lysine (Lys) residues or its N-terminal methionine (Met1) by a cascade of enzymatic reactions carried out by E1, E2, and E3 enzymes sequentially (Rape 2018). E1 initially binds to Ub for activation and then transfers activated Ub to E2. Subsequently, E3 simultaneously interacts with a Ub-loaded E2 and the target protein, mediating isopeptide bond formation between Ub substrate. According to the number of Ub coupled to one lysine residue in protein, ubiquitination can be divided into mono-ubiquitination and polyubiquitination. Conversely, ubiquitination can be reversed by deubiquitination, which is carried out by deubiquitinating enzymes (DUBs, also known as deubiquitylases or deubiquitinases). The human genome encodes >600 E3s, >40 E2s, 2 E1s and more than 100 DUBs (Clague et al. 2019) (Fig. 4.1). Therefore, understanding the mechanism and regulation of ubiquitination may provide new opportunities for disease diagnosis and treatment.

The cellular functions of ubiquitination span a broad spectrum of proteolytic and nonproteolytic roles, such as proteasomal degradation, DNA repair, intracellular trafficking, assembly of multiprotein complex and governance of enzymic activity, and so on (Komander and Rape 2012). Therefore, the disruption of ubiquitination has broad consequences, which may be harmful to cell functioning. For instance, it could contribute to aberrant activation or deactivation of signaling pathways (such as, those involved in cellular metabolism or tumorigenesis), incorrect or inadequate assembly of protein complexes (for instance, during the regulation of inflammatory responses), or mislocalization of proteins.

Deregulation of cellular protein homeostasis has been established as a significant mechanism contributing to many diseases, including metabolic and neurodegenerative diseases, cancers, etc. (Pohl and Dikic 2019). Cellular protein homeostasis is delicately balanced by de novo synthesis, posttranslational modification and degradation (Hershko and Ciechanover 1992). Eukaryotic cells operate two major proteolytic systems: the UPS and autophagy. The UPS is a selective proteolytic system in which the conjugation of ubiquitin to substrates induces degradation by the proteasome.

4.2.1 Type of Ubiquitination

As stated above, cellular Ub modifications occur in various forms, which are usually referred to as "Ub code" (Fig. 4.2). Mono-ubiquitination is the most abundant Ub modifications and has a role in DNA repair, transcription, signal transduction, viral budding, endocytosis, and even proteasomal degradation (Chen and Mallampalli 2009; Braten et al. 2016). Upon transfer of Ub to the α-amino group of the target Lys, any of the eight Ub amino groups (Met1, Lys6, Lys11, Lys27, Lys29, Lys33,

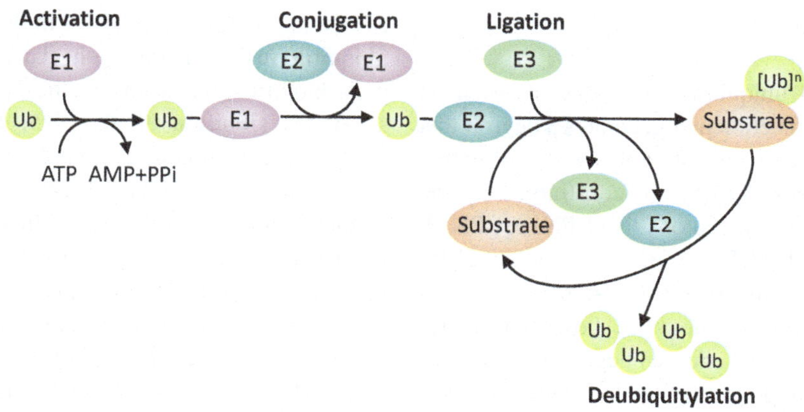

Fig. 4.1 The ubiquitylation pathway. E1 ubiquitin-activating enzymes use ATP to generate a high-energy thioester between ubiquitin and an active site Cys in the enzyme. Charged E1 enzymes transfer ubiquitin to E2 ubiquitin-conjugating enzymes, which then cooperate with E3 ubiquitin ligases to produce a ubiquitylated substrate. E3 ligases can transfer ubiquitin directly from E2 enzymes or undergo charging of their reactive Cys from which ubiquitin is transferred to the substrate. Approximately ~100 human DUBs cleave ubiquitin off its target to terminate this signaling event

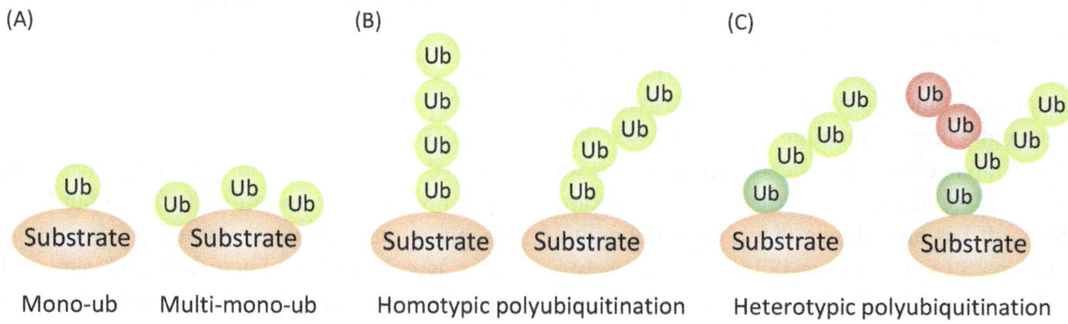

Fig. 4.2 Variety of ubiquitin signals. The different types of ubiquitin (Ub) linkages. (**a**) Mono-ub and multi-mono-ub. (**b**) Homotypic polyubiquitination. Ubiquitin can be conjugated by another ubiquitin through any of its seven Lys residues (Lys6, Lys11, Lys27, Lys29, Lys33, Lys48, and Lys63) or, alternatively, Met1. (**c**) Heterotypic polyubiquitination. Ubiquitin can be ubiquitinated through two or more different linkages in a single polymerization reaction, generating a mixed ubiquitin chain

Lys48, Lys63) can be attached to the C terminus of another Ub to form Ub chains of variable length, linkage type, and configuration (homo- and heterotypic/branched Ub chains). The functional significance of some Ub modifications (such as Lys48- and Lys63-linked ubiquitination) is widely recognized while the biological importance of the majority of Ub modifications is little understood. Lys48-linked Ub polymers are first identified and one of the most predominant among homotypic polyUb chains (Peng et al. 2003; Swatek and Komander 2016). These ubiquitinated signal proteins for proteolytic degradation, thereby regulating signal transduction, cell division, stress response, adaptive immune system, and development (Hershko and Ciechanover 1998; Wang and Maldonado 2006). The remaining homotypic polyUb chains are collectively called atypical (Swatek and Komander 2016), and their physiological roles have been comprehensively summarized elsewhere (Kulathu and Komander 2012; Akutsu et al. 2016).

4.2.2 UPS

Lys48-linked ubiquitination is one of the most abundant connection and generates a degron that induces proteasomal proteolysis (Ciechanover 2015). Lys48 polymers are constrained by the UBDs of specific receptors, like RPN10, RPN13, RAD23, and UBQLNs (Grice and Nathan 2016; Collins and Goldberg 2017). Numerous studies over the past few decades have shown that Lys48-linked proteasome based degradation targets many cellular proteins, thereby regulating various biological processes.

Among atypical types, Lys11 linkage is relatively abundant, even if its exact portion varies according to physiological states and thus remains controversial (Swatek and Komander 2016; Ohtake and Tsuchiya 2017; Xu et al. 2009; Ziv et al. 2011; Tsuchiya et al. 2017). Lys11 linkage, which is found in a mixed or branched chain with Lys48 and Lys63 typically, enable to promote proteasomal degradation. For instance, the anaphase promoting complex (APC/C) assembles a mixed chain containing Lys11, Lys48, and Lys63 linkages by a two-step mechanism (Brown et al. 2016; Lu et al. 2015; Meyer and Rape 2014). It has also shown that a mixed ubiquitin chain with Lys11 and Lys48 linkages synergistically mediates proteasomal degradation of misfolded proteins in the ER lumen via ER-associated degradation (ERAD) (Locke et al. 2014). One of the examples of Lys11 linkages is the oxygen sensor hypoxia-inducible factor 1α (HIF-1α) in hypoxia responses (Bremm et al. 2014) and STING in response to viral infection during innate immunity (Qin et al. 2014). Although most studies support the crucial roles of Lys11 linkages in proteasomal degradation, emerging evidence suggests that homotypic Lys11-linked chains do not facilitate proteasomal degradation (Grice et al. 2015; Martínez-Fonts and Matouschek 2016). All in all, the precise role of Lys11 linkages in proteasomal degradation remains unclear.

Similarly, Lys29 linkage likely signals to the proteasome, as this chain type is enriched upon proteasome inhibition (Wagner et al. 2011; Kim et al. 2011). Lys29-linked ubiquitin chains are assembled by E3 ligase UFD4 in the pathway of ubiquitin fusion degradation (UFD), in which polyubiquitination occurs at the N terminus of substrates (Johnson et al. 1995; Koegl et al. 1999). Notably, ubiquitination of UFD substrates is mediated by two E3s: UFD4 and UFD2. The former is specific to the Lys29 linkage, and the latter is specific to the Lys48 linkage. Moreover, there are other E3 ligases that may assemble Lys29 linkages, including a set of homologous to E6-AP carboxyl terminus (HECT) proteins such as ITCH, UBR5, UBE3C, and Smurf1 (Ohtake and Tsuchiya 2017; Dwane et al. 2017). However, Lys29-linked chains seem to have many roles in cells, which require further characterization in the future.

4.3 Pathophysiological Roles of Ubiquitination

4.3.1 Obesity

The obesity epidemic challenges health policymakers due to its far-reaching impact on health of populations and the enormous financial burden it can place on health-care systems. There is a strong association between obesity and increased risk of developing acute and chronic diseases, such as diabetes, cardiovascular disease, cancer, hypertension, stroke, and myocardial infarction. Obesity is characterized by the excessive accumulation of white adipose tissue in the body. Fundamentally, it is a state of energy imbalance caused by calorie intake exceeding its consumption. As the obesity epidemic spreads globally, how obesity develops and contributes to associated diseases is an imperative public health objective will be better understood.

As mentioned above, increased fat-cell number (hyperplasia) and increased fat-cell size (hypertrophy) result in increased body fat mass, which is the characteristic of obesity (Couillard et al. 2000). The number of adipocytes in an organism depends to a large extent on the process of adipocyte differentiation, which produces mature adipocytes from fibroblast-like preadipocytes. Briefly, both adipogenesis and proliferation of adipocytes play crucial roles in obesity.

The role of inflammation in the progression of obesity and obesity-associated complications has long been established. More than two decades ago, the first molecular connection between inflammation and obesity-associated tumor necrosis factor α (TNFα) was discovered, as evidenced by overexpression of TNFα in obese rodents and humans (Kern et al. 1995; Saghizadeh et al. 1996; Hotamisligil et al. 1993, 1995; Sethi and Hotamisligil 1999). Absence of TNFα led to improved insulin sensitivity in both genetic (ob/ob) and high-fat diet (HFD)-induced obesity (Uysal et al. 1997). In addition to inflammatory mediators including TNFα, certain molecules of adipocytes can regulate the immune response. For example, deficiency of leptin, an important adipose-derived hormone, resulted in impaired immunity in both humans and mice. Adiponectin, resistin, and visfatin, which are secreted by adipocytes, also have critical roles in immunological modulation.

4.3.1.1 Ubiquitination and Metabolic Signaling Pathways

Ubiquitination of FTO

Human fat mass and obesity-associated (FTO) gene has been robustly associated with an abnormal body mass index and a predisposition to obesity, and it is the first GWAS-identified obesity gene (Dina et al. 2007; Frayling et al. 2007). FTO is an RNA demethylase that links amino acid availability and mammalian target of the rapamycin complex 1 (mTORC1) signaling to regulate growth and mRNA translation (Cheung et al. 2013; Gulati et al. 2013). Evidence from epidemiological and functional studies suggests that FTO confers an increased risk of obesity by subtly changing food intake and preference (Speakman 2010). It has found that FTO undergoes post-translational ubiquitination on Lys-216. K216R mutation reduced the level of nuclear FTO and completely abolished its nuclear translocation in response to amino acid starvation, which are crucial for body mass and composition (Zhu et al. 2018a).

Ubiquitination of AKT

Insulin signaling plays a unique role in the regulation of energy homeostasis by promoting nutrient storage during the fed state via enhancing glucose uptake and triglyceride synthesis, and inhibiting triglyceride breakdown and fatty acids (FAs) release through lipolysis. AKT, the prominent molecule of insulin signaling pathway, could be K63-linked ubiquitinated by Ubc13, but this process was counter-regulated by GPS2 through inhibition of Ubc13 enzymatic activity. Moreover, adipocyte-specific deletion of GPS2 results in increased adiposity and altered lipid flux in adipocytes (Cederquist et al. 2017).

Ubiquitination of NF-κB

It is widely appreciated that chronic low-grade inflammation plays a key role in the development of obesity. Consistent with this, numerous recent studies have implicated the central role of nuclear factor κB (NF-κB), one of primary inflammatory signaling pathways, in obesity pathogenesis.

NF-κB can be activated by various inflammatory cytokines, among which the most well studied are interleukin-1 (IL-1) and TNFα. These cytokines bind to their receptors respectively and activate different signaling cascades that converge on TGF-β activated kinases 1 (TAK1), which then phosphorylates IκB kinase (IKK) and triggers the activation of NF-κB (Hayden and Ghosh 2008). Ubiquitination plays an essential role in the regulation of NF-κB pathway. IL-1β could trigger the activation of ubiquitin ligase TRAF6, which then mediates the K63-linked polyubiquitination of TAB2, a subunit of TAK1 kinase complex, to activate TAK1 (Cao et al. 1996; Wang et al. 2001; Deng et al. 2000). While in the TNFα cascade, under stimuli, cells could recruit several E3 ligases, including cIAP1, cIAP2, TRAF2, TRAF5, and LUBAC (HOIP, HOIL-1L and Sharpin), all of which cooperate with UBCH5 to mediate different polyubiquitin chains, including M1 liner, K11, and K63.

Among these molecules, IκB kinase β (IKKβ), a central coordinator of inflammatory responses through activation of NF-κB, has been implicated in obesity. Deficiency of IKKβ in smooth muscle cells (SMCs) protected low density lipoprotein receptor (LDLR)-knockout mice from HFD-induced vascular inflammation and were also resistant to HFD-induced obesity and its associated metabolic disorders, predominantly due to inhibiting proteasome-mediated β-catenin ubiquitination and degradation (Sui et al. 2014).

Certain pro-inflammatory stimuli, such as free fatty acids (FFAs), may result in the phosphorylation and subsequent ubiquitination of IκBs, which contribute to proteasome-degradation of IκB (Hayden and Ghosh 2012). The IκB kinase complex (IKK complex) is responsible for IκB phosphorylation in the classical NF-κB pathway. The phosphopeptide can be recognized by β-TrCP, also known as CRL1-β-TrCP that is comprised of Skp1, Cul1, and Rbx1, for ubiquitination and degradation. The degradation of IκB then facilitates the translocation of the NF-κB, thereby promoting the transcription of pro-inflammatory genes (Fig. 4.3).

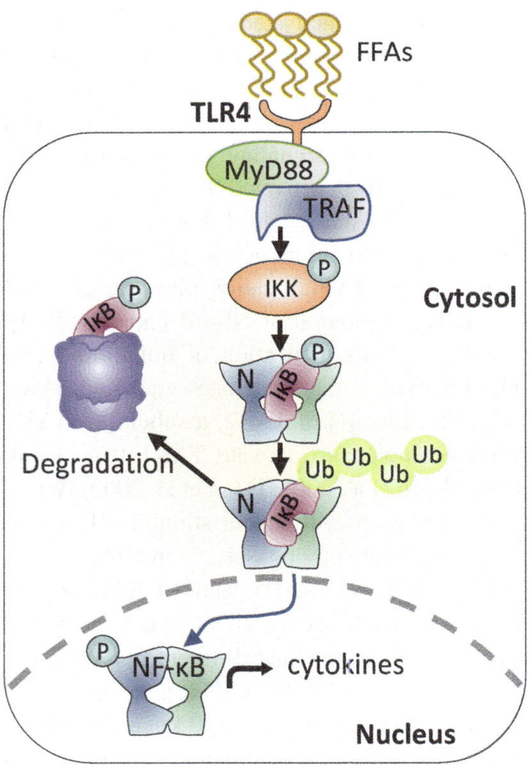

Fig. 4.3 Ubiquitination in inflammation. Free fatty acids (FFAs) can trigger the expression of pro-inflammatory genes through IKK and NF-kB

Unlike IKKα and IKKβ, the non-canonical IKKs, IKKε, and TANK-binding kinase 1 (TBK1) are induced downstream of NF-kB activation (Reilly et al. 2013). TBK1 can also be activated by phosphorylation in response to pro-inflammatory cytokines or other substances that bind to Toll-like receptors (TLRs) 3 and 4 (Perry et al. 2004). Moreover, TBK1 expression and activity are significantly increased in mature adipocytes of HFD-fed mice (Reilly et al. 2013). Adipose-specific TBK1 deficiency is resistant to HFD-induced obesity, due to dramatically increased oxygen consumption and energy expenditure in white adipose tissue (WAT) of obese, but not control-diet fed mice (Zhao et al. 2018a).

4.3.1.2 Ubiquitination and Transcriptional Factors

Ubiquitination of PPARγ

Peroxisome proliferator-activated receptor γ (PPARγ), a transcription factor that is highly expressed in the adipose tissues, regulates expression of numerous genes involved in FAs uptake, transport, and lipogenesis. PPARγ has been demonstrated to be a necessary and sufficient regulator of adipogenesis, glucose homeostasis, and insulin sensitivity (He et al. 2003; Imai et al. 2004; Sugii et al. 2009). Dysregulation of PPARγ contributes to obesity and obesity-associated diseases. PPARγ expression and activity are regulated at different levels from transcription to post-translational modification. The protein stability and transcriptional activity of PPARγ could be affected by phosphorylation, ubiquitination, and SUMOylation (Eeckhoute et al. 2012; Wadosky and Willis 2012). For instance, CUL4B-RING E3 ligase (CRL4B) negatively regulates PPARγ by promoting its polyubiquitination and proteasomal degradation. Deficiency of CUL4B resulted in upregulation of PPARγ-targeted genes, thus facilitating adipogenesis. In contrast, adipocyte-specific CUL4B deficiency aggravated HFD-induced body fat accumulation (Li et al. 2017). Additionally, other ubiquitin ligases, such as F-box only protein 4 (FBXO4) and C-terminus of HSC70-interacting protein (CHIP), are also

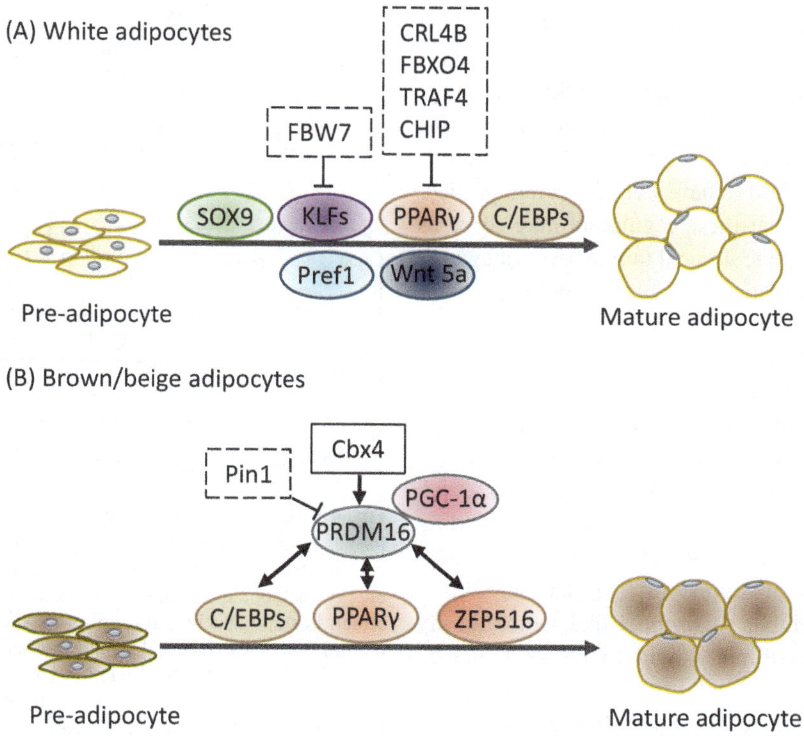

Fig. 4.4 Ubiquitination in adipogenesis. Many transcription factors are induced during adipocyte differentiation. (**a**) PPARγ and C/EBPs are crucial for adipocyte maturation. CUL4B-RING E3 ligase (CRL4B), F-box only protein 4 (FBXO4), CHIP, and TRAF4 could mediate ubiquitination of PPARγ by the ubiquitin-proteasome pathway. (**b**) PRDM16 is expressed in both brown and white adipocytes and functions as an important factor in transforming white adipocytes to beige adipocytes. Prolyl isomerase (Pin1) interacts with PRDM16 to promote its degradation through the ubiquitin-proteasome system, although specific E3 ligase remains unknown. Additionally, Cbx4, a SUMO E3 ligase for PRDM16, promotes its SUMOylating to impede the ubiquitination-mediated degradation, therefore improving PRDM16 stability

able to regulate ubiquitin-dependent degradation of PPARγ and mediate β-adrenergic signaling in adipocytes (Peng et al. 2018) and differentiation of adipocytes (Kim et al. 2017a). TRAF4, an E3 RING ubiquitin ligase, could also promote the degradation of PPARγ by a ubiquitin-proteasome pathway, thereby inhibiting adipogenesis during obesity (Chen et al. 2021) (Fig. 4.4a).

Ubiquitination of PRDM16

In mammals, WAT stores energy when nutrition is abundant, while brown adipose tissue (BAT) dissipates energy for thermogenesis primarily through uncoupled respiration mediated by mitochondrial uncoupling protein 1 (UCP1). There is another subpopulation of adipocytes called beige adipocytes, which express UCP1 and other thermogenic proteins to produce heat when responded to a cold environment or specific stimuli (Wu et al. 2012; Wang and Seale 2016). Adaptive thermogenesis by classical brown adipocytes and newly identified beige adipocytes can increase whole-body energy expenditure and therefore hold the potential for protecting against obesity. The zinc-finger transcriptional factor PRD1-BF1-RIZ1 homologous domain-containing protein 16 (PRDM16) is a powerful driver of brown and beige adipocyte identity (Kajimura et al. 2009; Seale et al. 2011; Ohno et al. 2013). It has been reported that Prolyl isomerase (Pin1) interacts with PRDM16 to promote its degradation through the UPS (Nakatsu et al. 2019). Moreover, Cbx4, a polycomb group protein, can function as a SUMO E3 ligase for

Prdm16, and sumoylation of Prdm16 at lysine 917 impedes its ubiquitination-mediated degradation, therefore improving its stability and pro-thermogenic function (Chen et al. 2018) (Fig. 4.4b). Although specific E3 ubiquitin-ligases for PRDM16 have not yet been identified, it is possible that these regulators may serve as plausible pharmacological targets to fight against obesity by activating browning of WAT.

Ubiquitination of KLF5

As mentioned above, the expansion of fat mass occurs either by an increase in the number of adipocytes (hyperplasia) or by an increase in the size of adipocytes (hypertrophy) (Rutkowski et al. 2015). PPARγ and CCAAT enhancer-binding protein α (CEBPα) are considered to be key determinants of adipocyte fate (Farmer 2006). KLF5, a member of kruppel-like factor family (KLF), could directly bind to the endogenous PPARγ promoter, and stimulate modest PPARγ promoter activity. Moreover, this effect can be enhanced in concert with C/EBPβ and C/EBPδ (Wu and Wang 2013). Clusterin (CLU) can inhibit the interaction between KLF5 and F-box/WD repeat-containing protein 7 (FBW7), thereby targeting KLF5 for ubiquitination and then enhancing KLF5-mediated transcriptional activation of both the CEBPα and the PPARγ promoter (Oh et al. 2020) (Fig. 4.4a).

Ubiquitination of IRX3

In addition to adipocytes, macrophages residing in adipose tissue work coordinately with adipocytes to orchestrate whole-body energy metabolism. Pro- or anti-inflammatory cytokines caused by macrophages have been shown to be key contributors in the pathogenesis of obesity (Mowers et al. 2013; Chawla et al. 2011). Recent studies showed that the iroquois homeobox 3 (IRX3), a mediator of FTO effects on body weight, promotes metabolic inflammation and inhibits lipolysis and adaptive thermogenesis, thereby accelerating the development of obesity. Lipopolysaccharide (LPS) treatment considerably reduced ubiquitination of IRX3 for degradation, which then facilitates the transcriptional capacity of IRX3 (Yao et al. 2021) (Fig. 4.5).

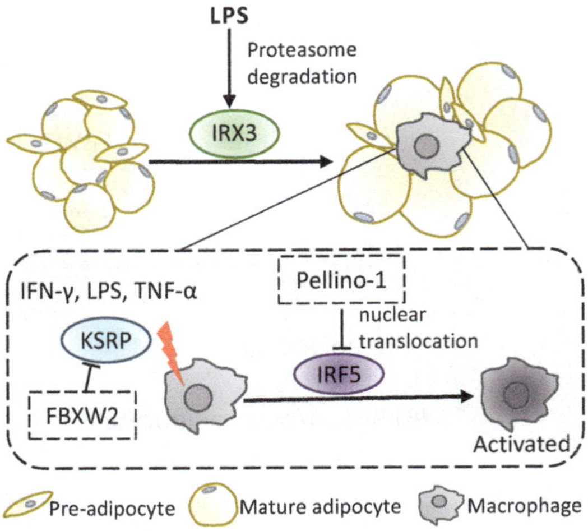

Fig. 4.5 Ubiquitination in macrophages of the adipose tissues. Macrophages reside in the adipose tissues work coordinately with adipocytes to orchestrate whole-body energy metabolism. Iroquois homeobox gene 3 (IRX3), which is stabilized by LPS, promotes metabolic inflammation and obesity. In addition, another transcriptional factor IRF5, ubiquitinated by Pellino-1, induces M1 macrophage polarization during obesity development. KSRP that regulates the response to LPS treatment can be degraded by FBXW2

Ubiquitination of IRF5

The turnover of M1 (classically activated) and M2 (alternatively activated) macrophage polarization is important for regulating pathological processes in obesity (Chavez-Galan et al. 2015). M1 and M2 represent two extreme dynamic activation states of macrophages (Murray et al. 2014). M1 macrophages produce pro-inflammatory cytokines such as IL-1β, TNF, IL-12, and IL-18, as well as reactive oxygen species (ROS) to promote inflammation, whereas M2 macrophages exhibit their effects of anti-inflammation by secreting anti-inflammatory cytokines. The signature transcription factor IRF5 induces M1 macrophage polarization. It has shown that the ubiquitin E3 ligase, Pellino-1, increased nuclear translocation of IRF5 through K63-linked ubiquitination in human and mouse M1 macrophages. The M1 polarization deficiency of Pellino-1-deficient macrophages enhances glucose intolerance in obese mice. In addition, compared with the control group, macrophages in the adipose tissues from obese individuals showed higher Pellino-1 expression and IRF5 nuclear translocation (Kim et al. 2017b) (Fig. 4.5).

4.3.1.3 Ubiquitination of Metabolic Enzymes

Deubiquitination of DNA-PKcs and ACLY

DNA-dependent protein kinase, catalytic subunit (DNA-PKcs) plays an important role in DNA double-stranded break responses (Blackford and Jackson 2017), and the induction of DNA-PK decreased monophosphate (AMP)-activated protein kinase (AMPK) activity and mitochondrial function in aged muscle, whereas the inhibition of DNA-PK protected against physical decline and improved insulin sensitivity (Park et al. 2017). ACLY is an enzyme responsible for lipid biosynthesis (Pinkosky et al. 2017). Ubiquitin-specific peptidase 21 (USP21) was found to interact with and deubiquitinates DNA-PKcs and ACLY, thereby controlling AMPK activity. Consequently, USP21 deficiency decreased obesity and diabetes induced by a HFD feeding (Kim et al. 2021).

Ubiquitination of ACC

Acetyl-coenzyme A carboxylase (ACC), the rate-limiting enzyme in FA synthesis, is phosphorylated and inactivated during fasting. More interestingly, there is a parallel pathway, in which the pseudokinase Tribbles 3 (TRB3), whose level is upregulated during fasting, stimulates lipolysis by triggering the degradation of ACC in the adipose tissues. TRB3 promoted ACC ubiquitination through an association with the E3 ubiquitin ligase constitutive photomorphogenic protein 1 (COP1). Indeed, deficiency of adipocyte TRB3 accumulated larger amounts of ACC protein than did wild-type cells (Qi et al. 2006).

Ubiquitination of KSRP

Previous studies have shown that the number of macrophages is increased in the adipose tissues in both obese mice and humans, and their content correlates positively with the level of obesity (Weisberg et al. 2003). Infiltration of macrophages, neutrophils, and lymphocytes into the adipose tissues is a characteristic of chronic inflammation. Accumulating evidence supports macrophages as important immune cells triggering chronic inflammatory responses, due to the observations that macrophages can express and secrete various pro-inflammatory chemokines and cytokines, including TNFα, IL-1β, and IL-6, thus impeding the normal functions of multiple cells and metabolic homeostasis. F-box and WD repeat domain containing 2 (FBXW2), a Skp1-Cul1-F-box-protein (SCF), is in charge of recognizing, binding, and polyubiquitinating substrates for subsequent degradation by the proteasome. FBXW2 is upregulated in macrophages of obese adipose tissues, and deficiency of myeloid FBXW2 alleviates obesity and related symptom by regulating degradation of KH-type splicing regulatory protein (KSRP) (Wang et al. 2020) (Fig. 4.5). KSRP is an RNA binding protein that destabilizes mRNAs via adenylate/uridylate-rich elements (AREs), and can negatively modulate a subset of cytokines and chemokines by reducing mRNA stability and translational efficiency (Ruggiero et al. 2009).

4.3.2 T2DM

Diabetes mellitus (DM) is a chronic metabolic disease with complex pathogenesis involving systemic disorders, and closely related to the obesity epidemic (DeFronzo et al. 2015). Over the past decades, the prevalence of DM has risen dramatically and gradually become a pandemic health concern. Surprisingly, the number of patients with DM is predicted to increase to 693 million by 2045 (Cho et al. 2018). T2DM is the most common diabetes, accounting for more than 90% among the three primary types of diabetes including T1DM and gestational DM.

Mounting evidence has proved that T2DM is caused by insulin resistance and dysfunction in insulin secretion that accompanies with pancreatic β cell failure. With the increasing number of diabetic patients all over the word, it is necessary to understand the molecular mechanisms underlying the development of insulin resistance and β cell failure, which is conducive to prevent and treat T2DM. Intriguingly, ubiquitination has been shown to be involved in the pathological process of T2DM (Yang et al. 2016). In this section, we will describe the reciprocal regulation of ubiquitination in insulin resistance and β cell dysfunction.

4.3.2.1 Ubiquitination and Metabolic Signaling in Insulin Resistance

Although many pathophysiological abnormalities can result in the dysregulation of glucose level, insulin resistance has long been considered as the core defect in T2DM. It is well-known that glucose homeostasis is modulated by insulin levels and insulin-associated signaling pathway. In insulin-associated signaling pathway, insulin binds to its tyrosine kinase receptors and leads to the recruitment and phosphorylation of insulin receptor (IR). Insulin resistance results from defective of key insulin signaling molecules, including IR, insulin receptors substrate (IRS), Phosphoinositide 3-Kinase/AKT (PI3K/AKT), and mitogen-activated protein kinase (MAPK), which are fundamental pathogenic factors of T2DM (Xiao et al. 2019).

A great deal of E3 ubiquitin ligases can regulate specific protein through ubiquitination, playing key roles in the development of insulin resistance. This section will review the different actions of "Ub code" in insulin resistance and major signaling molecules, thus elucidating the function of ubiquitination in T2DM.

Ubiquitination of IRS

As one of the proximal components of key insulin signaling pathway, IRS is considered as a crucial target of insulin receptor tyrosine kinase (Copps and White 2012). Mechanistically, IRS proteins exert profound influence on insulin signaling via multi-site S/T phosphorylation. Additionally, many intracellular Ser/Thr kinases can phosphorylate target proteins, including PKCs, mTOR, c-Jun N-terminal kinase (JNK), and S6 kinase (S6K) (James et al. 2021), thereby modulating insulin sensitivity. Therefore, defects in both IRS1 and IRS2 have been implicated in insulin resistance, further affecting a series metabolic disease (Fig. 4.6). Several E3 ubiquitin ligases that can interact with IRS proteins have been shown to participate in the development of T2DM (Yang et al. 2016).

First discovered as cytokine signaling suppressor, SOCS ubiquitin ligases function as potent inhibitors of insulin signaling in the regulation of body weight and glucose homeostasis. There are eight members in the SOCS family, among which SOCS-1, -3, -6 and -7 are capable of suppressing the insulin signaling pathway through three mechanisms, including inhibition of tyrosine phosphorylation of IRS, induction of the proteasomal degradation of the IRS, and inhibition of the IR kinase (Lebrun and Van Obberghen 2008). Importantly, the SOCS-1 and -3 could act as E3 ubiquitin ligases to degrade IRS1 and IRS2 proteins, and then lead to insulin resistance. Previous studies have shown that the expression level of SOCS-3 is significantly upregulated in animal models and patients with insulin resistance and T2DM. SOCS-1 inhibits insulin signaling through serving as part of an elongin BC ubiquitin ligase complex, indicating that it can't promote the subsequent ubiquitination and degradation of IRS (Dumpati et al. 2018; Howard and Flier 2006). In a word, the further discoveries relevant to SOCS properties and functions are expedient in the treatment for T2DM.

F-box proteins (FBPs), which belong to the SCF E3 ligases family in mammals, have been reported to participate in many biological and physiological events. Among which, FBXO2 serves as a functional E3 ligase in T2DM through enhancing ubiquitination-mediated IR degradation in the liver. Ablation of FBXO2 in obese mice improved insulin resistance and aberrant glucose tolerance (Liu et al. 2017). In addition, FBXO40, a muscle-specific E3 ligase, could also stimulate ubiquitin conjugation and lead to degradation of IRS1 (Zhang et al. 2020), similar as the effect of FBXO2.

IRS1 was recently found to be a proteolytic target of CRL7, an E3 ligase. During the pathogenesis of T2DM, the interplay between the mTORC1 and its effector kinase S6K1 enables IRS1 multi-site seryl phosphorylation, promoting IRS1 ubiquitination and degradation (Xu et al.

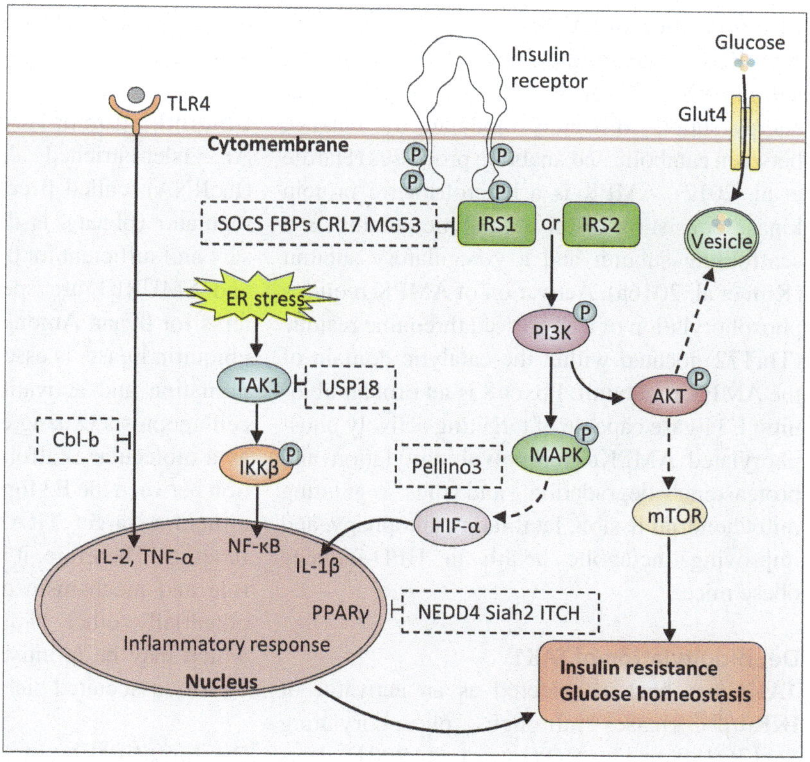

Fig. 4.6 The reciprocal relationship between ubiquitination and insulin signaling pathways. Some E3 ubiquitin ligases, including SOCS, FBPs, CRL7, MG53, Cbl-b, Pellion3, NEDD4, Siah2, and ITCH, play key roles in inflammatory responses and insulin resistance. In addition, USP18 is a deubiquitinating enzyme and can regulate insulin-associated molecules thereby modulating insulin resistance and glucose homeostasis. Both of them may contribute to the onset of T2DM

2012). Previous studies have shown that this negative feedback loop is able to restrain PI3 kinase/AKT signaling pathway through activating the insulin-like growth factor 1 (IGF-1) receptor (Scheufele et al. 2014). Therefore, CRL7 functions as a key factor in regulating IRS1 for mTORC1/S6K1-dependent degradation, and plays a pivotal role in the insulin-mediated glucose homeostasis and progression of T2DM.

MG53 (also called TRIM72), a muscle-specific E3 ligase, is significantly upregulated in HFD-induced obese mice and db/db diabetic mice (Wu et al. 2019), and can induce whole-body insulin resistance and metabolic syndrome. Mechanistically, MG53 mediates a simultaneous downregulation of the IR and IRS1 through ubiquitin-dependent degradation (Song et al. 2013), indicating that it is a powerful negative regulator of skeletal muscle insulin sensitivity and a new therapeutic target for T2DM and its complications.

Ubiquitination of TLR4

In T2DM, TLRs can elicit chronic inflammatory responses in respond to hyperglycemia and potential metabolic stimuli (Lancaster et al. 2018). Among which, TLR4 is involved in inflammatory responses in pancreatic islets and FA-mediated activation of TLR4 is considered as a key cause of insulin resistance (Pal et al. 2012). Thus, suppressing the TLR4 signaling is protective for pancreatic β cells and effective for improving insulin sensitivity by inhibiting inflammatory responses. As mentioned above, Cbl plays a key role in energy metabolism and insulin resistance. At the same time, Casitas B-cell lymphoma-b (Cbl-b), a unique E3 ubiquitin ligase involved in the maturation and activation of macrophages, is a critical regulator of diet-induced obesity and insulin resistance. In macrophages, Cbl-b inhibits saturated FAs-induced TLR4 signaling by ubiquitination and degradation of TLR4, as well as regulating the secretion of pro-inflammatory cytokines, such as TNFα and IL-6 (Abe et al. 2013). Therefore, Cbl-b-deficient mice exhibit activated macrophages due to saturated FAs, further leading to worse HFD-induced insulin resistance.

Ubiquitination of AMPK

AMPK is located at the center of metabolic signaling pathways, sensing fluctuations in cellular energy supply to acutely modulate the balance between catabolic and anabolic processes (Hardie et al. 2012). AMPK is a heterotrimeric protein kinase consisting of an α catalytic subunit, a β scaffolding subunit and a γ regulatory subunit (Kim et al. 2016a). Activation of AMPK requires phosphorylation of a conserved threonine residue (Thr172) located within the catalytic domain of the AMPK α-subunit. Fbxo48 is an orphan ubiquitin E3 ligase capable of targeting actively phosphorylated AMPKα for polyubiquitylation and proteasomal degradation, and thus regulating mitochondrial fission, facilitating autophagy, and improving metabolic health in HFD-induced obese mice.

Deubiquitylation of TAK1

TAK1 has been considered as an activator of IKKα/β kinases through phosphorylating Ser176/180 residues (Wang et al. 2001). It is reported that TAK1 indirectly participates in pathologic pathways related to diabetes (Jing et al. 2015), but the precise association remains to be elucidated. Several USPs that belong to a huge class of DUBs are involved in the amelioration of metabolic dysfunctions (Lim et al. 2020). For example, in HFD-induced or genetically obese mice, USP18 deubiquitinates TAK1 to inhibit its activation through the DUB activity (An et al. 2017), which constrains the downstream JNK and NF-κB signaling pathways. Therefore, USP18, expressed specifically in the liver, could improve lipid metabolism and insulin sensitivity, protecting liver steatosis and insulin resistance. Instead, another DUB, USP14 enables hepatic triglycerides to accumulate aberrantly, while loss of USP14 resulted from genetic ablation or pharmacological inhibition is capable to ameliorate hepatosteatosis, hyperglycemia, and insulin resistance in obese mice (Liu et al. 2018). Additionally, USP2 (Molusky et al. 2012), USP19 (Coyne et al. 2019), and USP7 (Hall et al. 2014; Forand et al. 2016) also function as regulators in metabolic disorders, suggesting that these USPs may provide efficient targets for intervention of hepatic glucose output for the patients with T2DM.

Ubiquitination of TRAF3IP2

An islet-enriched long non-coding RNA (lncRNA), called β cell function and apoptosis regulator (βFaar), is demonstrated to be necessary and sufficient for β cell apoptosis. TRAF3IP2 and SMURF1 are specifically interacting proteins for βFaar. Among which, SMURF1, an E3 ubiquitin ligase, is essential for TRAF3IP2 ubiquitination and activation of NF-κB-mediated β cell apoptosis (Zhang et al. 2021a). βFaar serves as a molecular scaffold to facilitate the interaction between the E3 ligase SMURF1 and its ubiquitination target TRAF3IP2 to improve β cell function. Therefore, it is necessary to explore the role and mechanism of βFaar, TRAF3IP2, and potentially other βFaar interacting molecules, which may be promising therapeutic targets for obesity-associated diabetes (Zhang et al. 2021a).

Deubiquitination in β Cell Apoptosis

Pancreatic β cell mass is reduced by 30–40% in patients with T2DM compared with non-diabetic subjects, and impaired insulin secretion caused by high loss rate of β cell mass occurs in both obese and lean diabetic patients (Matveyenko and Butler 2008) (Fig. 4.7), highlighting a fundamental role of β cell failure in T2DM pathogenesis.

The islet of T2DM is characterized by deposits of amyloid derived from the peptide islet amyloid polypeptide (IAPP) (Butler et al. 2003). Ubiquitin carboxyl-terminal esterase L1 (UCHL1), a specific DUB involved in the processing of ubiquitin precursors and the polyubiquitin chains, is abundantly expressed in pancreatic β cells in both humans and mice. Under stress conditions, such as high expression of IAPP (Costes et al. 2014) and lipotoxicity (Chu et al. 2012), UCHL1 plays a crucial role in pancreatic β cell survival. UCHL1-deficient islets exhibited decreased ER stress markers including CHOP and BiP, as well as reduced caspase-3-dependent β cell apoptosis (Chu et al. 2012). Moreover, UCHL1 acts as a brake against in β cell proliferation that is regulated by polyubiquitinating pro-

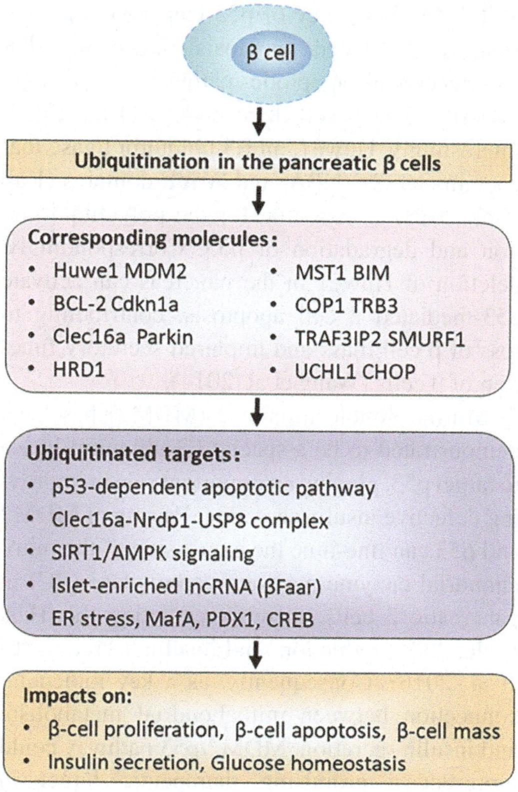

Fig. 4.7 Ubiquitination in pancreatic β cells. The ubiquitination system in pancreatic β cells is influenced by numerous molecules involved in β cell growth, including cell proliferation, cell mass, cell apoptosis, and cell death, which further affect insulin secretion and glucose homeostasis

teins with lysine 63-specific ubiquitin chains (Costes et al. 2014). Together, defective UCHL1 can be characterized as an early risk factor to damage pancreatic β cells, thus decreasing β cell viability and promoting the loss of β cell mass, which ultimately accelerate the onset of T2DM. Up to date, the direct substrates of UCHL1 are unknown in development of T2DM.

4.3.2.2 Ubiquitination and Transcriptional Factors

Ubiquitination of PPARγ

As described above, PPARs can modulate many biological processes, including glucose metabolism, insulin resistance, and energy homeostasis. The expression level of PPARγ is highest in the adipose tissues. Although the precise mechanism is unknown, PPARγ can enhance the expression of glucose transporter type 4 (Glut4) and CAP. Moreover, PPARγ is capable to regulate the expression of numerous factors, such as TNFα, adiponectin, and leptin, which play important roles in insulin sensitivity (Ahmadian et al. 2013). Notably, thiazolidinediones (TZDs), the pharmacological activators of PPARγ, can be used as an insulin-sensitizer to treat T2DM (Gross et al. 2017), indicating the significance of PPARγ in metabolic diseases.

Siah2, a RING-type E3 ubiquitin ligase, is reported to play a critical role in obesity-induced inflammation and insulin resistance (Kilroy et al. 2015). In the adipose tissues of Siah2-deficient mice, pro-inflammatory cytokines and chemokines are decreased. Siah2 can regulate PPARγ ubiquitination and ligand-dependent activation of PPARγ in adipocytes, ameliorating insulin resistance (Kilroy et al. 2012). Besides, Siah2 also affects hypoxia responses via MAPK signaling pathway that indirectly impairs glucose tolerance, supporting a crucial role of Siah2 in glucose metabolism (Segarra-Mondejar et al. 2018). Collectively, Siah2 serves as a key bridge between adipose tissue inflammation and insulin resistance.

ITCH is a ubiquitously expressed E3 ubiquitin ligase involved in T cell differentiation and a wide range of inflammatory pathways, including macrophage-associated signaling pathways (Shembade et al. 2008). Studies supported that ITCH-mediated inflammatory pathways may play a prominent role in protection against obesity and insulin resistance. Interestingly, ITCH deficiency is negatively associated with body weight and blood glucose levels in mice. Taken together, ITCH-mediated pathways have a great influence on the interplay between innate inflammation and metabolism disorders, such as T2DM.

Ubiquitination of FOXO1

Forkhead transcription factor FKHR (also known as FOXO1) is a key regulator of glucose homeostasis. Studies have shown that FKHR is phosphorylated through insulin signaling cascades, resulting in its cytoplasmic retention and the suppression of target gene expression. FKHR, FKHRL1 (FOXO3a), and AFX (FOXO4) (Burgering and Kops 2002) belong to FOXO subfamily of Forkhead transcription factors and can regulate the transcription of gluconeogenic key factors, including glucose-6-phosphatase (G6Pase), phosphoenolpyruvate carboxy kinase (PCK) and peroxisome proliferator-activated receptor-γ coactivator-1 (PGC-1) (Ayala et al. 1999; Schmoll et al. 2000; Hall et al. 2000; Daitoku et al. 2003). Moreover, hepatic glucose production and β cell compensation for insulin resistance is controlled by FKHR in T2DM in mouse models (Nakae et al. 2002). Insulin signaling can trigger FKHR polyubiquitination, resulting in its proteasome-dependent degradation. Currently, the E3 ligases mediating FKHR polyubiquitination remain unknown.

Ubiquitination of HIF-1α

Hypoxia-induced signaling pathways potently modulate cellular metabolism, in which hypoxia-inducible factors (HIFs) have key roles. Increasing HIF-1α protein can improve glucose control in patients with T2DM, resulting in wide-ranging benefits. For example, insulin signaling upregulates the expression of HIF-1α by the PI3K/MAPK pathway. Destabilization of HIF-1α is closely associated with insulin resistance and the onset of T2DM in return (Gunton 2020). TNF receptor associated factor 6 (TRAF6) is found to promote the ubiquitination of HIF-1α, and increased stability of TRAF6 could induce IL-1β expression under inflammatory and hypoxic conditions. In response to this pathway, Pellino3 (Medvedev et al. 2015), an essential factor in mediating immunity homeostasis, is shown to interfere the interaction between TRAF6 and HIF-1α, ultimately restraining the expression of IL-1β and protecting against insulin resistance (Yang et al. 2014). Collectively, Pellino3 may be a promising target for regulating insulin homeostasis and the development of T2DM.

Ubiquitination of p53

In the adipose tissues, the expression of tumor suppressor p53 is crucially involved in the development of β cell apoptosis and insulin resistance (Minamino et al. 2009). During the pathogenesis of T2DM, deficiency of p53 can protect β cells from drug-induced apoptosis, suggesting that p53-dependent apoptotic pathway is of great importance in β cell demise (Kon et al. 2012). Interestingly, Huwe1, an E3 ubiquitin ligase that contains HECT, UBA, and WWE domains (Kao et al. 2018), is essential for the polyubiquitination and degradation of p53. Correspondingly, deletion of Huwe1 in the pancreas can activate p53-mediated β cell apoptosis, contributing to loss of β cell mass and impaired secretory function of β cells (Wang et al. 2014).

Murine double minute 2 (MDM2) has been demonstrated to be a special E3 ubiquitin ligase to target p53, playing an important role in improving defective insulin secretion. Pairing of MDM2 and p53 can fine-tune the expression of the mitochondrial enzyme pyruvate carboxylase (PC) in pancreatic β cells, thereby regulating the TCA cycle, ATP production, and insulin secretion (Li et al. 2016). Consequently, as a key molecular connection between mitochondrial metabolism and insulin secretion, MDM2/p53 pathway could represent a promising therapeutic target in T2DM.

Ubiquitination of CREB

cAMP-responsive element-binding protein (CREB) is an essential transcription factor that binds to the cAMP response element within the promoter of many genes (Qi et al. 2009), which is of great importance in glucose homeostasis and β cell survival. Notably, CREB possesses a potential proteasomal targeting sequence that can be specifically recognized and induced for proteasomal degradation. Previous studies demonstrated that blocking CREB polyubiquitination and its subsequent degradation could regulate the expression of antiapoptotic B-cell lymphoma 2 (BCL-2) and protect β cells from apoptotic death (Costes et al. 2009). Besides, CREB can also regulate the pro-apoptotic gene p21 in β cells (Shin et al. 2014), thus mediating glucagon-like peptide-1 (GLP-1) receptor signaling and preserving β cell mass and insulin secretion.

Ubiquitination of MafA

Musculoaponeurotic fibrosarcoma oncogene homolog A (MafA), a β cell-specific transcription factor, is required for β cell normal development, maturation, and function. HMG-CoA reductase degradation 1 (HRD1), the best characterized ER-associated E3 ubiquitin ligase, plays a pivotal role in β cell function. Misfolded proteins are ubiquitinated by the cytoplasmic RING domain of HRD1 in the cytosol and ultimately destroyed via the UPS (Okuda-Shimizu and Hendershot 2007). Recent evidence showed that HRD1 can effectively degrade misfolded proinsulin to improve pancreatic β cell function, and thus absence of HRD1 results in a decrease in glucose-stimulated insulin secretion (GSIS) (Hu et al. 2019). Indeed, HRD1 expression level is markedly increased in both human and mouse diabetic islets, which facilitates the ubiquitination and degradation of MafA (Wu et al. 2020). Therefore, absence of MafA caused by ubiquitination of HRD1 could impair insulin secretion and ultimately lead to β cell dysfunction.

Ubiquitination of PDX1

The transcription factor pancreatic duodenal homeobox-1 (PDX1, previously called IPF1) (Jonsson et al. 1994) is identified as a crucial regulator in insulin transcription and β cell phenotype. In humans, mutations in the PDX1 gene can predispose individuals to develop maturity onset diabetes of adult. Downregulated PDX1 affects insulin production and secretion, and predisposes β cells to apoptosis (Claiborn et al. 2010). Mammalian sterile 20-like kinase-1 (MST1), a ubiquitously expressed serine-threonine kinase, is involved in multiple apoptotic signaling pathways (Ardestani et al. 2014). Under diabetogenic conditions, MST1 is markedly activated in β cells and specifically induces apoptosis through upregulation of the BCL-2 homology-3 (BH3)-only protein BIM. Furthermore, MST1 can directly phosphorylate PDX1, contributing to PDX1 ubiquitination and degradation by the E3 ubiquitin ligase cullin 3, ultimately impairing glucose homeostasis and insulin secretion (Brissova et al. 2002).

Ubiquitination of Xbp1

ER stress is a central pathology for insulin resistance and T2DM. X-box binding protein 1 spliced (XBP1s) is a transcription factor playing a crucial part in improving insulin resistance and maintaining glucose homeostasis (Liu et al. 2016). It has been demonstrated that in obese mice, the causes of ER stress and insulin resistance include reduced levels of XBP1s and impaired nuclear import of XBP1s in the liver, whereas in obese and diabetic mice, liver-specific overexpression of XBP1s improved ER and glucose homeostasis (Zhou et al. 2011; Lee et al. 2011). It was reported that two pivotal ubiquitin sites K60 and K77 of XBP1s are involved in proteasome-dependent degradation of XBP1s (Sun et al. 2020). Moreover, FBW7 and Cullin3-SPOP have been identified as essential regulators for ubiquitination of XBP1s (Chae et al. 2019; Sun et al. 2019).

4.3.2.3 Ubiquitination and Metabolic Enzymes

Ubiquitination of Fetuin-A

Fetuin-A, a hepatokine that promotes the production of inflammatory cytokines, has been recognized as a marker of inflammatory diseases. In particular, fetuin-A could be an endogenous ligand of TLR4 to regulate insulin sensitivity and manage insulin resistance in T2DM (Pal et al. 2012). The F-box and WD repeat domain containing 7 (FBXW7), a tumor suppressor (Yumimoto and Nakayama 2020), is also a central E3 ubiquitin ligase that mediates recognition of phosphorylated substrates for proteolysis. A recent report exhibited that depletion of FBXW7 in the liver leads to accumulation of fetuin-A and dysregulated glucose homeostasis. Mechanistically, FBXW7 directly binds to fetuin-A to induce its ubiquitination and subsequent proteasomal degradation, ultimately preventing the development of HFD-induced insulin resistance (Yan et al. 2021). Additionally, a functional FBXW7 variant (A204T) is positively correlated with blood glucose level and T2DM risk in animal models (Zhao et al. 2018b), highlighting a protective role of FBXW7 in the pathobiology of T2DM.

Ubiquitination of Parkin

It has been shown that dysfunction of parkin, a ubiquitin-protein isopeptide ligase, could lead to accumulation of its substrates, thus contributing to death of dopaminergic neurons (Zhong et al. 2005). Intriguingly, parkin is a fundamental regulator of mitophagy in diabetes and indirectly involved in the pathogenesis of T2DM (Soleimanpour et al. 2014). C-type lectin domain family 16, member A (Clec16a) has been identified as a diabetes gene (Soleimanpour et al. 2014) that encodes an E3 ligase, which can ubiquitinate neuregulin receptor degradation protein (Nrdp1) and control mitophagy flux in pancreatic β cells. Loss of Clec16a can facilitate an increase in parkin, one of Nrdp1 targets (Hoshino et al. 2014). On the contrary, overexpression of Clec16a reduced the level of cleaved caspase-3 via assembly of the Clec16a-Nrdp1-USP8 mitophagy complex (Pearson et al. 2018), indicating that restoration of the Clec16a pathway partially improves the negative effects of glucolipotoxicity on β cell survival.

Ubiquitination of SIRT1

Sirtuin-1 (SIRT1) plays an essential role in the development of metabolic diseases. The activation of SIRT1 can induce ketogenesis through promoting FAs oxidation and gluconeogenesis. Additionally, with working in concert with AMPK, SIRT1 can directly inhibit inflammatory responses (Packer 2020). Inhibition of SIRT1/AMPK signaling has a key role in the pathogenesis of T2DM. The COP1 protein is an evolutionarily conserved E3 ubiquitin ligase that belongs to the largest family of E3 enzymes CRL. Mice lacking COP1 in β cells are more likely to develop diabetes due to insulin granule docking defects, indicating the importance of COP1 in regulating β cell insulin secretion (Suriben et al. 2015). Notably, the COP1 can protect β cell function and improve insulin secretion through ubiquitinating the ETV family, which are crucial transcription factors strongly associated with obesity and T2DM (Lin et al. 2021). Additionally, TRB3-COP1-SIRT1 pathway plays an important role in lipotoxicity, which can lead to insulin resistance in hepatocytes by targeting SIRT1 for degradation (Ren et al. 2019).

4.3.3 NAFLD

NAFLD has become the most prevalent liver disease with noneffective pharmacological therapies to prevent or treat. NAFLD encompasses a range of severity, from simple steatosis to non-alcoholic steatohepatitis (NASH). Generally, the "two hits" hypothesis can be used to explain the progression of NAFLD to NASH. The "first hit" is represented by the accumulation of lipids in hepatocytes (liver steatosis), which then sensitizes the liver to injury induced by "second hits," such as oxidative and ER stress. Second hits is followed by histological hallmarks of NASH such as the development of insulin resistance, inflammation, ballooned hepatocytes, and the formation of inclusion bodies in the liver (Khan and Newsome 2018).

4.3.3.1 Ubiquitination and Metabolic Signaling Pathways

Ubiquitination of AMPK

AMPK is a critical sensor of cellular energy and nutrient levels, which, under the energy stress condition, instigates the cellular energy-generating systems along with suppression of the anabolic signaling pathways. Previous studies have expanded the functions of AMPK in a variety of metabolic syndromes, such as obesity, T2DM, fatty liver syndrome, and cardiovascular diseases. It was reported that the ubiquitination and proteasome-dependent degradation of AMPKα1 and AMPKα2 can be mediated by Makorin Ring Finger Protein 1 (MKRN1) to regulate their protein homeostasis (Lee et al. 2018). MKRN1 depletion promotes glucose consumption and suppresses lipid accumulation due to AMPK stabilization and activation. In addition, the E3 ubiquitin-ligase CHIP was found to specifically target AMPKα1, but not AMPKα2, and knockout of liver CHIP could delay the pathogenic progression into NASH by activating adiponectin-AMPK-FOXO pathway (Kim et al. 2016b).

Ubiquitination of ASK1

Apoptosis Signal-regulating Kinase 1 (ASK1), a member of the mitogen-activated protein kinase kinase kinase (MAP3K) family and an upstream activator of the JNK and p38 MAPK signaling cascades (Ichijo et al. 1997; Ichijo 1999), is activated by various stressors such as ROS, TNFα, ER stress, and LPS (Imoto et al. 2006; Tobiume et al. 2001). Liver-specific ASK1 expression blunts obesity-associated hepatic steatosis and fibrosis potentially through induction of autophagy (Challa et al. 2019). E3 ligase protein F-box and WD repeat domain containing protein 5 (FBXW5) is the central component of the SCF complex that directly interacts with and mediates K63-linked polyubiquitination of ASK1 in hepatocytes and thus exacerbating ASK1-JNK/p38 MAPK signaling, inflammation, and lipid accumulation. Thus, diet-induced systemic and hepatic metabolic disorders is exacerbated due to hepatocyte-specific overexpression of FBXW5, as well as the activation of ASK1-related MAPK signaling in vivo (Bai et al. 2019).

Additionally, TNF receptor-associated factor 6 (TRAF6) can also catalyze polyubiquitination activation of ASK1 upon metabolic challenge, thereby inhibiting the downstream JNK and p38 signaling pathway, whereas TNFα-induced protein 8-like 1 (TIPE1), a new member of the TNFα-induced protein 8 family, could directly interact with ASK1 to suppress the ubiquitination by TRAF6 and protect against hepatic steatosis, inflammation, and fibrosis during the development of NASH (Challa et al. 2019). Intriguingly, ovarian tumor domain, ubiquitin aldehyde binding 1 (OTUB1), the founding member of deubiquitinases, directly bound to TRAF6 and suppressed its lysine 63-linked polyubiquitination, thus inhibiting the activation of ASK1 and its downstream pathway (Zhang et al. 2021b).

Moreover, deubiquitinase tumor necrosis factor alpha-induced protein 3 (TNFAIP3) can also directly interact with and deubiquitinate ASK1 in hepatocytes. Evidence showed that hepatocyte-specific ablation of TNFAIP3 exacerbated NAFLD- and NASH-related phenotypes in mice, including aberrant glucose metabolism, lipid accumulation, and enhanced inflammation, and these phenomena are dependent on the presence of ASK1. On the contrary, in mouse and non-human primate NASH models, the transfer of the TNFAIP3 gene in the liver by transgenic or adenovirus-mediated vectors greatly prevents the occurrence and development of the disease (Zhang et al. 2018).

Furthermore, melatonin could activate a scaffold protein β-arrestin-1 and enable it to bind to ASK1, which antagonized the TRAFs-mediated ASK1 deubiquitination, and thus reduced ASK1 protein stability, although the specific E3 ligase was unclear (Li et al. 2019).

Ubiquitination of IκBα

As mentioned above, IκB is an important molecule of NF-κB signaling pathway that is involved in regulation of NAFLD. Tumor necrosis factor receptor-associated ubiquitous scaffolding and signaling protein (TRUSS) is characterized as a tumor necrosis factor receptor 1 (TNF-R1)-interacting protein (Soond et al. 2003). The TRUSS mRNA is widely expressed in mouse tissues but enriched in the brain, heart, liver, and testis (Soond et al. 2003, 2006). Functional studies indicated that TRUSS could activate many transcription factors, including NF-κB and activator protein 1, and serve as a scaffolding protein that links TNF-R1 to its downstream components. Moreover, TRUSS physically interacts with the inhibitor of nuclear factor κB α (IκBα) and promotes its ubiquitination and degradation, leading to aberrant activation of NF-κB, therefore regulating NAFLD development (Yu et al. 2018) (Fig. 4.8). However, the direct upstream E3 ligases for IκBα in NAFLD need further investigation.

Ubiquitination of p-Smad3

Increased apoptosis of hepatocytes is an important characteristic of NASH, as demonstrated in both clinical and animal studies (Feldstein et al. 2003; Hatting et al. 2013), and hepatocellular apoptosis then activates the abnormal liver's regenerative responses, including fibrosis and

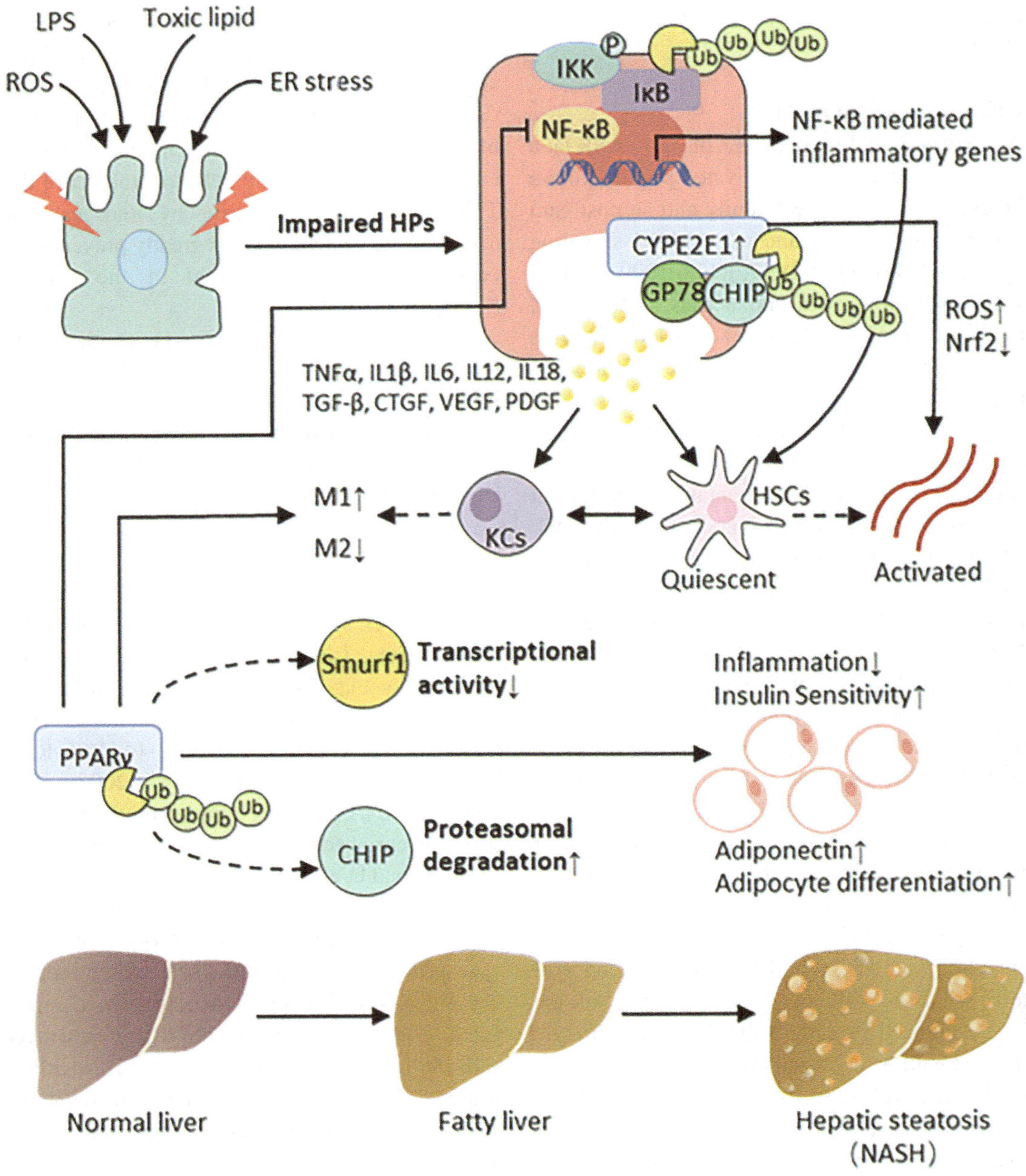

Fig. 4.8 The relationship between ubiquitination and progression of NAFLD. During development of NAFLD, various stimuli (e.g., ROS, LPS, ER stress, toxic lipids) accelerate damages of hepatocytes. These stressed and stimulated hepatocytes release NF-κB-mediated inflammatory genes that activate both Kupffer cells (KCs) and hepatic stellate cells (HSCs) by ubiquitin-mediated degradation of IκB domain. Also, apoptosis/necroptosis of hepatocytes secrete different pro-inflammatory cytokines that induce M1 polarization of KCs and myofibroblast phenotype in HSCs. PPARγ is an important regulator of adipocyte biogenesis, insulin sensitivity, and inflammation in the pathogenesis of NAFLD. It is notable that PPARγ could suppress NF-κB-mediated inflammatory signaling and induce M2 polarization of macrophages. Furthermore, E3 ligase Smurf1 induces non-proteolytic lysine 63 (K63)-linked ubiquitin modification that suppresses PPARγ transcriptional activity, while E3 ligase CHIP induces proteasome-mediated degradation of PPARγ, thereby regulating haptic lipid accumulation and insulin sensitivity

inflammation (Alkhouri et al. 2011). Moreover, chronic inflammation in the liver stimulates macrophages to produce cytokines, which further promotes hepatocyte apoptosis, thus aggravating NASH (Guicciardi and Gores 2010). Among the cytokines, transforming growth factor beta1 (TGFβ1) is strongly associated with fibrosis and apoptosis, and plays a major role in the progression of NASH (Fabregat et al. 2016). Activation of TGFβ1 signaling increases the expression level of hepatic v-ets avian erythroblastosis virus E26 oncogene homolog 1 (Ets-1) in NASH mice. In addition, Ets-1 can directly interact with phosphorylated Smad3 and prevent its ubiquitination and proteasomal degradation, thus enhancing the activity of TGFβ1/Smad3 signaling and accelerating the development of NASH (Liu et al. 2019).

Ubiquitination of TRAF6

Leukocyte immunoglobulin-like receptor B4 (LILRB4), an immunoreceptor, primarily expressed on the membrane of immune cells, where it transduces a negative signal during immune response. During NAFLD progression, LILRB4 expression in hepatocytes increases dramatically. Mechanistically, LILRB4 enable to recruit SHP1 to inhibit the ubiquitination of TRAF6 and subsequent inactivation of NF-κB and MAPK signaling pathways (Lu et al. 2018), but the direct upstream E3 ligases are unclear and need further investigation.

Ubiquitination in ERAD

Endoplasmic reticulum (ER) is a membranous network responsible for synthesis, maturation, and sorting of proteins to the plasma membrane or extracellular milieu (Hotamisligil 2010). The unfolded protein response (UPR) is activated in response to pathophysiological agents or conditions that trigger ER stress by reducing protein synthesis, facilitating protein degradation, and increasing production of chaperones and foldases that guide nascent or misfolded protein to fold correctly (Hetz 2012). ERAD refers to the process by which misfolded proteins that cannot be functionally resolved are transported from the ER to the cytosol and degraded by the ubiquitin-proteasome complexes (Vembar and Brodsky 2008; Tsai et al. 2002). ERAD has been identified as a cellular pathway of the protein quality control system in ER. In HFD mice, liver proteasome activity is reduced, which causes ER stress, UPR activation, and insulin resistance (Otoda et al. 2013). This suggests that defects in proteasome function may play an important role in NAFLD, consistent with the notion that ER stress plays a crucial role in the development of NAFLD (Pagliassotti 2012; Malhi and Kaufman 2011) (Fig. 4.8).

Gp78, an E3 ligase, can regulate ERAD by ubiquitination of misfolded ER proteins. Gp78-deficient mice developed obesity, recapitulating age-related human NASH. Accordingly, liver histology of Gp78-deficient mice revealed typical steatosis, hepatic inflammation, and fibrosis, followed by progression to hepatocellular tumors (Zhang et al. 2015) (Fig. 4.8).

In addition, CHIP, a cytoplasmic Hsc70-cochaperone, functions with its cognate UbcH5a E2 and Hsc70/Hsp40 co-chaperones in substrate ubiquitination. CHIP comprises a catalytic U-Box with a cross-brace structure, and a classical N-terminal tetratricopeptide (TPR) domain for Hsc70/Hsp40 or Hsp90 recruitment. It was reported that CHIP ubiquitinates native P450 proteins in vitro, and knockdown or knockout of CHIP results in the stabilization of functionally active P450s, thereby regulating hepatic oxidative stress, JNK1-activation, insulin resistance, FA- and TG-accumulation, and liver injury, and eventually NASH (Kim et al. 2016b; Kwon et al. 2020) (Fig. 4.8).

The ER-localized E3 ubiquitin-protein ligase, ring finger protein5 (RNF5), efficiently promotes the degradation of misfolded proteins as well as regulates the stability and clearance of proteins functioning in multiple cellular activities. Interestingly, RNF5 was significantly decreased in NASH livers across multiple species. Hepatocyte-specific RNF5 knockout markedly aggravated hepatic steatosis, inflammatory response, and fibrosis in mice challenged with diet-induced NASH, by interacting with HRD1 and then promoting its lysine 48 (K48)-linked and K33-linked ubiquitination and subsequent proteasomal degradation (Yang et al. 2021).

Ubiquitination of c-FLIPL

NASH is characterized by progressive liver injury, which is a result of hepatocyte death. Accumulating studies have revealed that hepatocyte death is a critical driver of liver disease progression (Brenner et al. 2013; Ding and Yin 2004; Luedde et al. 2014). Aberrantly increased liver cell death is linked to activation of liver regeneration and wound repair, contributing to persistent inflammation and liver fibrosis. c-FLIP$_L$ is a negative regulator of cell death, and its ubiquitination can mediate the function of Neuregulin 4 (Nrg4) on steatosis-to-NASH progression. Nrg4 is a key component of an endocrine checkpoint that preserves hepatocyte health and counters NASH. Transgenic expression of Nrg4 in the adipose tissues alleviated diet-induced NASH by blocking ubiquitination and proteasomal degradation of c-FLIP$_L$ in a cell-autonomous manner. Moreover, Itchy E3 ubiquitin protein ligase (Itch) was previously identified as a JNK-regulated E3 ubiquitin ligase for c-FLIP$_L$ (Chang et al. 2006), implicating a role in NASH progression (Guo et al. 2017).

Ubiquitination of TAK1

TAK1, a member of MAP3K family, is an important factor on IKKβ-NF-kB and MKK-JNK pathways that are dominant molecular signalings orchestrating inflammation and insulin resistance (Ajibade et al. 2013). TRAF3, a tumor necrosis factor receptor (TNFR)-associated factor (TRAF) family member, can function as signaling adaptors in various pathophysiologic processes (Bradley and Pober 2001). In NAFLD, TRAF3 expression is increased, and it can bind to and ubiquitinate TAK1, leading to TAK1 autophosphorylation and activation. Activated TAK1 subsequently results in an activation of downstream MKK-JNK-IRS1307 and IKKβ-NF-kB signaling, as well as an inhibition of AKT-GSK3β/FOXO1 phosphorylation cascades, which induce insulin resistance, inflammatory responses and glucose metabolic disorder in the liver, thus exacerbating FAs accumulation and hepatic steatosis (Wang et al. 2016).

Additionally, the E3 ligase TRIM8 could ubiquitinate TAK1 and their interaction was inhibited by the TNFAIP3 interacting protein 3 (TNIP3) can inhibit the interaction between the E3 ligase TRIM8 and TAK1, thereby attenuating TAK1 ubiquitination and suppressing NASH (Liu et al. 2020). TRAF6 could also ubiquitinate TAK1 and inhibit the TAK1 phosphorylation and activation of TAK1-JNK/p38 signaling, which mediates the function of Tumor necrosis factor α-induced protein 8-like 2 (TIPE2) during NAFLD development (Liu et al. 2021a). The E3 ligase tripartite motif 16 (TRIM16) could preferentially interact with phospho-TAK1 to promote its degradation, and thus suppress NASH progression by attenuating the MAPK signaling pathway (Wang et al. 2021). TRAF3 binds to TAK1 to induce TAK1 ubiquitination and subsequent autophosphorylation, thereby enhancing the activation of downstream IKKβ-NF-κB and MKK-JNK-IRS1 signaling cascades (Wang et al. 2016).

Moreover, DUBs are also involved in the regulation of TAK1. For example, USP8 can regulate TAK1 ubiquitination and deteriorate its protein stability, which in turn induce pyroptosis and M1 polarization of macrophages, thus contributing to liver fibrosis (Shu et al. 2021).

Ubiquitination of Raptor

Raptor, a regulatory protein of mTORC1, maintains the correct subcellular localization of mTORC1 and allows the binding of mTORC1 with substrates (Hussain et al. 2013). CUL4B-DDB1, an E3 ligase complex, interacts with and ubiquitinates Raptor for degradation. This interaction was interfered by pancreatic progenitor cell differentiation and proliferation factor (PPDPF) (Ma et al. 2021), therefore negatively regulating mTORC1-S6K-SREBP1 signaling in NAFLD.

4.3.3.2 Ubiquitination and Transcription Factors

Ubiquitination of PPARs

As described above, PPARs control lipid metabolism, glucose metabolism, and overall energy homeostasis. Among them, PPARα (encoded by PPARA) has mainly been characterized in the liver, where it regulates the adaptive response to fasting by controlling FAs transport, β-oxidation, and ketogenesis (Pawlak et al. 2015). Despite PPARα is one of the most abundantly expressed nuclear receptors in the liver, its mRNA transcripts are progressively decreased during NASH progression in humans (Francque et al. 2015). It has been demonstrated that both whole body and hepatocyte-specific PPARα-deficient mice develop aggravated NASH upon HFD and methionine and choline-deficient diet (MCD) feeding (Abdelmegeed et al. 2011; Pawlak et al. 2014; Montagner et al. 2016), and pharmacological activation of PPARα has preventive and curative effects on NASH in preclinical models (Ip et al. 2003, 2004). Mechanistically studies have showed that in addition to epigenetic and post-transcriptional regulation involved in the functions of PPARα in NAFLD (Zheng et al. 2010; Loyer et al. 2016), certain protein modifications are implicated in this process. Serine 73 of PPARα could be phosphorylated by GSK3β, facilitating its ubiquitination and protein turnover, and eventually hepatic steatosis (Hinds et al. 2016). The expression of PPARγ is normally low but is drastically induced as hepatic steatosis develops in the liver (Gavrilova et al. 2003). Studies have shown that overexpression of PPARγ promotes the lipid droplets accumulation in the liver, while hepatic disruption of PPARγ improves the fatty liver condition in leptin-deficient obese mice or mice that were fed on a HFD (Matsusue et al. 2003; Moran-Salvador et al. 2011). It was elucidated that Smad ubiquitin regulatory factor 1 (Smurf1) regulates its target genes engaged in lipid synthesis and FAs uptake by inducing non-proteolytic lysine 63 (K63)-linked ubiquitin modification (Zhu et al. 2018b). Moreover, as mentioned above, CHIP was also demonstrated to facilitate the ubiquitination and degradation of PPARγ, thus impairing insulin sensitivity (Fig. 4.8).

Ubiquitination of Nrf2

ROS is important to the development of liver fibrosis, and it could activate the hepatic stellate cells (HSCs) and stimulate the production of extracellular matrix proteins, including type I collagen. Nuclear factor erythroid 2-related factor 2 (Nrf2) is essential for cellular protective mechanisms against ROS by activating the transcriptional of antioxidant-responsive element-dependent genes. Watanabe et al. found that MIYAIRI 588, a butyrate-producing probiotic, reduced CDAA-diet-induced hepatic lipid deposition and significantly improved the triglyceride content, insulin resistance, serum endotoxin levels, and hepatic inflammatory indexes by precluding ubiquitination of Nrf2 and increasing the half-life of the Nrf2 protein (Endo et al. 2013), although the upstream E3 ligase remained unknown. Of note, KEAP1-CUL3-RBX1, a E3 ubiquitin ligase complex, was reported to mediate the Nrf2 for proteasome degradation in cancer (Cullinan et al. 2004; Martinez et al. 2013; Di Gregorio et al. 2021). Moreover, deubiquitinase USP11 directly reverses Nrf2 polyubiquitination and stabilizes Nrf2 in non-small cell lung carcinoma (NSCLC) (Meng et al. 2021). It is of interest to explore whether these regulations are also existed in NAFLD.

Ubiquitination of SREBP1

Sterol regulatory element binding protein (SREBP) family of transcription factors controls cholesterol and lipid metabolism (Osborne 2000). There are two genes, srebf1 and srebf2, which encode three different SREBP proteins, including SREBP1a, SREBP1c, and SREBP2. Among which, SREBP1c is an important regulator of insulin-dependent gene expression. It is induced in hepatic steatosis and its hyperactivation leads to liver triglyceride accumulation (Shimomura et al. 1999). FBW7, an F-box protein with seven tandem WD40 repeats, is a key substrate recognition subunit of the SCF E3 ubiquitin ligase, and mediates recognition of phosphorylated substrates, such as phosphorylated cyclin E, c-Myc, and c-Jun for degradation (Koepp et al. 2001; Nateri et al. 2004; Strohmaier et al. 2001; Welcker et al. 2004). In addition, Ericsson and his colleagues found that FBW7 enhances nuclear SREBP1a ubiquitination and degradation in the way that GSK-3 phosphorylates T426 and S430. FBW7 also degrades nuclear SREBP1c and SREBP2, and inactivating endogenous FBW7 can stabilize nuclear SREBP1c and -2, increase expression of SREBP target genes, improve synthesis of cholesterol and FAs, and enhance receptor-mediated uptake of LDL (Sundqvist et al. 2005).

Besides, homolog of mouse double minute 2 (HDM2), an E3 neddylation ligase of SREBP1c, could promote neddylation of SREBP1c, which competes with its ubiquitination and stabilizes SREBP1c protein level, eventually promoting hepatic steatosis.

4.3.3.3 Ubiquitination and Metabolic Enzymes

Ubiquitination of ATGL

Adipose triglyceride lipase (ATGL) plays a compelling role in hepatic lipid turnover and in the pathophysiology of NAFLD. Hepatic ATGL is post-transcriptionally regulated by E3 ubiquitin ligase COP1 through polyubiquitylation and proteasomal degradation (Ghosh et al. 2016). Pigment epithelium-derived factor (PEDF) facilitates nuclear import of predominantly cytosolic ATGL protein and promotes its subsequent proteasomal degradation of ATGL in the nucleus in COP1-dependent manner, thus controlling hepatocyte lipid accumulation and mobilization during NAFLD development (Niyogi et al. 2019). In addition, ubiquitin/ISG15-conjugating enzyme E2 L6 (UBE2L6/Ube2l6) could also promote ubiquitination of ATGL, thereby aggravating HFD-induced obesity, insulin resistance, and hepatic steatosis (Wei et al. 2021).

Moreover, ubiquitination of ATGL has a significant role in sensing intracellular FAs. Excessive FA release, as a result of uncontrolled lipolysis, results in lipotoxicity, which can in turn promote the progression of metabolic disorders. It has recently showed that changes in ROS levels are sensed by peroxin 2 (PEX2), which modulates K48-linked polyubiquitination of ATGL at the K92 site for proteasomal degradation (Ding et al. 2021).

Ubiquitination of ACLY

ACLY, a lipogenic enzyme in the process of de novo lipogenesis (DNL), catalyzes the conversion of citrate to acetyl-CoA (Bai and Li 2019; Icard et al. 2021). Both the ACLY expression level and acetyl-CoA content are closely associated with NAFLD (Guo et al. 2019). Specifically, liver-specific deletion of ACLY can inhibit hepatic DNL and protect db/db mice from hepatic steatosis (Wang et al. 2009). Furthermore, HRD1, a subunit of the ERAD complex, could interact with and ubiquitinate ACLY, thereby degrading ACLY. Correspondingly, HRD1 suppressed the Acetyl-CoA level and inhibited lipogenesis via an ACLY-dependent pathway. In addition, the expression of hepatic HRD1 was negatively correlated with NAFLD. In db/db mice and in separated mouse primary hepatocytes overexpression of HRD1 enable to ameliorate hepatic steatosis and enhanced insulin sensitivity (Li et al. 2021).

Ubiquitination of FASN

DNL is a fundamental biosynthetic process driving FAs synthesis and contributes to 20–25% of triglyceride deposition in the liver of individuals with NAFLD (Chakravarthy et al. 2009). Fatty acid synthase (FASN) is a key regulator of hepatic DNL and a promising therapeutic target for NAFLD. FASN inhibitors, such as TVB-2640 and FT-101, have supported the therapeutic potential of FASN for NAFLD treatment (Beysen et al. 2021; Syed-Abdul et al. 2020). The E3 ligase tripartite motif containing 28 (TRIM28) could interact with and ubiquitinate FASN for degradation, which mediates the effect of sorting nexin 8 (SNX8) on NAFLD development. SNX8 interacts with FASN and recruits TRIM28 to facilitate UPS-mediated degradation. Moreover, FASN and SH2 domain containing protein-tyrosine phosphatase Shp2/PTPN11 can be recruited by the E3 ligase COPI and form a complex which induces FASN degradation (Yu et al. 2013), while the USP14 and USP2a promote hepato-steatosis through binding to FASN and increase its stability directly (Liu et al. 2018; Graner et al. 2004).

Ubiquitination of Sirt1

SIRT1, a nicotinamide adenine dinucleotide (NAD)-dependent deacetylase, serves as an energy sensor to modulate insulin sensitivity through its substrates. MST1, a member of the mammalian sterile 20-like kinases (MSTs), shares the highest degree of homology with MST2, and is ubiquitously expressed. MST1 is part of the Hippo signaling pathway and involved in multiple cellular processes. Liver-specific knockout of MST1 in mice showed more severe hepatic metabolic damage under fasting and HFD than that of control mice. In contrast, MST1 overexpression inhibited SREBP1c expression and increased the expression of antioxidant genes in primary hepatocytes by inhibiting SIRT1 ubiquitination (Geng et al. 2016). Although it is unknown which E3 ligase regulates SIRT1 ubiquitination in MST1-dependent process, many E3 ligases have been identified as direct E3 ligases for SIRT1 under some conditions. For instance, GRAIL is a type I transmembrane protein that localizes to the endocytic pathway and bears homology to RING zinc-finger proteins. It was upregulated in the livers of obese mice and palmitate acid-treated liver cells. Importantly, GRAIL deficiency remarkably alleviated the HFD-induced hepatic fat accumulation in vitro and in vivo by ubiquitinating SIRT1 and regulating its protein stability (Liu et al. 2021b). Similarly, COP1 is an evolutionarily conserved E3 ligase for UPS. It was found that COP1 can be recruited by its adaptor protein TRB3, and then targeted SIRT1 for degradation in hepatocytes, as described above (Ren et al. 2019). Besides, the E3 ubiquitin ligase SMURF2 was found to interact with SIRT1 and mediate its ubiquitination and degradation, during colorectal cancer (CRC) cell proliferation and tumorigenesis (Liu et al. 2021b). The E3 ligase CHFR, responsible for SIRT1 proteasomal degradation, leads to an increase of p53 acetylation and its target gene transcription under oxidative stress conditions (Kim et al. 2016c).

4.4 Conclusions

Deregulation of ubiquitination results in diverse types of cellular dysfunction that exacerbate the development of metabolic diseases, whereas appropriate ubiquitination generally maintains cellular homeostasis and functions to improve metabolic health. As aberrant ubiquitination is implicated in numerous obesity-related diseases, such as T2DM and NAFLD, therapeutic regulation of ubiquitination to adequate levels is a potential treatment strategy for these diseases.

UPS, a selective proteolytic system in which the conjugation of ubiquitin to substrates induces degradation by the proteasome, is the most important pathway to mediate protein degradation and responsible for approximately 80–90% of cellular proteolysis. In the 20 years since the concept of a proteolysis targeting chimera (PROTAC) molecule harnessing the UPS to degrade a target protein was reported, targeted protein degradation (TPD) has moved from academia to industry, where numerous companies have disclosed programs in preclinical and early clinical development. It is possible that targeted protein degradation will become a key therapeutic modality for the treatment of metabolic diseases in the future, and further studies are necessary to identify more applicable targets and then design PROTAC molecules to fight against obesity and obesity-related diseases.

In addition to UPS, eukaryotic cells operate another major proteolytic system: autophagy, which is a process by which cytoplasmic constituents are degraded by the lysosome and responsible for approximately 10–20% of cellular proteolysis, although the ratios may vary depending on physiological states and cell types. Emerging evidence shows that there are many crosslinks between UPS and autophagy. Firstly, both autophagy and UPS could use ubiquitin as a label to degrade ubiquitinated substrates. Secondly, ubiquitinated proteasomes could be cleared by autophagy, and autophagy-related proteins could also be degraded through the proteasome. Thirdly, these two pathways can synergistically degrade the same substrate within the cell. Finally, they can regulate each other's activity. Consistently, either autophagy disability or ubiquitination results in diverse types of cellular dysfunction, such as metabolic diseases, whereas enhancement of autophagy promotes cellular homeostasis generally. Hence, therapeutic co-upregulation of ubiquitination and autophagy to appropriate levels may be a potential strategy for treating these diseases, while further research on establishing non-invasive methods for measuring ubiquitin and autophagy is beneficial for detecting the effects of drug treatment on ubiquitin and autophagy in humans.

Acknowledgements This work was supported by the National Natural Science Foundation of China (92157205, 81970561, 82172896, and 82103128), the China National Postdoctoral Program for Innovative Talents (BX2021201), the 1.3.5 Project for Disciplines Excellence, West China Hospital, Sichuan University (ZYJD23005), and the Post-Doctor Research Project, West China Hospital, Sichuan University (2020HXBH121).

Competing Interests The authors have declared that no competing interest exists.

References

Abdelmegeed MA et al (2011) PPARalpha expression protects male mice from high fat-induced nonalcoholic fatty liver. J Nutr 141:603–610

Abe T et al (2013) Cbl-b is a critical regulator of macrophage activation associated with obesity-induced insulin resistance in mice. Diabetes 62:1957–1969

Ahmadian M et al (2013) PPARgamma signaling and metabolism: the good, the bad and the future. Nat Med 19:557–566

Ajibade AA, Wang HY, Wang RF (2013) Cell type-specific function of TAK1 in innate immune signaling. Trends Immunol 34:307–316

Akutsu M, Dikic I, Bremm A (2016) Ubiquitin chain diversity at a glance. J Cell Sci 129:875–880

Alkhouri N, Carter-Kent C, Feldstein AE (2011) Apoptosis in nonalcoholic fatty liver disease: diagnostic and therapeutic implications. Expert Rev Gastroenterol Hepatol 5:201–212

An S et al (2017) USP18 protects against hepatic steatosis and insulin resistance through its deubiquitinating activity. Hepatology 66:1866–1884

Ardestani A et al (2014) MST1 is a key regulator of beta cell apoptosis and dysfunction in diabetes. Nat Med 20:385–397

Ayala JE et al (1999) Conservation of an insulin response unit between mouse and human glucose-6-phosphatase catalytic subunit gene promoters: transcription factor FKHR binds the insulin response sequence. Diabetes 48:1885–1889

Bai L, Li H (2019) Innate immune regulatory networks in hepatic lipid metabolism. J Mol Med (Berl) 97:593–604

Bai L et al (2019) F-box/WD repeat-containing protein 5 mediates the ubiquitination of apoptosis signal-regulating kinase 1 and exacerbates nonalcoholic steatohepatitis in mice. Hepatology 70:1942–1957

Beysen C et al (2021) Inhibition of fatty acid synthase with FT-4101 safely reduces hepatic de novo lipogenesis and steatosis in obese subjects with non-alcoholic fatty liver disease: results from two early-phase randomized trials. Diabetes Obes Metab 23:700–710

Blackford AN, Jackson SP (2017) ATM, ATR, and DNA-PK: the trinity at the heart of the DNA damage response. Mol Cell 66:801–817

Bluher M (2019) Obesity: global epidemiology and pathogenesis. Nat Rev Endocrinol 15:288–298

Bradley JR, Pober JS (2001) Tumor necrosis factor receptor-associated factors (TRAFs). Oncogene 20:6482–6491

Braten O et al (2016) Numerous proteins with unique characteristics are degraded by the 26S proteasome following monoubiquitination. Proc Natl Acad Sci U S A 113:E4639–E4647

Bremm A, Moniz S, Mader J, Rocha S, Komander D (2014) Cezanne (OTUD7B) regulates HIF-1alpha homeostasis in a proteasome-independent manner. EMBO Rep 15:1268–1277

Brenner C, Galluzzi L, Kepp O, Kroemer G (2013) Decoding cell death signals in liver inflammation. J Hepatol 59:583–594

Brissova M et al (2002) Reduction in pancreatic transcription factor PDX-1 impairs glucose-stimulated insulin secretion. J Biol Chem 277:11225–11232

Brown NG et al (2016) Dual RING E3 architectures regulate multiubiquitination and ubiquitin chain elongation by APC/C. Cell 165:1440–1453

Burgering BM, Kops GJ (2002) Cell cycle and death control: long live Forkheads. Trends Biochem Sci 27:352–360

Butler AE et al (2003) Beta-cell deficit and increased beta-cell apoptosis in humans with type 2 diabetes. Diabetes 52:102–110

Cao Z, Xiong J, Takeuchi M, Kurama T, Goeddel DV (1996) TRAF6 is a signal transducer for interleukin-1. Nature 383:443–446

Cederquist CT et al (2017) Systemic insulin sensitivity is regulated by GPS2 inhibition of AKT ubiquitination and activation in adipose tissue. Mol Metab 6:125–137

Chae U et al (2019) A negative feedback loop between XBP1 and Fbw7 regulates cancer development. Oncogenesis 8:12

Chakravarthy MV et al (2009) Identification of a physiologically relevant endogenous ligand for PPARalpha in liver. Cell 138:476–488

Challa TD et al (2019) Liver ASK1 protects from non-alcoholic fatty liver disease and fibrosis. EMBO Mol Med 11:e10124

Chang L et al (2006) The E3 ubiquitin ligase itch couples JNK activation to TNFalpha-induced cell death by inducing c-FLIP(L) turnover. Cell 124:601–613

Chavez-Galan L, Olleros ML, Vesin D, Garcia I (2015) Much more than M1 and M2 macrophages, there

are also CD169(+) and TCR(+) macrophages. Front Immunol 6:263
Chawla A, Nguyen KD, Goh YP (2011) Macrophage-mediated inflammation in metabolic disease. Nat Rev Immunol 11:738–749
Chen BB, Mallampalli RK (2009) Masking of a nuclear signal motif by monoubiquitination leads to mislocalization and degradation of the regulatory enzyme cytidylyltransferase. Mol Cell Biol 29:3062–3075
Chen Q, Huang L, Pan D, Zhu LJ, Wang YX (2018) Cbx4 sumoylates Prdm16 to regulate adipose tissue thermogenesis. Cell Rep 22:2860–2872
Chen Y et al (2021) Curcumin prevents obesity by targeting TRAF4-induced ubiquitylation in m(6) A-dependent manner. EMBO Rep 22:e52146
Cheung MK, Gulati P, O'Rahilly S, Yeo GS (2013) FTO expression is regulated by availability of essential amino acids. Int J Obes 37:744–747
Cho NH et al (2018) IDF diabetes atlas: global estimates of diabetes prevalence for 2017 and projections for 2045. Diabetes Res Clin Pract 138:271–281
Chu KY, Li H, Wada K, Johnson JD (2012) Ubiquitin C-terminal hydrolase L1 is required for pancreatic beta cell survival and function in lipotoxic conditions. Diabetologia 55:128–140
Ciechanover A (2015) The unravelling of the ubiquitin system. Nat Rev Mol Cell Biol 16:322–324
Clague MJ, Urbe S, Komander D (2019) Breaking the chains: deubiquitylating enzyme specificity begets function. Nat Rev Mol Cell Biol 20:338–352
Claiborn KC et al (2010) Pcif1 modulates Pdx1 protein stability and pancreatic beta cell function and survival in mice. J Clin Invest 120:3713–3721
Cnop M et al (2005) Mechanisms of pancreatic beta-cell death in type 1 and type 2 diabetes: many differences, few similarities. Diabetes 54(Suppl 2):S97–S107
Collins GA, Goldberg AL (2017) The logic of the 26S proteasome. Cell 169:792–806
Copps KD, White MF (2012) Regulation of insulin sensitivity by serine/threonine phosphorylation of insulin receptor substrate proteins IRS1 and IRS2. Diabetologia 55:2565–2582
Costes S et al (2009) Degradation of cAMP-responsive element-binding protein by the ubiquitin-proteasome pathway contributes to glucotoxicity in beta-cells and human pancreatic islets. Diabetes 58:1105–1115
Costes S, Gurlo T, Rivera JF, Butler PC (2014) UCHL1 deficiency exacerbates human islet amyloid polypeptide toxicity in beta-cells: evidence of interplay between the ubiquitin/proteasome system and autophagy. Autophagy 10:1004–1014
Couillard C et al (2000) Hyperleptinemia is more closely associated with adipose cell hypertrophy than with adipose tissue hyperplasia. Int J Obes Relat Metab Disord 24:782–788
Coyne ES et al (2019) The deubiquitinating enzyme USP19 modulates adipogenesis and potentiates high-fat-diet-induced obesity and glucose intolerance in mice. Diabetologia 62:136–146
Cullinan SB, Gordan JD, Jin J, Harper JW, Diehl JA (2004) The Keap1-BTB protein is an adaptor that bridges Nrf2 to a Cul3-based E3 ligase: oxidative stress sensing by a Cul3-Keap1 ligase. Mol Cell Biol 24:8477–8486
Daitoku H, Yamagata K, Matsuzaki H, Hatta M, Fukamizu A (2003) Regulation of PGC-1 promoter activity by protein kinase B and the forkhead transcription factor FKHR. Diabetes 52:642–649
DeFronzo RA et al (2015) Type 2 diabetes mellitus. Nat Rev Dis Primers 1:15019
Deng L et al (2000) Activation of the IkappaB kinase complex by TRAF6 requires a dimeric ubiquitin-conjugating enzyme complex and a unique polyubiquitin chain. Cell 103:351–361
Di Gregorio J et al (2021) UBXN7 cofactor of CRL3(KEAP1) and CRL2(VHL) ubiquitin ligase complexes mediates reciprocal regulation of NRF2 and HIF-1alpha proteins. Biochim Biophys Acta Mol Cell Res 1868:118963
Dina C et al (2007) Variation in FTO contributes to childhood obesity and severe adult obesity. Nat Genet 39:724–726
Ding WX, Yin XM (2004) Dissection of the multiple mechanisms of TNF-alpha-induced apoptosis in liver injury. J Cell Mol Med 8:445–454
Ding L et al (2021) Peroxisomal beta-oxidation acts as a sensor for intracellular fatty acids and regulates lipolysis. Nat Metab 3:1648–1661
Dumpati R, Ramatenki V, Vadija R, Vellanki S, Vuruputuri U (2018) Structural insights into suppressor of cytokine signaling 1 protein-identification of new leads for type 2 diabetes mellitus. J Mol Recognit 31:e2706
Dwane L, Gallagher WM, Ni Chonghaile T, O'Connor DP (2017) The emerging role of non-traditional ubiquitination in oncogenic pathways. J Biol Chem 292:3543–3551
Eeckhoute J, Oger F, Staels B, Lefebvre P (2012) Coordinated regulation of PPARgamma expression and activity through control of chromatin structure in adipogenesis and obesity. PPAR Res 2012:164140
Endo H, Niioka M, Kobayashi N, Tanaka M, Watanabe T (2013) Butyrate-producing probiotics reduce nonalcoholic fatty liver disease progression in rats: new insight into the probiotics for the gut-liver axis. PLoS One 8:e63388
Fabregat I et al (2016) TGF-beta signalling and liver disease. FEBS J 283:2219–2232
Farmer SR (2006) Transcriptional control of adipocyte formation. Cell Metab 4:263–273
Feldstein AE et al (2003) Hepatocyte apoptosis and fas expression are prominent features of human nonalcoholic steatohepatitis. Gastroenterology 125:437–443
Forand A et al (2016) Disruption of the phosphate transporter Pit1 in hepatocytes improves glucose metabolism and insulin signaling by modulating the USP7/IRS1 interaction. Cell Rep 16:2736–2748
Francque S et al (2015) PPARalpha gene expression correlates with severity and histological treatment

response in patients with non-alcoholic steatohepatitis. J Hepatol 63:164–173

Frayling TM et al (2007) A common variant in the FTO gene is associated with body mass index and predisposes to childhood and adult obesity. Science 316:889–894

Gavrilova O et al (2003) Liver peroxisome proliferator-activated receptor gamma contributes to hepatic steatosis, triglyceride clearance, and regulation of body fat mass. J Biol Chem 278:34268–34276

Geng C et al (2016) Mst1 regulates hepatic lipid metabolism by inhibiting Sirt1 ubiquitination in mice. Biochem Biophys Res Commun 471:444–449

Ghosh M et al (2016) Ubiquitin ligase COP1 controls hepatic fat metabolism by targeting ATGL for degradation. Diabetes 65:3561–3572

Graner E et al (2004) The isopeptidase USP2a regulates the stability of fatty acid synthase in prostate cancer. Cancer Cell 5:253–261

Grice GL, Nathan JA (2016) The recognition of ubiquitinated proteins by the proteasome. Cell Mol Life Sci 73:3497–3506

Grice GL et al (2015) The proteasome distinguishes between heterotypic and homotypic lysine-11-linked polyubiquitin chains. Cell Rep 12:545–553

Gross B, Pawlak M, Lefebvre P, Staels B (2017) PPARs in obesity-induced T2DM, dyslipidaemia and NAFLD. Nat Rev Endocrinol 13:36–49

Guicciardi ME, Gores GJ (2010) Apoptosis as a mechanism for liver disease progression. Semin Liver Dis 30:402–410

Gulati P et al (2013) Role for the obesity-related FTO gene in the cellular sensing of amino acids. Proc Natl Acad Sci U S A 110:2557–2562

Gunton JE (2020) Hypoxia-inducible factors and diabetes. J Clin Invest 130:5063–5073

Guo L et al (2017) Hepatic neuregulin 4 signaling defines an endocrine checkpoint for steatosis-to-NASH progression. J Clin Invest 127:4449–4461

Guo L et al (2019) Enhanced acetylation of ATP-citrate lyase promotes the progression of nonalcoholic fatty liver disease. J Biol Chem 294:11805–11816

Hall RK et al (2000) Regulation of phosphoenolpyruvate carboxykinase and insulin-like growth factor-binding protein-1 gene expression by insulin. The role of winged helix/forkhead proteins. J Biol Chem 275:30169–30175

Hall JA, Tabata M, Rodgers JT, Puigserver P (2014) USP7 attenuates hepatic gluconeogenesis through modulation of FoxO1 gene promoter occupancy. Mol Endocrinol 28:912–924

Hardie DG, Ross FA, Hawley SA (2012) AMPK: a nutrient and energy sensor that maintains energy homeostasis. Nat Rev Mol Cell Biol 13:251–262

Hatting M et al (2013) Hepatocyte caspase-8 is an essential modulator of steatohepatitis in rodents. Hepatology 57:2189–2201

Hayden MS, Ghosh S (2008) Shared principles in NF-kappaB signaling. Cell 132:344–362

Hayden MS, Ghosh S (2012) NF-kappaB, the first quarter-century: remarkable progress and outstanding questions. Genes Dev 26:203–234

He W et al (2003) Adipose-specific peroxisome proliferator-activated receptor gamma knockout causes insulin resistance in fat and liver but not in muscle. Proc Natl Acad Sci U S A 100:15712–15717

Hershko A, Ciechanover A (1992) The ubiquitin system for protein degradation. Annu Rev Biochem 61:761–807

Hershko A, Ciechanover A (1998) The ubiquitin system. Annu Rev Biochem 67:425–479

Hetz C (2012) The unfolded protein response: controlling cell fate decisions under ER stress and beyond. Nat Rev Mol Cell Biol 13:89–102

Hinds TD Jr et al (2016) Biliverdin reductase A attenuates hepatic steatosis by inhibition of glycogen synthase kinase (GSK) 3beta phosphorylation of serine 73 of peroxisome proliferator-activated receptor (PPAR) alpha. J Biol Chem 291:25179–25191

Hoshino A et al (2014) Inhibition of p53 preserves parkin-mediated mitophagy and pancreatic beta-cell function in diabetes. Proc Natl Acad Sci U S A 111:3116–3121

Hotamisligil GS (2010) Endoplasmic reticulum stress and the inflammatory basis of metabolic disease. Cell 140:900–917

Hotamisligil GS, Shargill NS, Spiegelman BM (1993) Adipose expression of tumor necrosis factor-alpha: direct role in obesity-linked insulin resistance. Science 259:87–91

Hotamisligil GS, Arner P, Caro JF, Atkinson RL, Spiegelman BM (1995) Increased adipose tissue expression of tumor necrosis factor-alpha in human obesity and insulin resistance. J Clin Invest 95:2409–2415

Howard JK, Flier JS (2006) Attenuation of leptin and insulin signaling by SOCS proteins. Trends Endocrinol Metab 17:365–371

Hu Y et al (2019) Endoplasmic reticulum-associated degradation (ERAD) has a critical role in supporting glucose-stimulated insulin secretion in pancreatic beta-cells. Diabetes 68:733–746

Hussain S et al (2013) Ubiquitin hydrolase UCH-L1 destabilizes mTOR complex 1 by antagonizing DDB1-CUL4-mediated ubiquitination of raptor. Mol Cell Biol 33:1188–1197

Icard P et al (2021) Understanding the central role of citrate in the metabolism of cancer cells and tumors: an update. Int J Mol Sci 22:6587. https://doi.org/10.3390/ijms22126587

Ichijo H (1999) From receptors to stress-activated MAP kinases. Oncogene 18:6087–6093

Ichijo H et al (1997) Induction of apoptosis by ASK1, a mammalian MAPKKK that activates SAPK/JNK and p38 signaling pathways. Science 275:90–94

Imai T et al (2004) Peroxisome proliferator-activated receptor gamma is required in mature white and brown adipocytes for their survival in the mouse. Proc Natl Acad Sci U S A 101:4543–4547

Imoto K et al (2006) Impact of mitochondrial reactive oxygen species and apoptosis signal-regulating kinase 1 on insulin signaling. Diabetes 55:1197–1204
Ip E et al (2003) Central role of PPARalpha-dependent hepatic lipid turnover in dietary steatohepatitis in mice. Hepatology 38:123–132
Ip E, Farrell G, Hall P, Robertson G, Leclercq I (2004) Administration of the potent PPARalpha agonist, Wy-14,643, reverses nutritional fibrosis and steatohepatitis in mice. Hepatology 39:1286–1296
James DE, Stockli J, Birnbaum MJ (2021) The aetiology and molecular landscape of insulin resistance. Nat Rev Mol Cell Biol 22:751–771
Jing Y et al (2015) Hepatic p38alpha regulates gluconeogenesis by suppressing AMPK. J Hepatol 62:1319–1327
Johnson ES, Ma PC, Ota IM, Varshavsky A (1995) A proteolytic pathway that recognizes ubiquitin as a degradation signal. J Biol Chem 270:17442–17456
Jonsson J, Carlsson L, Edlund T, Edlund H (1994) Insulin-promoter-factor 1 is required for pancreas development in mice. Nature 371:606–609
Kajimura S et al (2009) Initiation of myoblast to brown fat switch by a PRDM16-C/EBP-beta transcriptional complex. Nature 460:1154–1158
Kamei N et al (2006) Overexpression of monocyte chemoattractant protein-1 in adipose tissues causes macrophage recruitment and insulin resistance. J Biol Chem 281:26602–26614
Kanda H et al (2006) MCP-1 contributes to macrophage infiltration into adipose tissue, insulin resistance, and hepatic steatosis in obesity. J Clin Invest 116:1494–1505
Kao SH, Wu HT, Wu KJ (2018) Ubiquitination by HUWE1 in tumorigenesis and beyond. J Biomed Sci 25:67
Kern PA et al (1995) The expression of tumor necrosis factor in human adipose tissue. Regulation by obesity, weight loss, and relationship to lipoprotein lipase. J Clin Invest 95:2111–2119
Khan RS, Newsome PN (2018) NAFLD in 2017: novel insights into mechanisms of disease progression. Nat Rev Gastroenterol Hepatol 15:71–72
Kilroy G, Kirk-Ballard H, Carter LE, Floyd ZE (2012) The ubiquitin ligase Siah2 regulates PPARgamma activity in adipocytes. Endocrinology 153:1206–1218
Kilroy G et al (2015) The ubiquitin ligase Siah2 regulates obesity-induced adipose tissue inflammation. Obesity (Silver Spring) 23:2223–2232
Kim W et al (2011) Systematic and quantitative assessment of the ubiquitin-modified proteome. Mol Cell 44:325–340
Kim J, Yang G, Kim Y, Kim J, Ha J (2016a) AMPK activators: mechanisms of action and physiological activities. Exp Mol Med 48:e224
Kim SM, Grenert JP, Patterson C, Correia MA (2016b) CHIP(−/−)-mouse liver: adiponectin-AMPK-FOXO-activation overrides CYP2E1-elicited JNK1-activation, delaying onset of NASH: therapeutic implications. Sci Rep 6:29423
Kim M, Kwon YE, Song JO, Bae SJ, Seol JH (2016c) CHFR negatively regulates SIRT1 activity upon oxidative stress. Sci Rep 6:37578
Kim JH et al (2017a) C-terminus of HSC70-interacting protein (CHIP) inhibits adipocyte differentiation via ubiquitin- and proteasome-mediated degradation of PPARgamma. Sci Rep 7:40023
Kim D et al (2017b) Cytosolic Pellino-1-mediated K63-linked ubiquitination of IRF5 in M1 macrophages regulates glucose intolerance in obesity. Cell Rep 20:832–845
Kim A et al (2021) Ablation of USP21 in skeletal muscle promotes oxidative fibre phenotype, inhibiting obesity and type 2 diabetes. J Cachexia Sarcopenia Muscle 12:1669–1689
Koegl M et al (1999) A novel ubiquitination factor, E4, is involved in multiubiquitin chain assembly. Cell 96:635–644
Koepp DM et al (2001) Phosphorylation-dependent ubiquitination of cyclin E by the SCFFbw7 ubiquitin ligase. Science 294:173–177
Komander D, Rape M (2012) The ubiquitin code. Annu Rev Biochem 81:203–229
Kon N, Zhong J, Qiang L, Accili D, Gu W (2012) Inactivation of arf-bp1 induces p53 activation and diabetic phenotypes in mice. J Biol Chem 287:5102–5111
Kulathu Y, Komander D (2012) Atypical ubiquitylation—the unexplored world of polyubiquitin beyond Lys48 and Lys63 linkages. Nat Rev Mol Cell Biol 13:508–523
Kwon D, Kim SM, Correia MA (2020) Cytochrome P450 endoplasmic reticulum-associated degradation (ERAD): therapeutic and pathophysiological implications. Acta Pharm Sin B 10:42–60
Lancaster GI et al (2018) Evidence that TLR4 is not a receptor for saturated fatty acids but mediates lipid-induced inflammation by reprogramming macrophage metabolism. Cell Metab 27:1096–1110.e5
Lebrun P, Van Obberghen E (2008) SOCS proteins causing trouble in insulin action. Acta Physiol (Oxf) 192:29–36
Lee J et al (2011) p38 MAPK-mediated regulation of Xbp1s is crucial for glucose homeostasis. Nat Med 17:1251–1260
Lee MS et al (2018) Loss of the E3 ubiquitin ligase MKRN1 represses diet-induced metabolic syndrome through AMPK activation. Nat Commun 9:3404
Li X et al (2016) The MDM2-p53-pyruvate carboxylase signalling axis couples mitochondrial metabolism to glucose-stimulated insulin secretion in pancreatic beta-cells. Nat Commun 7:11740
Li P et al (2017) Lack of CUL4B in adipocytes promotes PPARgamma-mediated adipose tissue expansion and insulin sensitivity. Diabetes 66:300–313
Li D et al (2019) Melatonin safeguards against fatty liver by antagonizing TRAFs-mediated ASK1 deubiquitination and stabilization in a β-arrestin-1 dependent manner. J Pineal Res 67:e12611

Li K et al (2021) Hrd1-mediated ACLY ubiquitination alleviate NAFLD in db/db mice. Metabolism 114:154349
Lim KH, Joo JY, Baek KH (2020) The potential roles of deubiquitinating enzymes in brain diseases. Ageing Res Rev 61:101088
Lin H et al (2021) IP6-assisted CSN-COP1 competition regulates a CRL4-ETV5 proteolytic checkpoint to safeguard glucose-induced insulin secretion. Nat Commun 12:2461
Liu J et al (2016) Inflammation improves glucose homeostasis through IKKbeta-XBP1s interaction. Cell 167:1052–1066
Liu B et al (2017) Aberrant expression of FBXO2 disrupts glucose homeostasis through ubiquitin-mediated degradation of insulin receptor in obese mice. Diabetes 66:689–698
Liu B et al (2018) Proteome-wide analysis of USP14 substrates revealed its role in hepatosteatosis via stabilization of FASN. Nat Commun 9:4770
Liu D et al (2019) Ets-1 deficiency alleviates nonalcoholic steatohepatitis via weakening TGF-beta1 signaling-mediated hepatocyte apoptosis. Cell Death Dis 10:458
Liu D et al (2020) TNFAIP3 interacting protein 3 overexpression suppresses nonalcoholic steatohepatitis by blocking TAK1 activation. Cell Metab 31:726–740
Liu Y et al (2021a) Tumor necrosis factor alpha-induced protein 8-like 2 alleviates nonalcoholic fatty liver disease through suppressing transforming growth factor beta-activated kinase 1 activation. Hepatology 74:1300–1318
Liu P et al (2021b) E3 ubiquitin ligase Grail promotes hepatic steatosis through Sirt1 inhibition. Cell Death Dis 12:323
Locke M, Toth JI, Petroski MD (2014) Lys11- and Lys48-linked ubiquitin chains interact with p97 during endoplasmic-reticulum-associated degradation. Biochem J 459:205–216
Loyer X et al (2016) Liver microRNA-21 is overexpressed in non-alcoholic steatohepatitis and contributes to the disease in experimental models by inhibiting PPARalpha expression. Gut 65:1882–1894
Lu Y, Wang W, Kirschner MW (2015) Specificity of the anaphase-promoting complex: a single-molecule study. Science 348:1248737
Lu Y et al (2018) Hepatic leukocyte immunoglobulin-like receptor B4 (LILRB4) attenuates nonalcoholic fatty liver disease via SHP1-TRAF6 pathway. Hepatology 67:1303–1319
Luedde T, Kaplowitz N, Schwabe RF (2014) Cell death and cell death responses in liver disease: mechanisms and clinical relevance. Gastroenterology 147:765–783
Lumeng CN, Saltiel AR (2011) Inflammatory links between obesity and metabolic disease. J Clin Invest 121:2111–2117
Ma N et al (2021) PPDPF alleviates hepatic steatosis through inhibition of mTOR signaling. Nat Commun 12:3059
Malhi H, Kaufman RJ (2011) Endoplasmic reticulum stress in liver disease. J Hepatol 54:795–809
Martinez VD et al (2013) Frequent concerted genetic mechanisms disrupt multiple components of the NRF2 inhibitor KEAP1/CUL3/RBX1 E3-ubiquitin ligase complex in thyroid cancer. Mol Cancer 12:124
Martinez-Fonts K, Matouschek A (2016) A rapid and versatile method for generating proteins with defined ubiquitin chains. Biochemistry 55:1898–1908
Matsusue K et al (2003) Liver-specific disruption of PPARgamma in leptin-deficient mice improves fatty liver but aggravates diabetic phenotypes. J Clin Invest 111:737–747
Matveyenko AV, Butler PC (2008) Relationship between beta-cell mass and diabetes onset. Diabetes Obes Metab 10(Suppl 4):23–31
Medvedev AE, Murphy M, Zhou H, Li X (2015) E3 ubiquitin ligases Pellinos as regulators of pattern recognition receptor signaling and immune responses. Immunol Rev 266:109–122
Meng C et al (2021) The deubiquitinase USP11 regulates cell proliferation and ferroptotic cell death via stabilization of NRF2 USP11 deubiquitinates and stabilizes NRF2. Oncogene 40:1706–1720
Meyer HJ, Rape M (2014) Enhanced protein degradation by branched ubiquitin chains. Cell 157:910–921
Minamino T et al (2009) A crucial role for adipose tissue p53 in the regulation of insulin resistance. Nat Med 15:1082–1087
Molusky MM, Li S, Ma D, Yu L, Lin JD (2012) Ubiquitin-specific protease 2 regulates hepatic gluconeogenesis and diurnal glucose metabolism through 11beta-hydroxysteroid dehydrogenase 1. Diabetes 61:1025–1035
Montagner A et al (2016) Liver PPARalpha is crucial for whole-body fatty acid homeostasis and is protective against NAFLD. Gut 65:1202–1214
Moran-Salvador E et al (2011) Role for PPARgamma in obesity-induced hepatic steatosis as determined by hepatocyte- and macrophage-specific conditional knockouts. FASEB J 25:2538–2550
Mowers J et al (2013) Inflammation produces catecholamine resistance in obesity via activation of PDE3B by the protein kinases IKKepsilon and TBK1. elife 2:e01119
Murray PJ et al (2014) Macrophage activation and polarization: nomenclature and experimental guidelines. Immunity 41:14–20
Nakae J et al (2002) Regulation of insulin action and pancreatic beta-cell function by mutated alleles of the gene encoding forkhead transcription factor Foxo1. Nat Genet 32:245–253
Nakatsu Y et al (2019) Prolyl isomerase Pin1 suppresses thermogenic programs in adipocytes by promoting degradation of transcriptional co-activator PRDM16. Cell Rep 26:3221–3230
Nateri AS, Riera-Sans L, Da Costa C, Behrens A (2004) The ubiquitin ligase SCFFbw7 antagonizes apoptotic JNK signaling. Science 303:1374–1378
Niyogi S, Ghosh M, Adak M, Chakrabarti P (2019) PEDF promotes nuclear degradation of ATGL through COP1. Biochem Biophys Res Commun 512:806–811

Oh YS (2015) Mechanistic insights into pancreatic beta-cell mass regulation by glucose and free fatty acids. Anat Cell Biol 48:16–24

Oh GS et al (2020) Regulation of adipocyte differentiation by clusterin-mediated Kruppel-like factor 5 stabilization. FASEB J 34:16276–16290

Ohno H, Shinoda K, Ohyama K, Sharp LZ, Kajimura S (2013) EHMT1 controls brown adipose cell fate and thermogenesis through the PRDM16 complex. Nature 504:163–167

Ohtake F, Tsuchiya H (2017) The emerging complexity of ubiquitin architecture. J Biochem 161:125–133

Okuda-Shimizu Y, Hendershot LM (2007) Characterization of an ERAD pathway for nonglycosylated BiP substrates, which require Herp. Mol Cell 28:544–554

Osborne TF (2000) Sterol regulatory element-binding proteins (SREBPs): key regulators of nutritional homeostasis and insulin action. J Biol Chem 275:32379–32382

Otoda T et al (2013) Proteasome dysfunction mediates obesity-induced endoplasmic reticulum stress and insulin resistance in the liver. Diabetes 62:811–824

Packer M (2020) Role of deranged energy deprivation signaling in the pathogenesis of cardiac and renal disease in states of perceived nutrient overabundance. Circulation 141:2095–2105

Pagliassotti MJ (2012) Endoplasmic reticulum stress in nonalcoholic fatty liver disease. Annu Rev Nutr 32:17–33

Pal D et al (2012) Fetuin-A acts as an endogenous ligand of TLR4 to promote lipid-induced insulin resistance. Nat Med 18:1279–1285

Park SJ et al (2017) DNA-PK promotes the mitochondrial, metabolic, and physical decline that occurs during aging. Cell Metab 26:447

Pawlak M et al (2014) The transrepressive activity of peroxisome proliferator-activated receptor alpha is necessary and sufficient to prevent liver fibrosis in mice. Hepatology 60:1593–1606

Pawlak M, Lefebvre P, Staels B (2015) Molecular mechanism of PPARalpha action and its impact on lipid metabolism, inflammation and fibrosis in non-alcoholic fatty liver disease. J Hepatol 62:720–733

Pearson G et al (2018) Clec16a, Nrdp1, and USP8 form a ubiquitin-dependent tripartite complex that regulates beta-cell mitophagy. Diabetes 67:265–277

Peng J et al (2003) A proteomics approach to understanding protein ubiquitination. Nat Biotechnol 21:921–926

Peng J et al (2018) An Hsp20-FBXO4 axis regulates adipocyte function through modulating PPARgamma ubiquitination. Cell Rep 23:3607–3620

Perry AK, Chow EK, Goodnough JB, Yeh WC, Cheng G (2004) Differential requirement for TANK-binding kinase-1 in type I interferon responses to toll-like receptor activation and viral infection. J Exp Med 199:1651–1658

Pinkosky SL, Groot PHE, Lalwani ND, Steinberg GR (2017) Targeting ATP-citrate lyase in hyperlipidemia and metabolic disorders. Trends Mol Med 23:1047–1063

Pohl C, Dikic I (2019) Cellular quality control by the ubiquitin-proteasome system and autophagy. Science 366:818–822

Qi L et al (2006) TRB3 links the E3 ubiquitin ligase COP1 to lipid metabolism. Science 312:1763–1766

Qi L et al (2009) Adipocyte CREB promotes insulin resistance in obesity. Cell Metab 9:277–286

Qin Y et al (2014) RNF26 temporally regulates virus-triggered type I interferon induction by two distinct mechanisms. PLoS Pathog 10:e1004358

Rape M (2018) Ubiquitylation at the crossroads of development and disease. Nat Rev Mol Cell Biol 19:59–70

Reilly SM, Saltiel AR (2017) Adapting to obesity with adipose tissue inflammation. Nat Rev Endocrinol 13:633–643

Reilly SM et al (2013) An inhibitor of the protein kinases TBK1 and IKK-varepsilon improves obesity-related metabolic dysfunctions in mice. Nat Med 19:313–321

Ren X, Chen N, Chen Y, Liu W, Hu Y (2019) TRB3 stimulates SIRT1 degradation and induces insulin resistance by lipotoxicity via COP1. Exp Cell Res 382:111428

Ruggiero T et al (2009) LPS induces KH-type splicing regulatory protein-dependent processing of microRNA-155 precursors in macrophages. FASEB J 23:2898–2908

Rutkowski JM, Stern JH, Scherer PE (2015) The cell biology of fat expansion. J Cell Biol 208:501–512

Saghizadeh M, Ong JM, Garvey WT, Henry RR, Kern PA (1996) The expression of TNF alpha by human muscle. Relationship to insulin resistance. J Clin Invest 97:1111–1116

Scheufele F et al (2014) Evidence for a regulatory role of Cullin-RING E3 ubiquitin ligase 7 in insulin signaling. Cell Signal 26:233–239

Schmoll D et al (2000) Regulation of glucose-6-phosphatase gene expression by protein kinase Balpha and the forkhead transcription factor FKHR. Evidence for insulin response unit-dependent and -independent effects of insulin on promoter activity. J Biol Chem 275:36324–36333

Seale P et al (2011) Prdm16 determines the thermogenic program of subcutaneous white adipose tissue in mice. J Clin Invest 121:96–105

Segarra-Mondejar M et al (2018) Synaptic activity-induced glycolysis facilitates membrane lipid provision and neurite outgrowth. EMBO J 37:e97368. https://doi.org/10.15252/embj.201797368

Sethi JK, Hotamisligil GS (1999) The role of TNF alpha in adipocyte metabolism. Semin Cell Dev Biol 10:19–29

Shembade N et al (2008) The E3 ligase itch negatively regulates inflammatory signaling pathways by controlling the function of the ubiquitin-editing enzyme A20. Nat Immunol 9:254–262

Shimomura I et al (1999) Insulin selectively increases SREBP-1c mRNA in the livers of rats with streptozotocin-induced diabetes. Proc Natl Acad Sci U S A 96:13656–13661

Shin S et al (2014) CREB mediates the insulinotropic and anti-apoptotic effects of GLP-1 signaling in adult mouse beta-cells. Mol Metab 3:803–812

Shu B et al (2021) The METTL3/MALAT1/PTBP1/USP8/TAK1 axis promotes pyroptosis and M1 polarization of macrophages and contributes to liver fibrosis. Cell Death Discov 7:368
Soleimanpour SA et al (2014) The diabetes susceptibility gene Clec16a regulates mitophagy. Cell 157:1577–1590
Song R et al (2013) Central role of E3 ubiquitin ligase MG53 in insulin resistance and metabolic disorders. Nature 494:375–379
Soond SM, Terry JL, Colbert JD, Riches DW (2003) TRUSS, a novel tumor necrosis factor receptor 1 scaffolding protein that mediates activation of the transcription factor NF-kappaB. Mol Cell Biol 23:8334–8344
Soond SM, Terry JL, Riches DW (2006) TRUSS, a tumor necrosis factor receptor-1-interacting protein, activates c-Jun NH(2)-terminal kinase and transcription factor AP-1. FEBS Lett 580:4591–4596
Speakman JR (2010) FTO effect on energy demand versus food intake. Nature 464:E1; discussion E2
Strohmaier H et al (2001) Human F-box protein hCdc4 targets cyclin E for proteolysis and is mutated in a breast cancer cell line. Nature 413:316–322
Sugii S et al (2009) PPARgamma activation in adipocytes is sufficient for systemic insulin sensitization. Proc Natl Acad Sci U S A 106:22504–22509
Sui Y et al (2014) IKKbeta links vascular inflammation to obesity and atherosclerosis. J Exp Med 211:869–886
Sun S, Kelekar S, Kliewer SA, Mangelsdorf DJ (2019) The orphan nuclear receptor SHP regulates ER stress response by inhibiting XBP1s degradation. Genes Dev 33:1083–1094
Sun H et al (2020) Inhibition of XBP1s ubiquitination enhances its protein stability and improves glucose homeostasis. Metabolism 105:154046
Sundqvist A et al (2005) Control of lipid metabolism by phosphorylation-dependent degradation of the SREBP family of transcription factors by SCF(Fbw7). Cell Metab 1:379–391
Suriben R et al (2015) Beta-cell insulin secretion requires the ubiquitin ligase COP1. Cell 163:1457–1467
Swatek KN, Komander D (2016) Ubiquitin modifications. Cell Res 26:399–422
Syed-Abdul MM et al (2020) Fatty acid synthase inhibitor TVB-2640 reduces hepatic de novo lipogenesis in males with metabolic abnormalities. Hepatology 72:103–118
Tobiume K et al (2001) ASK1 is required for sustained activations of JNK/p38 MAP kinases and apoptosis. EMBO Rep 2:222–228
Tsai B, Ye Y, Rapoport TA (2002) Retro-translocation of proteins from the endoplasmic reticulum into the cytosol. Nat Rev Mol Cell Biol 3:246–255
Tsuchiya H et al (2017) In vivo ubiquitin linkage-type analysis reveals that the Cdc48-Rad23/Dsk2 axis contributes to K48-linked chain specificity of the proteasome. Mol Cell 66:488–502
Uysal KT, Wiesbrock SM, Marino MW, Hotamisligil GS (1997) Protection from obesity-induced insulin resistance in mice lacking TNF-alpha function. Nature 389:610–614
Vembar SS, Brodsky JL (2008) One step at a time: endoplasmic reticulum-associated degradation. Nat Rev Mol Cell Biol 9:944–957
Wadosky KM, Willis MS (2012) The story so far: post-translational regulation of peroxisome proliferator-activated receptors by ubiquitination and SUMOylation. Am J Physiol Heart Circ Physiol 302:H515–H526
Wagner SA et al (2011) A proteome-wide, quantitative survey of in vivo ubiquitylation sites reveals widespread regulatory roles. Mol Cell Proteomics 10:M111.013284
Wang J, Maldonado MA (2006) The ubiquitin-proteasome system and its role in inflammatory and autoimmune diseases. Cell Mol Immunol 3:255–261
Wang W, Seale P (2016) Control of brown and beige fat development. Nat Rev Mol Cell Biol 17:691–702
Wang C et al (2001) TAK1 is a ubiquitin-dependent kinase of MKK and IKK. Nature 412:346–351
Wang Q et al (2009) Abrogation of hepatic ATP-citrate lyase protects against fatty liver and ameliorates hyperglycemia in leptin receptor-deficient mice. Hepatology 49:1166–1175
Wang L et al (2014) Dichotomous role of pancreatic HUWE1/MULE/ARF-BP1 in modulating beta cell apoptosis in mice under physiological and genotoxic conditions. Diabetologia 57:1889–1898
Wang PX et al (2016) Hepatocyte TRAF3 promotes liver steatosis and systemic insulin resistance through targeting TAK1-dependent signalling. Nat Commun 7:10592
Wang C et al (2020) E3 ligase FBXW2 is a new therapeutic target in obesity and atherosclerosis. Adv Sci (Weinh) 7:2001800
Wang L et al (2021) Tripartite motif 16 ameliorates nonalcoholic steatohepatitis by promoting the degradation of phospho-TAK1. Cell Metab 33:1372–1388
Wei W, Li Y, Li Y, Li D (2021) Adipose-specific knockout of ubiquitin-conjugating enzyme E2L6 (Ube2l6) reduces diet-induced obesity, insulin resistance, and hepatic steatosis. J Pharmacol Sci 145:327–334
Weisberg SP et al (2003) Obesity is associated with macrophage accumulation in adipose tissue. J Clin Invest 112:1796–1808
Welcker M et al (2004) The Fbw7 tumor suppressor regulates glycogen synthase kinase 3 phosphorylation-dependent c-Myc protein degradation. Proc Natl Acad Sci U S A 101:9085–9090
Wu Z, Wang S (2013) Role of kruppel-like transcription factors in adipogenesis. Dev Biol 373:235–243
Wu J et al (2012) Beige adipocytes are a distinct type of thermogenic fat cell in mouse and human. Cell 150:366–376
Wu HK et al (2019) Glucose-sensitive myokine/cardiokine MG53 regulates systemic insulin response and metabolic homeostasis. Circulation 139:901–914

Wu T et al (2020) HRD1, an important player in pancreatic beta-cell failure and therapeutic target for type 2 diabetic mice. Diabetes 69:940–953

Xiao Y et al (2019) Extracellular vesicles in type 2 diabetes mellitus: key roles in pathogenesis, complications, and therapy. J Extracell Vesicles 8:1625677

Xu P et al (2009) Quantitative proteomics reveals the function of unconventional ubiquitin chains in proteasomal degradation. Cell 137:133–145

Xu X et al (2012) Identification of the degradation determinants of insulin receptor substrate 1 for signaling cullin-RING E3 ubiquitin ligase 7-mediated ubiquitination. J Biol Chem 287:40758–40766

Yan SQ et al (2021) FBXW7 gene polymorphism is associated with type 2 diabetes in the Uygur population in Xinjiang, China. Hereditas 158:27

Yang S et al (2014) The E3 ubiquitin ligase Pellino3 protects against obesity-induced inflammation and insulin resistance. Immunity 41:973–987

Yang XD, Xiang DX, Yang YY (2016) Role of E3 ubiquitin ligases in insulin resistance. Diabetes Obes Metab 18:747–754

Yang Q et al (2021) The E3 ubiquitin ligase ring finger protein 5 ameliorates NASH through ubiquitin-mediated degradation of 3-hydroxy-3-methylglutaryl CoA reductase degradation protein 1. Hepatology 74:3018–3036

Yao J et al (2021) Macrophage IRX3 promotes diet-induced obesity and metabolic inflammation. Nat Immunol 22:1268–1279

Yu J et al (2013) Modulation of fatty acid synthase degradation by concerted action of p38 MAP kinase, E3 ligase COP1, and SH2-tyrosine phosphatase Shp2. J Biol Chem 288:3823–3830

Yu CJ et al (2018) TRUSS exacerbates NAFLD development by promoting IkappaBalpha degradation in mice. Hepatology 68:1769–1785

Yumimoto K, Nakayama KI (2020) Recent insight into the role of FBXW7 as a tumor suppressor. Semin Cancer Biol 67:1–15

Zhang T et al (2015) Gp78, an E3 ubiquitin ligase acts as a gatekeeper suppressing nonalcoholic steatohepatitis (NASH) and liver cancer. PLoS One 10:e0118448

Zhang P et al (2018) The deubiquitinating enzyme TNFAIP3 mediates inactivation of hepatic ASK1 and ameliorates nonalcoholic steatohepatitis. Nat Med 24:84–94

Zhang L, Chen Z, Wang Y, Tweardy DJ, Mitch WE (2020) Stat3 activation induces insulin resistance via a muscle-specific E3 ubiquitin ligase Fbxo40. Am J Physiol Endocrinol Metab 318:E625–E635

Zhang F et al (2021a) The long non-coding RNA betaFaar regulates islet beta-cell function and survival during obesity in mice. Nat Commun 12:3997

Zhang JL et al (2021b) OTUB1 alleviates NASH through inhibition of the TRAF6-ASK1 signaling pathways. Hepatology 75:1218. https://doi.org/10.1002/hep.32179

Zhao P et al (2018a) TBK1 at the crossroads of inflammation and energy homeostasis in adipose tissue. Cell 172:731–743

Zhao J et al (2018b) Hepatic F-box protein FBXW7 maintains glucose homeostasis through degradation of Fetuin-A. Diabetes 67:818–830

Zheng L, Lv GC, Sheng J, Yang YD (2010) Effect of miRNA-10b in regulating cellular steatosis level by targeting PPAR-alpha expression, a novel mechanism for the pathogenesis of NAFLD. J Gastroenterol Hepatol 25:156–163

Zhong L, Tan Y, Zhou A, Yu Q, Zhou J (2005) RING finger ubiquitin-protein isopeptide ligase Nrdp1/FLRF regulates parkin stability and activity. J Biol Chem 280:9425–9430

Zhou Y et al (2011) Regulation of glucose homeostasis through a XBP-1-FoxO1 interaction. Nat Med 17:356–365

Zhu T, Yong XLH, Xia D, Widagdo J, Anggono V (2018a) Ubiquitination regulates the proteasomal degradation and nuclear translocation of the fat mass and obesity-associated (FTO) protein. J Mol Biol 430:363–371

Zhu K et al (2018b) Non-proteolytic ubiquitin modification of PPARgamma by Smurf1 protects the liver from steatosis. PLoS Biol 16:e3000091

Ziv I et al (2011) A perturbed ubiquitin landscape distinguishes between ubiquitin in trafficking and in proteolysis. Mol Cell Proteomics 10:M111.009753

The Cullin-RING Ligase Family in Immune Regulation

5

Di Wu

Abstract

The Cullin-RING ligase (CRL) E3 family is the largest ubiquitin ligase family and consists of a catalytic subunit, a scaffold protein, an adaptor, and a substrate receptor. Each component is different in the different members of the CRL family. CRLs affect the ubiquitination and stability of multiple substrates and regulate many biological processes. In immune regulation, components of CRLs also play critical roles in different types of immune cells and under different physical or pathological conditions. In this review, we summarized the recent advances in research about the functions of CRLs in immune regulation. Different components of the CRL family are associated with the regulation of different immunological processes.

Keywords

Cullin-RING ligase · Ubiquitination
Cellular immunity · Humoral immunity

D. Wu (✉)
Cancer Institute of the Second Affiliated Hospital and Institute of Translational Medicine, School of Medicine, Zhejiang University,
Hangzhou, Zhejiang, China
e-mail: diwu@zju.edu.cn

Cullin-RING ligases (CRLs) are a family of ubiquitin E3 ligases that consist of a catalytic subunit, a scaffold protein, an adaptor, and the substrate receptor (Fig. 5.1). Each component is distinct from others in the different CRLs (Zhao and Sun 2013).

The catalytic subunit of a CRL is a single small peptide known as the RING component. To date, only two family members have been found in mammals: RING-Box 1 (Rbx1/Roc1) and RING-Box 2 (Rnf7/Rbx2/Roc2/Sag). Rbx1 and Rbx2 are similar small peptides with 53% sequence identity in humans (Wei and Sun 2010). Despite the similarity between Rbx1 and Rbx2, it seems that these proteins are not redundant. Biochemically, Rbx1 and Rbx2 serve different roles as CRL family members. In one CRL E3 complex, only one member was present at a time (Huang et al. 2009). Functionally, complete

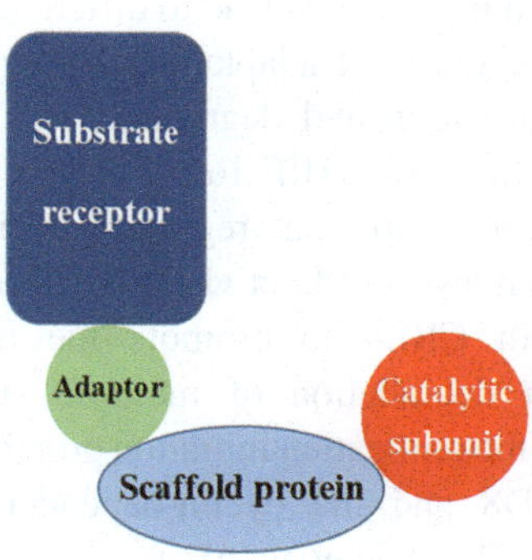

Fig. 5.1 Diagram of Cullin-RING ligases family

H. Hu, X. Fu (eds.), *Ubiquitination in Immune System and Immune Response*, Advances in Experimental Medicine and Biology 1466, https://doi.org/10.1007/978-981-97-7288-9_5

knockout of Rbx1 or Rnf7 results in embryonic lethality, which is associated with different survival times and mechanisms (Tan et al. 2009, 2011). Cutting edge research about conditional knockout of Rbx1 or Rbx2 in Treg cells further suggests the nonredundant relationship between Rbx1 and Rbx2 (Unpublished), which will be introduced in detail in this paper.

The scaffold proteins of CRLs are Cullin proteins. The most well-known members of the Cullin family include Cullin1, Cullin2, Cullin3, Cullin4a, Cullin4b, and Cullin5. Cullin proteins associate with the catalytic subunit (Rbx1 or Rbx2). In general, it is believed that Cullin1, 2, 3, 4a, and 4b tend to associate with Rbx1 as their catalytic subunit, and Cullin5 tends to associate with the catalytic subunit Rbx2. The Cullin proteins also connect with the adaptor protein, and the adaptor protein connects with the substrate receptor (Zhao and Sun 2013).

The substrate receptor directly interacts with the substrates and determines the specificity of the CRL. Up to several hundred types of substrate receptors have been reported to date. Among them, F-box and WD repeat domain-containing 7 (Fbxw7), von Hippel-Lindau (Vhl) and DDB1-cullin4-associated factor 2 (Dcaf2) are three frequently reported members. Fbxw7 is one of the substrate receptors of the CRL family, and it is well known as an oncogene involved in the generation and metastasis of multiple types of cancers (Yeh et al. 2018). The substrate receptor Vhl forms a complex with CRL2 and has been identified as a tumor suppressor gene whose inherited mutation in humans can lead to different cancers. Vhl is a regulator of adaptations to hypoxia via the ubiquitination and degradation of hypoxia-inducible factor-1α (HIF-1α), which is a master transcription factor that regulates responses to low oxygen levels (Aki et al. 2019). Dcaf2 associates with CRL4 to promote the ubiquitin-dependent degradation of multiple substrates, including the replication initiation proteins CDT1 and SETD8 and the cyclin-dependent kinase inhibitor P21 (Zhou et al. 2020).

As the largest ubiquitin E3 family, CRLs regulate the degradation of approximately 20% of cellular proteins (Soucy et al. 2009) and regulate numerous biological processes, including tumorigenesis, tumor metastasis, development, and tissue repair. Recently, emerging evidence has revealed the importance of CRLs in immune regulation.

5.1 CRLs in Macrophages

Macrophages are a type of immune cell that participate in the innate immune response. The main function of macrophages is to phagocytize pathogens and cell fragments, and macrophages can also activate lymphocytes or other immune cells to respond to pathogens.

Cullin proteins, which are the scaffold structures of CRLs, are involved in the regulation of macrophages. Cullin family members in macrophages are dynamically regulated under stress conditions. *Entamoeba histolytica* (*Eh*)-induced proinflammatory responses are critical in disease pathogenesis; after live *Eh* makes contact with human macrophages, Cullin-1/4a/4b/5 but not cullin-2/3 are degraded within 10 min, and Caspase 1 is involved in Cullin-1/4a/5 degradation. This finding suggests that Cullin proteins might regulate macrophage function (Chadha et al. 2021).

One member of the Cullin family, Cullin 4b (Cul4b), has been researched in detail and has been shown to play critical roles in macrophages. Hung et al. generated myeloid-specific Cul4b-knockout mice (*LyzM*^{cre};*Cul4b*$^{fl/fl}$ mice) and challenged them with an intraperitoneal injection of lipopolysaccharide (LPS). Myeloid-specific Cul4b deficiency worsens LPS-induced peritonitis, enhances DNA replication and proliferation, increases the production of chemokines, and decreases the production of proinflammatory cytokines in macrophages (Hung et al. 2014). Song et al. further systematically investigated the response of Cul4b-deficient macrophages to multiple PAMPs, including LPS (a ligand of TLP4), poly(I:C) (a ligand of TLR3), R848/CL097 (ligands of TLR7 and TLR8), and PGN (a ligand of TLR2). Cul4b-deficiency potentiated LPS-, poly(I:C)- and PGN-induced responses, as revealed by the increased synthesis of inflammatory cytokines, including TNF-α, IL-1β, and IL-6.

However, the response against ligands of TLR7/8 (R848 and CL097, respectively) was similar in *wt* and *Cul4b*-deficient macrophages. These results suggest that Cul4b negatively regulates TLR4-, TLR3-, and TLR2-mediated inflammatory responses. Cul4b forms a complex with EZH2, H2AK119ub1, and H3K27me3 that is enriched at the promoter of the *Pten* gene. Pten is a negative regulator of Akt. Akt can promote the activity of GSK3β by phosphorylating of GSK3β at serine 9. GSK3β is critically involved in immune responses triggered by TLRs. Increased production of proinflammatory cytokines in Cul4b-deficient macrophages is mediated, at least partially, by enhanced GSK3β activity (Song et al. 2021).

Other CRL components also have been shown to regulate macrophage function. Fbxw7, a substrate receptor of CRLs, participates in the stress responses of macrophages in different contexts by degrading different substrates. Balamurugan et al. showed that Fbxw7 interacts with and degrades C/EBPδ, which is essential for Tlr4 expression, and suppresses the inflammatory response in macrophages (Balamurugan et al. 2013). Song et al. showed that Fbxw7 mediates the ubiquitination and degradation of Shp2, which can couple with c-Cbl to degrade RIG-1; thus, Fbxw7 stabilizes RIG-1 in macrophages to establish antiviral immunity (Song et al. 2017). He et al. showed that Fbxw7 suppressed H3K27me3 modification and promoted Ccl2 and Ccl7 expression by degrading the histone-lysine N-methyltransferase enhancer of zeste homologue 2 (EZH2) in macrophages to promote intestinal inflammation (He et al. 2019). Overall, Fbxw7 regulates the stability of different substrates in macrophages in response to different stress conditions, indicating the importance and complexity of the role of Fbxw7 in macrophages.

5.2 CRLs in Dendritic Cells

Dendritic cells (DCs) are named for their dendritic pleomorphism and can efficiently absorb, process, and present antigens and activate lymphocytes, thus functioning as the most powerful professional antigen-presenting cells (APCs) in vivo.

Huang et al. showed that Dcaf2, a substrate receptor of CRLs, controls the function of dendritic cells (DCs). In mouse bone marrow–derived dendritic cells (BMDCs), Dcaf2 expression was rapidly reduced after LPS stimulation, indicating an association between Dcaf2 and the TLR4 pathway in DCs. Conditional knockout of Dcaf2 in DCs ($Cd11c^{cre}$;$Dcaf2^{fl/fl}$ mice) led to cell cycle arrest at the G2/M phase and an increase in the frequency and number of DCs in vivo, accompanied by an increase in the frequency of IL-17-producing CD4$^+$-T cells in the inguinal lymph nodes (iLNs) of ageing mice. Using an Aldara-induced psoriasis-like skin inflammation model and an experimental autoimmune encephalomyelitis model, $Cd11c^{cre}$;$Dcaf2^{fl/fl}$ mice were shown to suffer from a more severe inflammatory reaction than wild-type mice, indicating that Dcaf2-deficiency in DCs facilitates experimental autoimmunity. Biochemically, Dcaf2 induces the ubiquitination and degradation of NIK to inhibit noncanonical NF-κB activation and inhibit Il-23 expression in DCs (Huang et al. 2018).

5.3 CRLs in T Lymphocytes

T lymphocytes play pivotal roles in adaptive immune responses. The CRL family also regulates T lymphocytes. Mathewson et al. showed that Rbx2, a catalytic subunit of CRLs, was essential in supporting the activation of T lymphocytes. In Lck^{cre};$Rbx2^{fl/fl}$ mice with deficiencies in Rbx2 in T lymphocytes, T cell development in the thymus was largely normal, and peripheral T cells showed no obvious changes in the steady state. However, Rbx2-deficient T cells could not be activated by anti-CD3 stimulation in vitro, as revealed by decreased proliferation and impaired cytokine synthesis. In vivo, Rbx2-deficient T cells failed to induce graft-versus-host disease (GvHD) in a xenogeneic hematopoietic stem cell transplantation experiment. This evidence demonstrates the impaired functions of Rbx2-deficient T cells, indicating that Rbx2 is essential for T cell activation (Mathewson et al. 2016).

T lymphocytes can be divided into $CD4^+$ helper T cells and $CD8^+$ cytotoxic T cells in the periphery, both of which are regulated by CRL components.

$CD4^+$ helper T cells can establish and regulate numerous immune responses and are critical for in vivo immune regulation. The CRL scaffold protein Cullin 4b is essential for $CD4^+$-T cell activation. The activation of naive $CD4^+$-T cells by CD3/CD28 stimulation leads to robust upregulation of Cul4b in $CD4^+$ cells. Conditional knockout of *Cul4b* in T cells by *$CD4^{cre}$* does not affect thymus development. Adoptive transfer of *wt* naive $CD4^+$ T cells ($CD4^+CD25^-CD44^-CD62L^+$) to *$Rag1^{-/-}$* mice led to severe colon, while *Cul4b*-deficient naive $CD4^+$ T cells lost this ability, indicating the loss of Cul4b impaired the immune function of $CD4^+$ cells. Cul4b regulates homeostasis, proliferation, and survival in $CD4^+$-T cells by promoting the repair of DNA damaged during DNA replication. Biochemically, Cul4b degrades multiple proteins related to DNA damage responses and promotes the phosphorylation of Smc1a and changes in protein level (Dar et al. 2021).

Regulatory T cells (Treg cells) are a specialized subpopulation of $CD4^+$ helper T cells that play an immunosuppressive role to prevent immune overactivation. Foxp3 is the pivotal regulator of Treg cell function and is a marker to distinguish Treg cells. In addition to Foxp3, other regulators of Treg cells have been reported, and CRLs are involved in Treg cell regulation. Vhl, a substrate receptor in the CRL family, is important for the immunosuppressive function of Treg cells. The loss of Vhl in Treg cells in mice (*$Foxp3^{cre}$;$Vhl^{fl/fl}$* mice) results in fatal inflammatory responses and a shortened lifespan in *$Foxp3^{cre}$;$Vhl^{fl/fl}$* mice of approximately 70 days. Vhl-deficient Treg cells suffer from impaired suppressive functions, accompanied by increased expression of IFN-γ, and abrogating IFN-γ expression can rescue the functional impairment of Vhl-deficient Treg cells. Hif-1α, the classic substrate of Vhl, accumulated in Vhl-deficient Treg cells, and double knockout of Vhl and Hif-1α in Treg cells (*$Foxp3^{cre}$;$Vhl^{fl/fl}$;Hif-$1\alpha^{fl/fl}$* mice) largely abolished the inflammatory phenotype of *$Foxp3^{cre}$;$Vhl^{fl/fl}$* mice, indicating the significance of the classic Vhl-Hif-1α axis in Treg cells (Lee et al. 2015). Wu et al. (Wu et al. 2022; Wu et al. 2023) investigated the role of Rbx1 and Rbx2, the catalytic subunits of the CRL family, in Treg cells. Notably, conditional deletion of Rbx2 in Treg cells did not cause any obvious inflammatory phenotype in mice, while Rbx1-deficiency in Treg cells caused a disastrous inflammatory response, with a severity similar to that in *Scurfy* mutant mice (Gavin et al. 2007). This evidence suggests that Rbx1 plays significant function in Treg cells, and Rbx1 and Rbx2 are not functionally redundant but are relatively independent of each other. Rbx1-deficiency leads to impaired function and a reduced number of Treg cells. Rbx1-deficient Treg cells do not overexpress IFN-γ, indicating that the mechanisms by which Vhl and Rbx1 affect Treg cells are distinct from each other. Single-cell RNA-seq and transcriptome data revealed that Rbx1 maintains the effector subpopulations of Treg cells and supports the expression of many functional molecules in Treg cells. As a ubiquitin E3, Rbx1 specifically degrades numerous substrates in Treg cells. Bim is a potential substrate of Rbx1 and shows obvious accumulation in Rbx1-deficient Treg cells. The deletion of Bim partially rescues the phenotype induced by Rbx1-deficiency in Treg cells (Yeh et al. 2018). Notably, single-cell RNA-seq data suggested that Rbx1 was downregulated in Treg cells in the alveolar lavage fluid of patients with both mild and severe/critical COVID-19 (Liao et al. 2020), which suggests the potential pathological or physiological role of the dynamic regulation of Rbx1 in Treg cells.

$CD8^+$ cytotoxic T cells are executors of cellular immunity and are involved in antiviral responses and anti-tumor immunity. Vhl, a substrate receptor in the CRL family, is involved in regulating $CD8^+$-T cell functions, and the loss of Vhl in $CD8^+$-T cells enhances CD8-mediated cellular immunity. Doedens et al. reported that Vhl-deficient $CD8^+$-T cells respond to viral infection more efficiently than wild-type cells (Doedens et al. 2013). Liikanen et al. showed that Vhl-deficient $CD8^+$-T cells in tumors tended to enhance anti-tumor immunity and increased the efficacy of CAR-T therapy (Liikanen et al. 2021). In both antiviral and anti-tumor immune responses, the underlying mechanism of Vhl in CD8$^+$-T cells was mainly mediated by the classic substrate Hif-1α.

5.4 CRLs in Humoral Immunity

Components of the CRL E3 family also participate in regulating humoral immunity. In B lymphocytes, Rbx1, which is a catalytic subunit of CRLs, is dynamically regulated under stress conditions. When B cells are subjected to ER stress, the Rbx1 protein is cleaved at the N-terminus by Caspase 1, and ER stress is an early event in B cell activation. Although the physical implications of the observation are not clear, these findings suggest that RBX1 may be involved in B cell function (Shteingart et al. 2012).

In addition to the direct regulation of B cells, humoral immunity can also be regulated via Follicular helper T cells (Tfh cells), and components of the CRL family are also involved in this process. Zhu et al. showed that Vhl was essential for the regulation and function of Tfh cells, and *Vhl*-deficient CD4$^+$-T cells cannot efficiently differentiate into Tfh cells during lymphocytic chorio-meningitis virus (LCMV) infection. The substrate Hif-1α mediates the effects of Vhl on Tfh cells, and the Vhl-Hif-1α axis regulates RNA m^{6}A methylation and the expression of the *Icos* gene, which is a master regulator of Tfh cells (Zhu et al. 2019).

5.5 CRLs in Group 2 Innate Lymphoid Cells (ILC2s)

Innate lymphoid cells (ILCs) are immune cells that are characterized by lymphoid morphology and cytokine profiles similar to those of T cells but do not express clonally distributed diverse antigen receptors; these cells are mainly found at barrier surfaces such as the lungs, intestine, and skin, as well as adipose tissue and some mucosal-associated tissues. ILCs can be categorized into three subsets (Group 1 ILCs (ILC1), Group 2 ILCs (ILC2s), and Group 3 ILCs (ILC3s)) based on their transcriptional and functional similarities to T helper cell subsets. ILC2s express GATA3, RORa, and BCL11b and produce type 2 cytokines such as interleukin-4 (IL-4), IL-5, IL-9, IL-13, and the epidermal growth factor amphiregulin (AREG).

Li et al. crossed *Vhl$^{fl/fl}$* mice with *Zbtb16^{cre}* mice to generate *Zbtb16^{cre};Vhl$^{fl/fl}$* mice and knock out the *Vhl* gene in all ILCs in vivo. Interestingly, specific surface markers of ILC2 were changed after Vhl deletion, while ILC1 and ILC3 were largely normal, suggesting a specific role of Vhl in ILC2s. Functionally, *Zbtb16^{cre};Vhl$^{fl/fl}$* mice suffered from a more modest inflammatory response in the papain-induced lung inflammation model than wild-type mice, as revealed by the presence of multiple inflammatory markers. ILC2s undergo expansion and activation in the presence of IL33; the deletion of Vhl abolishes this ability of ILC2s to respond to IL-33 stimulation, indicating functional inertia in Vhl-deficient ILC2s. The classic substrate of Vhl is Hif-1α, which accumulates in Vhl-deficient ILC2s. Abrogates Hif-1α expression in *Zbtb16^{cre};Vhl$^{fl/fl}$;Hif-1α$^{fl/fl}$* mice rescues the phenotype associated with Vhl deficiency in ILC2s. The function of Vhl in ILC2s is mediated by its classic substrate Hif-1α (Li et al. 2018).

5.6 Prospectives

This review summarizes the findings about CRL E3 ligase components in immune regulation (Table 5.1). CRLs regulate many biological processes in vivo. Recently, increasing evidence has revealed the critical role of CRLs in immune regulation. Further work should focus on the specific mechanisms in cell types or particular processes. Given that CRL inhibition is used as an anticancer strategy for drug development, it is important to evaluate the effect of CRL inhibitors on the immune system.

Table 5.1 Summary of findings about CRL E3 ligase components in immune regulation

Proteins	Structure of CRLs	Immune cell type	Description	References
Cul-1/4a/4b/5	Scaffold protein	Macrophage	Cul-1/4a/4b/5 are degraded after live *Eh* contaction	Chadha et al. (2021)
Cul4b	Scaffold protein	Macrophage	Myeloid-specific Cul4b deficiency worsens LPS-induced peritonitis	Hung et al. (2014)
			Cul4b negatively regulates TLR4-, TLR3- and TLR2-mediated inflammatory responses	Song et al. (2021)
Fbxw7	Substrate receptor	Macrophage	Suppresses the inflammatory response	Balamurugan et al. (2013)
			Establish antiviral immunity	Song et al. (2017)
			Promote intestinal inflammation	He et al. (2019)
Dcaf2	Substrate receptor	Dendritic cell	Dcaf2-deficiency in DCs facilitates experimental autoimmunity	Huang et al. (2018)
Rbx2	Catalytic subunit	T lymphocyte	Rbx2 is essential for T cell activation	Mathewson et al. (2016)
Cul4b	Scaffold protein	$CD4^+$ helper T cell	Cul4b regulates homeostasis, proliferation, and survival in $CD4^+$-T cells	Dar et al. (2021)
Vhl	Substrate receptor	Regulatory T cells	Vhl-deficient Treg cells suffer from impaired suppressive functions, accompanied by increased expression of IFN-γ	Lee et al. (2015)
Rbx1	Catalytic subunit	Regulatory T cells	Rbx1-deficiency in Treg cells caused a disastrous inflammatory response, with a severity similar to that in *Scurfy* mutant mice	Wu et al. (2022); Wu et al. (2023)
Vhl	Substrate receptor	$CD8^+$ cytotoxic T cell	Vhl-deficient $CD8^+$-T cells respond to viral infection more efficiently than wild-type cells	Doedens et al. (2013)
			Vhl-deficient $CD8^+$-T cells in tumors tended to enhance anti-tumor immunity	Liikanen et al. (2021)
Rbx1	catalytic subunit	B lymphocyte	When B cells are subjected to ER stress, the Rbx1 protein is cleaved at the N-terminus by Caspase 1	Shteingart et al. (2012)
Vhl	substrate receptor	Follicular helper T cell	Vhl was essential for Tfh cells via regulates RNA m^6A methylation and the expression of the Icos gene	Zhu et al. (2019)
Vhl	substrate receptor	Group 2 innate lymphoid cells	Vhl specifically regulates development and function of ILC2	Li et al. (2018)

References

Aki D, Li Q, Li H, Liu YC, Lee JH (2019) Immune regulation by protein ubiquitination: roles of the E3 ligases VHL and itch. Protein Cell 10:395–404

Balamurugan K et al (2013) FBXW7α attenuates inflammatory signalling by downregulating C/EBPδ and its target gene Tlr4. Nat Commun 4:1662

Chadha A, Moreau F, Wang S, Dufour A, Chadee K (2021) Entamoeba histolytica activation of caspase-1 degrades cullin that attenuates NF-κB dependent signaling from macrophages. PLoS Pathog 17:e1009936

Dar AA et al (2021) The E3 ubiquitin ligase Cul4b promotes $CD4^+$ T cell expansion by aiding the repair of damaged DNA. PLoS Biol 19:e3001041

Doedens AL et al (2013) Hypoxia-inducible factors enhance the effector responses of $CD8^+$ T cells to persistent antigen. Nat Immunol 14:1173–1182

Gavin MA et al (2007) Foxp3-dependent programme of regulatory T-cell differentiation. Nature 445:771–775

He J et al (2019) Fbxw7 increases CCL2/7 in $CX3CR1^{hi}$ macrophages to promote intestinal inflammation. J Clin Invest 129:3877–3893

Huang DT et al (2009) E2-RING expansion of the NEDD8 cascade confers specificity to cullin modification. Mol Cell 33:483–495

Huang T et al (2018) $CRL4^{DCAF2}$ negatively regulates IL-23 production in dendritic cells and limits the development of psoriasis. J Exp Med 215:1999–2017

Hung MH, Jian YR, Tsao CC, Lin SW, Chuang YH (2014) Enhanced LPS-induced peritonitis in mice

deficiency of cullin 4B in macrophages. Genes Immun 15:404–412
Lee JH, Elly C, Park Y, Liu YC (2015) E3 ubiquitin ligase VHL regulates hypoxia-inducible factor-1α to maintain regulatory T cell stability and suppressive capacity. Immunity 42:1062–1074
Li Q et al (2018) E3 Ligase VHL Promotes Group 2 innate lymphoid cell maturation and function via glycolysis inhibition and induction of interleukin-33 receptor. Immunity 48:258–270
Liao M et al (2020) Single-cell landscape of bronchoalveolar immune cells in patients with COVID-19. Nat Med 26:842–844
Liikanen I et al (2021) Hypoxia-inducible factor activity promotes antitumor effector function and tissue residency by CD8+ T cells. J Clin Invest 131:e143729
Mathewson ND et al (2016) SAG/Rbx2-dependent neddylation regulates T-cell responses. Am J Pathol 186:2679–2691
Shteingart S, Hadar R, Cohen I, Ravid T, Tirosh B (2012) Endoplasmic reticulum stress induces a caspase-dependent N-terminal cleavage of RBX1 protein in B cells. J Biol Chem 287:31223–31232
Song Y et al (2017) E3 ligase FBXW7 is critical for RIG-I stabilization during antiviral responses. Nat Commun 8:14654
Song Y et al (2021) CUL4B negatively regulates Toll-like receptor-triggered proinflammatory responses by repressing Pten transcription. Cell Mol Immunol 18:339–349
Soucy TA et al (2009) An inhibitor of NEDD8-activating enzyme as a new approach to treat cancer. Nature 458:732–736
Tan M et al (2009) RBX1/ROC1 disruption results in early embryonic lethality due to proliferation failure, partially rescued by simultaneous loss of P27. Proc Natl Acad Sci USA 106:6203–6208
Tan M et al (2011) SAG/RBX2/ROC2 E3 ubiquitin ligase is essential for vascular and neural development by targeting NF1 for degradation. Dev Cell 21:1062–1076
Wei D, Sun Y (2010) Small RING finger proteins RBX1 and RBX2 of SCF E3 ubiquitin ligases: the role in cancer and as cancer targets. Genes Cancer 1:700–707
Wu D et al (2022) The Ube2m-Rbx1 neddylation-Cullin-RING-Ligase proteins are essential for the maintenance of Regulatory T cell fitness. Nat Commun 13:3021
Wu D, Sun Y (2023) The functional redundancy of neddylation E2s and E3s in modulating the fitness of regulatory T cells. Research 6:0212
Yeh CH, Bellon M, Nicot C (2018) FBXW7: a critical tumor suppressor of human cancers. Mol Cancer 17:115
Zhao Y, Sun Y (2013) Cullin-RING ligases as attractive anti-cancer targets. Curr Pharm Des 19:3215–3225
Zhou Z, Song X, Wavelet CM, Wan Y (2020) Cullin 4-DCAF proteins in tumorigenesis. Adv Exp Med Biol 1217:241–259
Zhu Y et al (2019) The E3 ligase VHL promotes follicular helper T cell differentiation via glycolytic-epigenetic control. J Exp Med 216:1664–1681

The Immune Modulatory Role of TIF1 Proteins

6

Qingchen Zhu and Yichuan Xiao

Abstract

The function of immune cells is delicately regulated under a variety of molecular networks. Transcriptional intermediary factor 1 (TIF1) family proteins, consisting of TRIM24, TRIM28 and TRIM33, share a highly conserved RING domain that is essential for the regulation of protein ubiquitination functioning as E3 ubiquitin ligases. TIF1 family proteins are diversely expressed in different types of immune cells, and participate in the regulation of various of cellular functions including chromosome modification, DNA repair, tumor progression, and immunity. In this review, we summarized current studies on TIF1 family proteins' functions in the modulation of immune cell development, anti-infection immunity, cancer immunology, inflammation, and autoimmune diseases.

Keywords

TRIM24 · TRIM28 · TRIM33 · Immune cells

Q. Zhu · Y. Xiao (✉)
CAS Key Laboratory of Tissue Microenvironment and Tumor, Shanghai Institute of Nutrition and Health, Chinese Academy of Sciences, University of Chinese Academy of Sciences, Shanghai, China
e-mail: ycxiao@sibs.ac.cn

6.1 Overview of TIF1 Proteins

The tripartite motif-containing (TRIM) protein family is a group of proteins with a conserved structure of RBBC motif and is composed of more than 80 members (Reymond et al. 2001). This motif constitutes the N terminal region of these proteins and has a highly conserved order consisting of a RING domain followed by one or two B-Box domains and then a coiled-coil domain. The structure of these proteins in the C-terminus is highly variable, which can be further divided into eleven subgroups owing to the diversity of C-terminus (Hatakeyama 2017; Short and Cox 2006).

The TIF1 family consists of three proteins including TRIM24, TRIM28, and TRIM33. Besides the RBBC motif, TIF1 family proteins contain a plant-homeodomain (PHD) and another bromodomain (BROMO) in the C-terminus (Fig. 6.1). The RING domain enables TIF1 family proteins to transfer ubiquitin or SUMO molecules to specific substrate proteins coordinating with other enzymes. The B-Box and coiled-coil domains may take part in interaction with other proteins. TIF1 proteins mostly participate in chromatin modification via PHD and BROMO domains, which read the N-terminal domain of methylated histone H3 via the PHD fingers and recruit acetylated histone through the BROMO domain (Ozato et al. 2008; Meroni and Diez-Roux 2005; Tsai et al. 2010).

H. Hu, X. Fu (eds.), *Ubiquitination in Immune System and Immune Response*, Advances in Experimental Medicine and Biology 1466, https://doi.org/10.1007/978-981-97-7288-9_6

Fig. 6.1 The structures of TIF1 family proteins. The N terminal regions contain RING domain, B-Box1 (B1), B-Box2 (B2), and coiled-coil domain (CC). The C-terminus is composed of plant-homeodomain (PHD) and bromodomain (BROMO)

6.2 TIF1 in Anti-infection Immunity

As the first line of defense against infection, innate immunity plays an important role in antiviral immunity. Upon infection, pathogen-associated molecular patterns (PAMPs) can be recognized by the host's pattern recognition receptors (PRRs) which are usually expressed in innate immune cells such as macrophages and dendritic cells (DCs). Sensing PAMPs can promote downstream signaling cascades, activate transcription factors such as NF-κB, IRF3, IRF7, and AP-1, then produce type I interferon and other cytokines to induce inflammation and activate adaptive immune response to inhibit pathogen reproduction (Iwasaki and Medzhitov 2004; Ivashkiv and Donlin 2014).

Over the past decade, many studies have reported that TRIM proteins positively or negatively regulate anti-infection immunity (Ozato et al. 2008; van Tol et al. 2017). For example, TRIM25 combines with the CARD domain of RIG-I and mediates its K63-liked ubiquitination, which is important for the production of IFN-β and antiviral immunity (Gack et al. 2007). Furthermore, TRIM31 enhances the antiviral effect by promoting the K63-linked ubiquitination of MAVS (Liu et al. 2017). In this section, we mainly discuss the role of TRIM24, TRIM28, and TRIM33 in anti-infection immunity.

6.2.1 TRIM24 and TRIM28 Modulates RLR Signaling

RLRs are the RNA receptors located in the cytoplasm. The RLRs are composed of three members: retinoic acid-inducible gene I (RIG-I), melanin differentiation-associated protein 5 (MDA5), and laboratory of genetics and physiology 2 (LGP2) (Loo and Gale Jr. 2011). All RLR proteins contain a helicase domain and a carboxy-terminal domain (CTD), which work together to recognize viral RNA. In addition, the N-terminus of RIG-I and MDA5 also have two caspase activation and recruitment domains (CARD) mediating downstream signaling transduction (Rehwinkel and Gack 2020; Jiang et al. 2011). Upon RIG-I and MDA5 recognizing RNA, the protein conformation changes to be activated, and its CARD domain binds to the CARD domain of the MAVS protein located in mitochondria, resulting in the activation of MAVS. On the one hand, activated MAVS can bind with TBK1 and IKKi, which directly catalyze the phosphorylation of IRF3 and IRF7. Activated IRF3/7 binds to the promoter region of type I IFN and promotes transcription of type I IFN. On the other hand, activated MAVS also binds to FADD, RIP1, and TRAF6 to form complexes, which then activate NF-κB and release pro-inflammatory factors (Yong and Luo 2018).

A published work has showed that TRIM24 overexpression facilitates RIG-I-induced IFN-β luciferase activity in a systematic analysis of all known 75 human TRIM family proteins (Versteeg et al. 2013). Furthermore, another global quantitative proteomic analysis revealed that five TRIM proteins were adjusted upon SeV infection, and found that knocking down TRIM24 and TRIM27 downregulated SeV-induced IFN-I production (Zhu et al. 2017). Our published work further confirmed that TRIM24 deletion dramatically inhibits VSV-induced type I IFN production and is more susceptible to VSV infection. TRIM24 in resting macrophages is mainly located in the nucleus; however, TRIM24 is transferred from the nucleus to the cytoplasm under interaction with CRM1 proteins upon VSV infection, and finally located in mitochondria. Next, TRIM24 interacts with TRAF3 and directly mediates its

K63-linked ubiquitination at K429 and K436, promoting the formation of the complexes of TRAF3 with MAVS and TBK1 (Fig. 6.2). Furthermore, activated TBK1 phosphorylates IRF3 and facilitates IFN-I expression to antagonize the replication of the RNA virus (Zhu et al. 2020).

Interestingly, loss of TRIM24 does not affect the activation of NF-κB and MAPK and the secretion of pro-inflammatory cytokines such as IL-6 and TNF-α. K63-linked ubiquitination of TRAF3 promotes the expression of IFN-I in RIG-I and TLR pathways, but we found that TRIM24 only facilitates the IFN-I secretion upon VSV infection. Many TRIM family proteins expression was found to be regulated upon infection. For example, TRIM26 expression is significantly upregulated upon the SeV infection (Wang et al. 2015). In our research, we found that virus-induced activated IRF3 directly bound to the promoter of *TRIM24* and triggered the transcriptional suppression of TRIM24, forming a negative feedback loop to restrain overactivation of antiviral signaling.

Besides TRIM24, TRIM28 was shown to inhibit IFN-I during Sendai virus (SeV) and VSV infection (Liang et al. 2011). TRIM28 is associated with the ID domain (aa 283–466) of IRF7 through the RBBC domain (aa 1–400) in vitro and in vivo. Moreover, TRIM28 specifically catalyzed SUMOylation of IRF7 at K444 and K446 sites depending on its RING domain, hardly affecting the SUMOylation of IRF3. The SUMOylation of IRF7 mediated by TRIM28 was enhanced after SeV infection. Consistent with previous research results, SUMOylation of transcription factors usually inhibited transcription, for instance, overexpression of TRIM28 suppressed IRF7-induced IFN-I transcription hence impairing the antiviral immunity (Gill 2003).

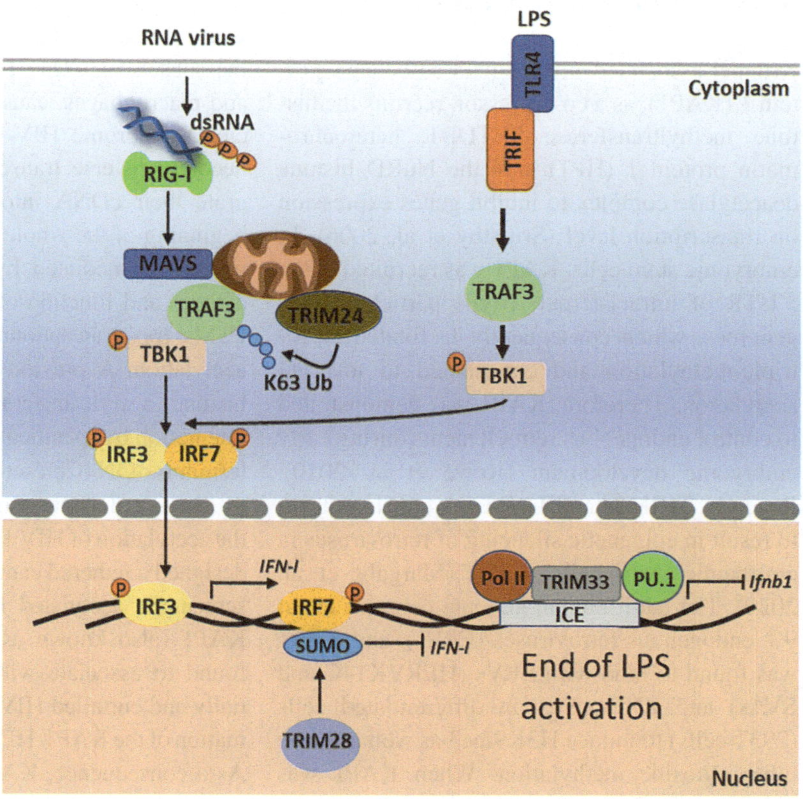

Fig. 6.2 The regulation of TIF1 proteins in IFN-I. Upon RNA virus infection (such as VSV and SeV), TRIM24 was translocated from the nucleus binding with TRAF3 and catalyzed its K63-linked ubiquitination, facilitating IFN-I expression. However, SeV infection promoted the SUMOylation of IRF7 catalyzed by TRIM28, which inhibited the IFN-I expression. At the end of LPS activation, TRIM33 bound with a regulatory element called ICE at 15kb upstream *Ifnb1* to disassociate PU.1 and CBP/P300, downregulating the *Ifnb1* expression. ICE, *Ifnb1* control element

6.2.2 TRIM33 in LPS-Induced IFN-β Expression

The transcriptional regulation of IFN-β is critical to eliminate invading pathogens. TRIM33 was found to inhibit *Ifnb1* mRNA expression at the late term of activated macrophages stimulated by LPS, without affecting the *Ifna4* mRNA expression. Mechanically, TRIM33 associated a regulatory element called ICE (*Ifnb1* control element) which is 15kb upstream of *Ifnb1* with the recruitment by PU.1 at the end of LPS stimulation, suppressing the binding of CBP/P300/PU.1 to *Ifnb1* promoter and lower the H3 acetylation which switches on *Ifnb1* transcription (Fig. 6.2). The suppression function to *Ifnb1* relied on the coiled-coil domain (CC) which was shown to act in the interaction among different proteins (Ferri et al. 2015).

6.2.3 TIF1 and Retroviruses

Krüppel-associated box domain-associated protein 1 (KAP1), as a co-repressor, recruits the histone methyltransferase SETDB1, heterochromatin protein 1 (HP1), and the NuRD histone deacetylase complex to inhibit genes expression on transcription level (Sripathy et al. 2006). In embryonic stem cells, KAP1 was recruited at the 5′UTR of intracisternal A-type particles (IAP) genomes, which consequently facilitated H3K9 triple-methylation and contributed to low H4 acetylation. Therefore, KAP1 was demonstrated to control endogenous retroelements during early embryonic development (Rowe et al. 2010). Recently, TRIM24 and TRIM33 were also shown to result in epigenetic silencing of retroviruses in embryonic stem cells (ESC) (Margalit et al. 2020). The adult human genome contains about 9% endogenous retroviruses (ERVs), and KAP1 was found to bind with ERVs (HERVK14C and SVAs) and ZNF genes in differentiated cells (293T cells) to induce H3K9me3 as well as some other cytosine methylation. When KAP1 was knocked out, HERVs and SVAs were reactivated as a result of low H3K9me3 level, accumulated dsRNA-inducing IFN-I transcription and some ISGs expression depending on RIG-I (Tie et al. 2018).

Interestingly, the influenza viruses (IAV and IBV) induced repression of SUMOylation of TRIM28 with quantitative proteomics SUMOylation identification, and the lower SUMOylation of TRIM28 inhibited the repression of ERVs, releasing the "self" dsRNA to initialize antiviral immunity depending on RIG-I-MAVS signaling pathways. However, the intact IAV failed to induce high ISGs expressions owing to the sequence-independent dsRNA-binding activity of IAV NS1, antagonizing the host antiviral immunity (Schmidt et al. 2019).

6.2.4 TRIM28 and TRIM33 in HIV-1 Infection

Human immunodeficiency virus type 1 (HIV-1) originated in central Africa and spread to Haiti, Europe, North America, and further to the whole world (Mendoza et al. 2014). It selectively invades the subsets of $CD4^+$ T lymphocytes, monocytes, and macrophages, causing acquired immunodeficiency syndrome. HIV-1 is a kind of retrovirus that needs to reverse transcribe their RNA and integrate their cDNA into the host genome at the beginning of the whole life cycle. The integration process is mediated by integrase (IN). Enzyme activity and function of IN could be adjusted by PTMs (posttranslational modifications), such as acetylation. A previous study demonstrated that histone acetyltransferase p300 facilitates HIV integration by specifically acetylating the carboxyl terminus of IN (Cereseto et al. 2005). In an attempt to identify delicate modification mechanisms in the acetylation of HIV IN, Cereseto and colleagues designed a tethered catalysis two-hybrid system to screen the acetylated HIV IN binding proteins. KAP1 (also known as TRIM28 or TIF1β) was found to associate with acetylated IN preferentially and curtailed HIV IN acetylation by the formation of the KAP1-HDAC1-IN proteins complex. As a consequence, KAP1 inhibited HIV integration and prevented the host from HIV infection (Allouch et al. 2011). Besides the adaptor function, TRIM28 can also restrain HIV expression as a SUMO E3 ligase. TRIM28 bound with P-TEFb catalytic subunit CDK9 through the RING domain and SUMOylated CDK9 in Lys44, Lys56, and Lys68, which led to the repression of CDK9 kinase activity. Therefore, CDK9 failed to super-phosphorylate the Ser2 resides of RNA polymerase II, which promotes HIV RNA transcribing, resulting in HIV-1 latency (Ma et al. 2019).

Recently, TRIM33 has also been reported to impair HIV-1 infection. In a screen of ubiquitin-conjugation siRNA library affecting HIV-1 IN protein degradation, TRIM33 was found to promote the poly-ubiquitin chains formation of IN as an E3 ligase, and ubiquitinated IN was then degraded by host proteasome (Ali et al. 2019), but TRIM24 and TRIM28, two proteins of TIF1 family proteins, don't catalyze the ubiquitination of IN protein.

6.2.5 TRIM28 in Other Virus Infection

Human cytomegalovirus (HCMV) is a universal pathogen that leads to life-long infection. The HCMV latency and lytic replication are regulated by complicated mechanisms (Sinclair 2008). TRIM28 phosphorylation switch was reported to mediate HCMV latency release. TRIM28 was recruited to the genome of HCMV, then enrolled HP1α and SETDB1 to induce H3K9 methylation and heterochromatin formation, which consequently contribute to transcriptional repression and HCMV latency in human hematopoietic stem cells (HSC). During lytic replication, TRIM28 was phosphorylated on Ser824 by mTOR, and it failed to recruit SETDB1 to silence viral genome transcription although remaining binding to the viral genome (Rauwel et al. 2015). A prior study revealed that TRIM28 was phosphorylated at Ser824 by ATM upon DNA damage (White et al. 2012), and this study found that chloroquine, an ATM activator, could also force HCMV out of latency by phosphorylating TRIM28. Furthermore, TNF-α stimulation strengthens HCMV reactivation by combining with chloroquine from latently infected HSC.

Highly pathogenic avian influenza viruses (HPAIV) such as H5N1 remain a global threat to public health. HPAIV usually induces high expression of type I IFN and pro-inflammatory cytokines to severely impair tissue. Upon infection by HPAIV such as H5N1, H7N7, and H7N9, the viral RNA was sensed by PKR, which induced phosphorylated at Ser473 via downstream signaling cascades of p38 and MSK1 which is independent of RIG-I sensing (Claas et al. 1998; Liu et al. 2016). Phosphorylation of TRIM28 lost function of transcriptional corepressor, resulting in pathological high levels of IFN-β and pro-inflammatory cytokines along with severe tissue damage. In contrast, human-adapted IAV strains such as H1N1 did not induce the phosphorylation of TRIM28 at Ser473, contributing to low levels of IFN-β and pro-inflammatory cytokines production (Krischuns et al. 2018).

Epstein-Barr virus (EBV) is a prevalent human virus. The ATM-KAP1 axis was also shown to regulate EBV replication. During EBV latency, KAP1 repressed EBV lytic genes by transcriptional silence, but KAP1 could be phosphorylated at Ser824 by ATM with lytic cycle activation signals, failing to silence the viral genome, and chloroquine could also activate EBV lytic cycles the same as HCMV (Li et al. 2017). Another research found that latent membrane protein-1 (LMP1), an EBV viral oncoprotein, utilized the host SUMO-conjugating enzyme Ubc9 to SUMOylate TRIM28, transcriptionally silencing lytic viral promoters and contributing to EBV latency (Bentz et al. 2015).

Adeno-associated virus (AAV) is widely used as a vector for gene therapy owing to its apparent safety. TRIM28 was also validated to be recruited at the *rep* ORF and enroll SETDB1 to form transcriptionally repressive heterochromatin, silencing the AAV2 genome. Moreover, the AAV replication proteins (Rep) mainly including Rep52 and Rep78 restrained the recruitment of the NIPP1-PP1 complex that dephosphorylated Ser824 of TRIM28, hence dramatically promoting the phosphorylation of TRIM28 at Ser824 (Smith-Moore et al. 2018). Consistent with previous studies, TRIM28 phosphorylation at Ser824 deprived its co-repression activity to the AAV genome, supporting lytic replication of AAV.

6.3 TIF1 in Cancer Immunology

The function of TIF1 family proteins in tumorigenesis has been discussed in numerous studies (Hatakeyama 2011). TRIM24 was found to specifically inhibit hepatocellular carcinoma (HCC) progression depending on RARα cooperating with TRIM33, while TRIM28 deletion alleviated the occurrence of HCC caused by TRIM24 deletion (Khetchoumian et al. 2007; Herquel et al. 2011). In another research, TRIM24 was shown to be stabilized in speckle-type POZ protein (SPOP) mutant PC and CRPC, inspiring AR signaling activation and promoting prostate cancer progression (Groner et al. 2016). In SPOP-wildtype PC, TRIM28 breaks the binding of TRIM24 and SPOP, stabilizing TRIM24 proteins to potentiate the prostate cancer progression (Fong et al. 2018). Moreover, TRIM33 was shown to regulate TGF-β signaling to affect many types of cancer progression (Yu et al. 2019a). Here, we focus on the function of TIF1 family proteins in cancer immunology despite a few published researches.

Macrophage exhibits the function of inhibiting or promoting tumor growth depending on the transformation of M1- and M2-type response to the tumor immune microenvironment factors. M2-type macrophages are usually educated by Th2 signature cytokines interleukin-4 (IL-4) and IL-13 and potentiate tumor growth mediated by STAT6 (Locati et al. 2020). In our study, we found that TRIM24 mediated K63-linked ubiquitination of CBP at K119 site, enhancing STAT6 acetylation at K383 site by potentiating the association of CBP and STAT6. Acetylated STAT6 inhibited the binding of STAT6 to the promoter region of M2 genes (*Arg1*, *Mrc1*, *Fizz1*, *Ym1*, *Ccl17* and *Ccl24*), downregulated those gene expressions, and finally suppressed tumor growth (Fig. 6.3) (Yu et al. 2019b).

In lung adenocarcinoma, TRIM28 was upregulated and the expression of TRIM28 was identified to be negatively correlated with prognosis. Mechanistically, TRIM28 might potentiate the SUMOylation of IRF5 and IRF8, weakening the tumor immune microenvironment to positively regulate tumor progression (Fig. 6.3) (Liu et al. 2020). Similarly, high level of TRIM28 might have a good prognosis on melanoma patients. TRIM28 could directly affect epigenetic silencing of IRFs, inhibiting the interferon signaling to further promote stem cell-associated molecular features and to downregulate immune cell infil-

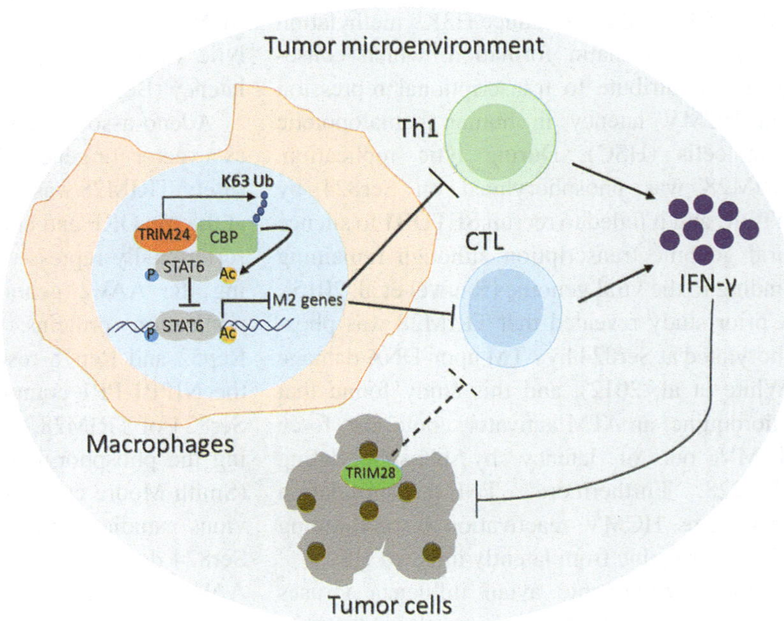

Fig. 6.3 TIF1 in cancer immunology. Macrophage TRIM24 catalyzes K63-linked ubiquitination of CBP and enhances STAT6 acetylation, inhibiting the M2 gene expressions, promoting the IFN-γ secretion of Th1 and CTL to kill tumor cells. The tumor TRIM28 expression correlates negatively with CTL infiltration, suggesting that TRIM28 may inhibit CTL infiltration

tration (Czerwinska et al. 2020). However, TRIM28 deficiency in immune cells such as macrophages and T cells is not yet reported to have a role in tumor progression.

6.4 TIF1 in Autoimmune and Inflammatory Diseases

Autoimmune and inflammatory diseases are common human diseases such as colitis and multiple sclerosis (MS), regulated by delicate mechanisms involved in epigenetic regulations, transcriptional regulations, and posttranslational modifications (Zhang et al. 2018; Liu et al. 2018). TIF1 family proteins have been reported to control these diseases through a variety of regulatory mechanisms (Fig. 6.4).

Th2 cells driving type 2 immune responses play an important regulatory role in a large proportion of atopic asthmatics, which is regulated by many factors. TRIM24 was screened through silico analyses using published transcriptional data and turned out to be a possible protein adjusting Th2-mediated immune responses in asthmatics owing to its functional activation in Th2 cells but not in Th1, Th17, or Treg cells. TRIM24 deficiency downregulated house dust mite (HDM) and helminth-induced IL-1-associated cytokine and chemokine expressions due to the suppressed IL-1R expression. Moreover, IL-1β did not exacerbate HDM-induced Th2 cytokine and chemokine expressions as well as airway allergy upon TRIM24 depletion in T cells (Perez-Lloret et al. 2016).

Unlike TRIM24, TRIM28 specific deletion in total T cells induced a spontaneous autoimmune phenotype and exacerbated EAE and colitis diseases by both regulating T cell activation and T cell tolerance. TCR signaling triggered phosphorylation of TRIM28 at Ser473 but not Ser824 which was phosphorylated upon DNA damage, and TRIM28 promoted T cell activation and development of $CD4^+$ and $CD8^+$ T cells in spleens, lymph nodes, and peripheral but not thymus by potentiating IL-2 production and cell cycle-related gene expressions (Chikuma et al. 2012). In addition, TRIM28 suppressed the

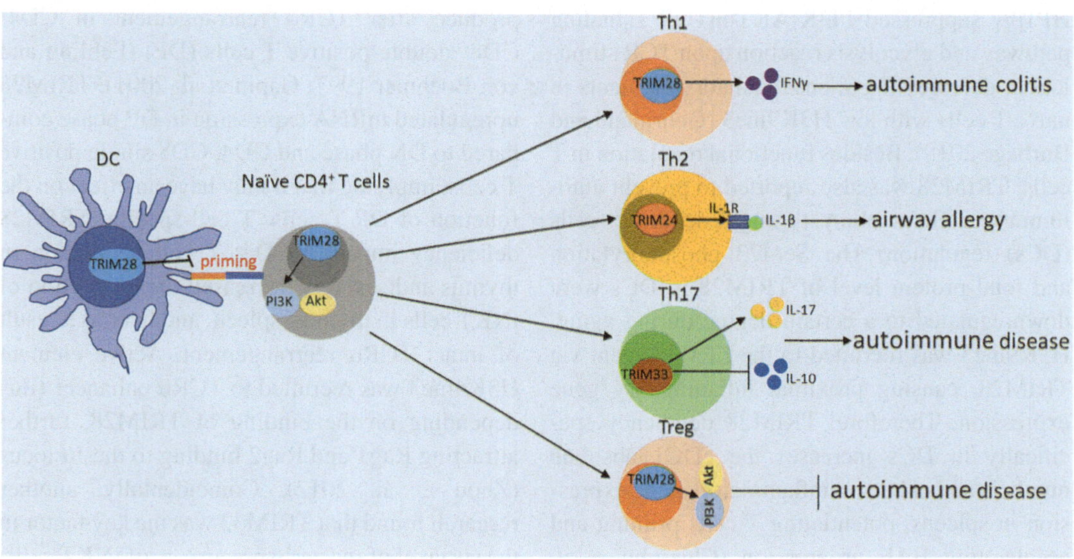

Fig. 6.4 TIF1 in autoimmune and inflammatory diseases. TRIM28 in DCs inhibited T cell priming and alleviated EAE progression. TRIM28 in naïve T cells was critical for effector T cells (Th1, Th2, and Th17) differentiation via PI3K-AKT signaling pathways. It also potentiated PMA/ionomycin-induced cytokines secretion in Th1, Th2, and Th17 cells, exacerbating auto-immune colitis. On the contrary, TRIM28 promoted Treg expansion to prevent autoimmune diseases. TRIM24 specifically upregulated IL-1R expression in Th2 cells and was important for Th2-dependent allergy and anti-helminth immunity. TRIM33 boosted IL-17 expression and inhibited IL-10 expression in Th17 cells to accelerate autoimmune disease progression

memory cells (CD62L^{lo} CD44^{hi}) proportion and differentiation of Treg and Th17 owing to its direct inhibition to *Tgfb3* and TGFβ signaling pathways. Meanwhile, TRIM28 was shown to associate with FOXP3 via the adaptor protein FIK in human T cells, downregulating FOXP3-target genes and inhibiting the activity of Treg cells (Huang et al. 2013).

Less consistent with those research, Chen Dong and colleagues found loss of TRIM28 specifically in Th17 cells contributed to low differentiation of Th17 and relieve of EAE and colitis progression. TRIM28 bound at *Il17-Il17f* locus co-localized with P300, H3K4me3, H3K4me1, H3K27ac, and DNA hydroxylmethylation (5hmc) which were normally associated with gene activation, and TRIM28 was recruited to the *Il17-Il17f* locus depending on IL-6-STAT3 signaling, resulting in RORγt binding to its target genes (Jiang et al. 2018). Moreover, after TRIM28-deficient naïve CD4^{+} T cells adoptively transferred to *Rag2*$^{-/-}$ mice, the mice survived longer and were relieved from colitis owing to defective effector T cell development and expansion. Mechanically, loss of TRIM28 or HP1β/γ suppressed PI3K-AKT-mTOR signaling pathway and glycolysis reaction upon TCR stimulation, derepressing silent regulatory elements in naïve T cells with low H3K9me3 (Gehrmann and Burbage 2019). Besides functional regulation in T cells, TRIM28 was also reported to prevent autoimmune inflammation through dendritic cells (DCs) regulation. The Ser473 phosphorylation and total protein level of TRIM28 in DCs were downregulated to a certain degree during aging. H3K9me3 was recruited to the ERV element via TRIM28, causing proximal inflammatory gene expression. Therefore, TRIM28 deficiency specifically in DCs increases the cDC1 absolute numbers and enhances inflammatory gene expression in spleens, potentiating T cells priming and accelerating EAE progression (Chikuma et al. 2021). Targeting TRIM28 may have an important role in autoimmune diseases.

TRIM33 also regulated autoimmune and inflammatory diseases via innate immunity and adaptive immunity. Firstly, T cell-specific TRIM33 deletion decreased IL-17 expression and promoted IL-10 expression, specifically affecting the Th17 differentiation in vitro and the disease scores of the EAE model in vivo. TRIM33 was recruited to the *Il17a* and *Il10* gene loci depending on Smad2, affecting suppressive marker H3K9me3 and H3K27me3 or activation marker H3K4me3 binding to the *Il17a* and *Il10* gene loci respectively (Tanaka et al. 2018). Moreover, TRIM33 was repressed on transcriptional level in Crohn's disease (CD), and TRIM33 in mature myeloid cells prevents severe DSS-induced colitis, accumulating less inflammatory monocytes and neutrophils in blood (Petit et al. 2019). Another research in zebrafish identified that TRIM33 was crucial for the migration of macrophages and neutrophils to inflammatory cues (Demy et al. 2017).

6.5 TIF1 and Immune Cells Development

T cells develop and mature in thymus under a series of regulatory mechanisms. Mature αβ T cells and invariant natural killer T cells (iNKT) produce after TCRα rearrangement in CD4^{+} CD8^{+} double positive T cells (DP) (Fehling and von Boehmer 1997; Gapin et al. 2001). TRIM28 upregulated mRNA expression in DP phase compared to DN phase and CD4, CD8 single positive T cells, implying that it may have an effect on the function of DP T cells. T cell-specific TRIM28 deficiency impaired CD4^{+} T cell generation in thymus and severely decreased the proportion of iNKT cells in thymus, spleen, and liver as a result of intact TCRα rearrangement. Active element H3K4me3 was recruited to TCRα enhancer (Eα) depending on the binding of TRIM28, further attracting Rag1 and Rag2 binding to the Jα locus (Zhou et al. 2012). Coincidentally, another research found that TRIM33 was the key factor in the survival of the early precursors of iNKT cells. TRIM33 deletion in T cells impaired the CD44^{low} NK1.1^{-} precursors (Doisne et al. 2009).

Besides T cell development, TRIM28 could also regulate mouse B lymphoid cell differentiation and function. B-cell progenitors (Pro-B cells) undergo Ig chain rearrangement in bone marrow and then migrate to the spleen for further maturation. Antibody production from mature B cells plays a broad function. TRIM28 deletion specifically in B-lymphoid cells impaired peripheral mature B cells including follicular (FO) and marginal zone (MZ) without affecting the total B cell numbers consisting of precursor, immature, and mature B cells in bone marrow, and TRIM28 knocked-out mice downregulated the antibody production induced by both T cell-dependent and T cell-independent immunization. TRIM28 deficiency derepressed histone marks H3K9me3 in *Pten* gene which participated in BCR signaling and B cell development via PI3K (Santoni de Sio et al. 2012). Another study revealed that TRIM28 destroyed class switch recombination (CSR) which contributed to antibody diversity by suppressing antigen-induced cytidine deaminase (AID) transcription and expression (Okazaki et al. 2011). This function of TRIM28 in B cell development and antibody production may play a crucial role in viral infection and autoimmune inflammatory diseases.

6.6 Conclusions and Future Perspective

TIF1 proteins participate in a variety of cellular functions including immune cell development, anti-infection immunity, cancer immunology, inflammation, and autoimmune diseases. TIF1 proteins ubiquitinate or SUMOylate the specific substrate to come up against different environments. Macrophage TRIM24 could act as an E3 ligase in TRAF3 and CBP to affect virus infection and tumor growth, while TRIM28 catalyzed SUMOylation of IRF7 and CDK9 to defend against different virus infections. Moreover, viral oncoproteins also can SUMOylate TRIM28 to affect infection. However, little is known about the ubiquitination and SUMOylation of TRIM33 in immune cells, and whether TIF1 proteins stability can be changed through ubiquitination or SUMOylation under immune-mediated diseases deserves further study.

Phosphorylation of TRIM28 at Ser824 and Ser473 due to different virus infections failed to recruit SETDB1 and to further silence transcription, supporting lytic replication of the virus. TCR signaling induces the phosphorylation of Ser473 but not Ser824. Phosphorylation at Ser473 was enhanced in DCs upon aging to positively induce aging-induced autoimmune diseases. Perhaps screening some compounds against these two phosphorylation sites would play a beneficial role in virus infection and autoimmune diseases. Moreover, phosphorylation of TRIM24 induced by DNA damage contributes to TRIM24 degradation. However, the phosphorylation of TRIM24 in immune cells is not clear.

The function of TIF1 family proteins in $CD4^+$ T cells has been widely revealed. TRIM24 specifically upregulated IL-1R expression in Th2 cells and was essential for Th2-dependent allergy and anti-helminth immunity. TRIM28 in naïve T cells was critical for effector T cells (Th1, Th2, and Th17) differentiation via PI3K-AKT signaling pathways. TRIM33 boosted IL-17 expression and inhibited IL-10 expression in Th17 cells to accelerate autoimmune disease progression. However, TIF1 proteins in $CD8^+$ T cells have not been discussed.

TIF1 proteins have an effect on chromatin modification to positively or negatively regulate gene transcriptional levels. TRIM33 depressed *Ifnb1* mRNA expression through impaired H3ac recruitment. TRIM28 potentiates the binding of H3K4me3 in Eα and further recruits the Rag1 and Rag2 to the Jα locus, enhancing the maturation of αβ T cells and iNKT. Therefore, the PHD and bromodomain are potential targets for some diseases.

References

Ali H et al (2019) Cellular TRIM33 restrains HIV-1 infection by targeting viral integrase for proteasomal degradation. Nat Commun 10(1):926

Allouch A et al (2011) The TRIM family protein KAP1 inhibits HIV-1 integration. Cell Host Microbe 9(6):484–495

Bentz GL et al (2015) LMP1-induced sumoylation influences the maintenance of Epstein-Barr virus latency through KAP1. J Virol 89(15):7465–7477

Cereseto A et al (2005) Acetylation of HIV-1 integrase by p300 regulates viral integration. EMBO J 24(17):3070–3081

Chikuma S et al (2012) TRIM28 prevents autoinflammatory T cell development in vivo. Nat Immunol 13(6):596–603

Chikuma S et al (2021) TRIM28 expression on dendritic cells prevents excessive T cell priming by silencing endogenous retrovirus. J Immunol 206(7):1528–1539

Claas EC et al (1998) Human influenza A H5N1 virus related to a highly pathogenic avian influenza virus. Lancet 351(9101):472–477

Czerwinska P et al (2020) Melanoma stem cell-like phenotype and significant suppression of immune response within a tumor are regulated by TRIM28 protein. Cancer 12(10):2998

Demy DL et al (2017) Trim33 is essential for macrophage and neutrophil mobilization to developmental or inflammatory cues. J Cell Sci 130(17):2797–2807

Doisne J-M et al (2009) iNKT cell development is orchestrated by different branches of TGF-β signaling. J Exp Med 206(6):1365–1378

Fehling HJ, von Boehmer H (1997) Early alpha beta T cell development in the thymus of normal and genetically altered mice. Curr Opin Immunol 9(2):263–275

Ferri F et al (2015) TRIM33 switches off Ifnb1 gene transcription during the late phase of macrophage activation. Nat Commun 6:8900

Fong KW et al (2018) TRIM28 protects TRIM24 from SPOP-mediated degradation and promotes prostate cancer progression. Nat Commun 9(1):5007

Gack MU et al (2007) TRIM25 RING-finger E3 ubiquitin ligase is essential for RIG-I-mediated antiviral activity. Nature 446(7138):916–920

Gapin L et al (2001) NKT cells derive from double-positive thymocytes that are positively selected by CD1d. Nat Immunol 2(10):971–978

Gehrmann U, Burbage M (2019) Critical role for TRIM28 and HP1β/γ in the epigenetic control of T cell metabolic reprograming and effector differentiation. Proc Natl Acad Sci USA 116(51):25839–25849

Gill G (2003) Post-translational modification by the small ubiquitin-related modifier SUMO has big effects on transcription factor activity. Curr Opin Genet Dev 13(2):108–113

Groner AC et al (2016) TRIM24 is an oncogenic transcriptional activator in prostate cancer. Cancer Cell 29(6):846–858

Hatakeyama S (2011) TRIM proteins and cancer. Nat Rev Cancer 11(11):792–804

Hatakeyama S (2017) TRIM family proteins: roles in autophagy, immunity, and carcinogenesis. Trends Biochem Sci 42(4):297–311

Herquel B et al (2011) Transcription cofactors TRIM24, TRIM28, and TRIM33 associate to form regulatory complexes that suppress murine hepatocellular carcinoma. Proc Natl Acad Sci USA 108(20):8212–8217

Huang C et al (2013) Cutting edge: a novel, human-specific interacting protein couples FOXP3 to a chromatin-remodeling complex that contains KAP1/TRIM28. J Immunol 190(9):4470–4473

Ivashkiv LB, Donlin LT (2014) Regulation of type I interferon responses. Nat Rev Immunol 14(1):36–49

Iwasaki A, Medzhitov R (2004) Toll-like receptor control of the adaptive immune responses. Nat Immunol 5(10):987–995

Jiang F et al (2011) Structural basis of RNA recognition and activation by innate immune receptor RIG-I. Nature 479(7373):423–427

Jiang Y et al (2018) Epigenetic activation during T helper 17 cell differentiation is mediated by tripartite motif containing 28. Nat Commun 9(1):1424

Khetchoumian K et al (2007) Loss of Trim24 (Tif1alpha) gene function confers oncogenic activity to retinoic acid receptor alpha. Nat Genet 39(12):1500–1506

Krischuns T et al (2018) Phosphorylation of TRIM28 enhances the expression of IFN-beta and proinflammatory cytokines during HPAIV infection of human lung epithelial cells. Front Immunol 9:2229

Li X, Burton EM, Bhaduri-McIntosh S (2017) Chloroquine triggers Epstein-Barr virus replication through phosphorylation of KAP1/TRIM28 in Burkitt lymphoma cells. PLoS Pathog 13(3):e1006249

Liang Q et al (2011) Tripartite motif-containing protein 28 is a small ubiquitin-related modifier E3 ligase and negative regulator of IFN regulatory factor 7. J Immunol 187(9):4754–4763

Liu Q, Zhou YH, Yang ZQ (2016) The cytokine storm of severe influenza and development of immunomodulatory therapy. Cell Mol Immunol 13(1):3–10

Liu B et al (2017) The ubiquitin E3 ligase TRIM31 promotes aggregation and activation of the signaling adaptor MAVS through Lys63-linked polyubiquitination. Nat Immunol 18(2):214–224

Liu J et al (2018) Peli1 negatively regulates noncanonical NF-κB signaling to restrain systemic lupus erythematosus. Nat Commun 9(1):1136

Liu J et al (2020) TRIM28 is a distinct prognostic biomarker that worsens the tumor immune microenvironment in lung adenocarcinoma. Aging 12(20):20308–20331

Locati M, Curtale G, Mantovani A (2020) Diversity, mechanisms, and significance of macrophage plasticity. Annu Rev Pathol 15:123–147

Loo YM, Gale M Jr (2011) Immune signaling by RIG-I-like receptors. Immunity 34(5):680–692

Ma X et al (2019) TRIM28 promotes HIV-1 latency by SUMOylating CDK9 and inhibiting P-TEFb. elife 8:e42426

Margalit L et al (2020) Trim24 and Trim33 play a role in epigenetic silencing of retroviruses in embryonic stem cells. Viruses 12(9):1015

Mendoza Y et al (2014) Human immunodeficiency virus type 1 (HIV-1) subtype B epidemic in Panama is mainly driven by dissemination of country-specific clades. PLoS One 9(4):e95360

Meroni G, Diez-Roux G (2005) TRIM/RBCC, a novel class of 'single protein RING finger' E3 ubiquitin ligases. Bioessays 27(11):1147–1157

Okazaki IM et al (2011) Histone chaperone Spt6 is required for class switch recombination but not somatic hypermutation. Proc Natl Acad Sci USA 108(19):7920–7925

Ozato K et al (2008) TRIM family proteins and their emerging roles in innate immunity. Nat Rev Immunol 8(11):849–860
Perez-Lloret J et al (2016) T-cell-intrinsic Tif1α/Trim24 regulates IL-1R expression on TH2 cells and TH2 cell-mediated airway allergy. Proc Natl Acad Sci USA 113(5):E568–E576
Petit V et al (2019) TRIM33 deficiency in monocytes and macrophages impairs resolution of colonic inflammation. EBioMedicine 44:60–70
Rauwel B et al (2015) Release of human cytomegalovirus from latency by a KAP1/TRIM28 phosphorylation switch. elife 4:e06068
Rehwinkel J, Gack MU (2020) RIG-I-like receptors: their regulation and roles in RNA sensing. Nat Rev Immunol 20(9):537–551
Reymond A et al (2001) The tripartite motif family identifies cell compartments. EMBO J 20(9):2140–2151
Rowe HM et al (2010) KAP1 controls endogenous retroviruses in embryonic stem cells. Nature 463(7278):237–240
Santoni de Sio FR et al (2012) KAP1 regulates gene networks controlling mouse B-lymphoid cell differentiation and function. Blood 119(20):4675–4685
Schmidt N et al (2019) An influenza virus-triggered SUMO switch orchestrates co-opted endogenous retroviruses to stimulate host antiviral immunity. Proc Natl Acad Sci USA 116(35):17399–17408
Short KM, Cox TC (2006) Subclassification of the RBCC/TRIM superfamily reveals a novel motif necessary for microtubule binding. J Biol Chem 281(13):8970–8980
Sinclair J (2008) Human cytomegalovirus: latency and reactivation in the myeloid lineage. J Clin Virol 41(3):180–185
Smith-Moore S et al (2018) Adeno-associated virus Rep proteins antagonize phosphatase PP1 to counteract KAP1 repression of the latent viral genome. Proc Natl Acad Sci USA 115(15):E3529–E3538
Sripathy SP, Stevens J, Schultz DC (2006) The KAP1 corepressor functions to coordinate the assembly of de novo HP1-demarcated microenvironments of heterochromatin required for KRAB zinc finger protein-mediated transcriptional repression. Mol Cell Biol 26(22):8623–8638
Tanaka S et al (2018) Trim33 mediates the proinflammatory function of Th17 cells. J Exp Med 215(7):1853–1868
Tie CH et al (2018) KAP1 regulates endogenous retroviruses in adult human cells and contributes to innate immune control. EMBO Rep 19(10):e45000
Tsai WW et al (2010) TRIM24 links a non-canonical histone signature to breast cancer. Nature 468(7326):927–932
van Tol S et al (2017) The TRIMendous role of TRIMs in virus-host interactions. Vaccine 5(3):23
Versteeg GA et al (2013) The E3-ligase TRIM family of proteins regulates signaling pathways triggered by innate immune pattern-recognition receptors. Immunity 38(2):384–398
Wang P et al (2015) TRIM26 negatively regulates interferon-beta production and antiviral response through polyubiquitination and degradation of nuclear IRF3. PLoS Pathog 11(3):e1004726
White D et al (2012) The ATM substrate KAP1 controls DNA repair in heterochromatin: regulation by HP1 proteins and serine 473/824 phosphorylation. Mol Cancer Res 10(3):401–414
Yong HY, Luo D (2018) RIG-I-like receptors as novel targets for pan-antivirals and vaccine adjuvants against emerging and re-emerging viral infections. Front Immunol 9:1379
Yu C et al (2019a) The roles of TIF1gamma in cancer. Front Oncol 9:979
Yu T et al (2019b) Modulation of M2 macrophage polarization by the crosstalk between Stat6 and Trim24. Nat Commun 10(1):4353
Zhang X et al (2018) Macrophage/microglial Ezh2 facilitates autoimmune inflammation through inhibition of Socs3. J Exp Med 215(5):1365–1382
Zhou XF et al (2012) TRIM28 mediates chromatin modifications at the TCRalpha enhancer and regulates the development of T and natural killer T cells. Proc Natl Acad Sci USA 109(49):20083–20088
Zhu SL et al (2017) Global quantitative proteomic analysis profiles host protein expression in response to Sendai virus infection. Proteomics 17(5):239
Zhu Q et al (2020) TRIM24 facilitates antiviral immunity through mediating K63-linked TRAF3 ubiquitination. J Exp Med 217(7):e20192083

Role of Ubiquitin Signaling in Modulating Dendritic Cell Function

7

Zuliang Jie

Abstract

As a professional antigen-presenting cell, dendritic cell (DC) plays an essential role in the connection of innate and adaptive immune responses. Ubiquitination is a post-translational mechanism of protein modification that plays a pivotal role in regulating DC maturation and function. To date, considerable progress has been made in understanding the underlying mechanisms of ubiquitination in modulating the function of DC in various diseases. Recent studies have emphasized that ubiquitin signaling in DCs plays crucial roles in regulating immune tolerance and functions, which can be promising targets for DC-based immunotherapy. In this chapter, we will focus on discussing the recent progress regarding the molecular mechanisms and functions of ubiquitination in DC-mediated immune homeostasis and responses.

Keywords

Dendritic cells · Ubiquitination · Immune responses · Post-translational modification

7.1 Introduction

Ubiquitination is a post-translational mechanism of protein modification that involves in various biological processes, including protein degradation, protein-protein interaction, receptor endocytosis, kinase activation, gene transcription, and assembly of signaling complexes (Sun 2008; Swatek and Komander 2016). Ubiquitination involves covalent conjugation of poly-ubiquitin or mono-ubiquitin chains onto lysin (K) resides of substrate proteins (Hershko and Ciechanover 1998). This process is catalyzed by sequential reactions of ubiquitin-activating (E1), ubiquitin-conjugating (E2), and ubiquitin-ligating (E3) enzymes (Hershko and Ciechanover 1998). The specificity of substrate protein is mainly determined by E3 ubiquitin ligase, which recognize and catalyze the conjugation of ubiquitin to the target proteins (Hershko and Ciechanover 1998). The E3 ubiquitin ligases are a diverse group of proteins, including the RING (Really Interesting New Gene)-type E3 ubiquitin ligase and the HECT (Homologous to the E6AP Carboxyl Terminus)-type E3 ubiquitin ligase (Berndsen and Wolberger 2014). During the formation of

Z. Jie (✉)
State Key Laboratory of Cellular Stress Biology, School of Life Sciences, Faculty of Medicine and Life Sciences, MOE Key Laboratory of Spectrochemical Analysis and Instrumentation, College of Chemistry and Chemical Engineering, State Key Laboratory of Vaccines for Infectious Diseases, Xiang An Biomedicine Laboratory, Xiamen University, Xiamen, Fujian, China
e-mail: jiezuliang@xmu.edu.cn

H. Hu, X. Fu (eds.), *Ubiquitination in Immune System and Immune Response*, Advances in Experimental Medicine and Biology 1466, https://doi.org/10.1007/978-981-97-7288-9_7

polyubiquitin chains, the C-terminal glycine residue of ubiquitin is connected to an internal K residue or the N-terminal methionine (M1) residue, which generate a large variety of ubiquitin chains with distinct biological functions, including K6, K11, K27, K29, K33, K48, K63, and M1 ubiquitin chains (Hu and Sun 2016). For example, K11- and K48-linked ubiquitin chains direct substrate proteins to the proteasome for degradation, whereas K63-linked and M1-linked ubiquitin chains are involved in mediating protein-protein interaction and enzymatic activation in signal transduction (Chen and Sun 2009). Ubiquitination is a reversible process that can be counteracted by deubiquitinase (DUBs), a large group of proteases that cleaves ubiquitin chains (Sun 2008). To date, more than 600 E3 ligases and about 100 DUBs have been reported in mammalian cells; these enzymes display substrate specificities and regulate distinct cellular functions (Berndsen and Wolberger 2014; Deshaies and Joazeiro 2009). Ubiquitination and deubiquitination play a crucial role in the modulation of diverse aspects of immune system functions (Sun 2008; Hu and Sun 2016; Zhou and Sun 2021; Wang et al. 2019). The role of ubiquitination in immune system including lymphocytes has been reviewed on other chapters. Herein, we will focus on discussing the role of ubiquitination in regulating dendritic cells' (DCs') maturation and function.

Dendritic cells (DCs) are a diverse group of professional antigen-presenting cells (APCs) with essential roles in initiating innate and adaptive immune responses (Steinman 2012). The primary function of DC is to phagocytose, process, and present antigens on the surface of MHC molecules to T cells (Cabeza-Cabrerizo et al. 2021). DCs recognize foreign antigens via pattern-recognition receptors (PRRs) on their surface, including TLRs, C-type lectins (CLRs), and intracellular retinoic acid-inducible gene-I (RIG-I) (Schlitzer et al. 2015). After binding to pathogen-associated molecular pattern (PAMP) on the pathogens, DCs uptake the foreign antigens, migrate, and present the antigen to T cells in the draining lymph nodes. Meanwhile, DCs upregulate MHC molecules and costimulatory molecules on their surface, and secret plenty of soluble immune-stimulatory cytokines and chemokines.

Although DCs constitute a small population of immune cells within tumor site and lymphoid organs, they are crucial for the mediation of antigen-specific immunity. DCs regulate immune responses or tolerance by presenting antigens to T cells, providing immunomodulatory signals and cytokine secretion. For example, in tumor microenvironment, DCs uptake antigens from dying tumor cells and migrate to draining lymph nodes, where DCs present the tumor antigens to T cells (Bottcher and Reis Sousa 2018; Wculek et al. 2020). As such, there is growing interest in modulating DC function to induce efficient antiviral or antitumor immunity. A better understanding of the diversity and functions of DC subpopulations and the underlying mechanisms that regulate DC functions could lead to improved immunotherapy. Recent studies have demonstrated that ubiquitination plays an essential role in regulating DC-mediate immunity and tolerance. In this chapter, we will focus on discussing the progress of ubiquitination in regulating DC's function.

7.2 The Role of Dendritic Cells in Immunity

DCs process and present pathogen-associated antigens on MHC molecules, provide costimulatory molecules and soluble cytokines to shape T cell responses. Derived from a hematopoietic lineage, DCs consist of developmentally and phenotypically distinct subpopulations (Table 7.1). Among these DC subsets, conventional DCs (cDCs) are specialized in antigen presentation for T cell activation. cDCs populations comprise two main subsets, the type 1 cDCs (cDC1s) and type 2 cDCs (cDC2s) (Merad et al. 2013; Mildner and Jung 2014). The cDC1s (CD8α^+ and/or CD103$^+$ cDC1) are specialized in initiating cellular immunity against intracellular pathogens through processing and cross-presentation of exogenous antigens to prime CD8$^+$ T cell and T_H1 cell responses (Mildner and Jung 2014; Chiang et al. 2016; Hildner et al. 2008). On the other hand, cDC2s (CD11b$^+$ cDC2) are efficient in inducing

Table 7.1 Transcription factors, surface markers, and specific functions of DC subsets

DC subsets		Transcription factors	Surface markers	Functional specialization
cDCs	cDC1	BATF3, IRF8, BCL6, ID2, ZBTB46, NFIL3	$CD11c^+$, $MHCII^+$, $CD8a^+$ (resident), $CD103^+$ (migratory), $CD24^+$, $XCR1^+$, $CLEC9A^+$, $DEC205^+$	Specialized in cross-presentation, secretion of IL-12 and type I interferon, involved in cellular immunity against intracellular pathogens and tumors, T_H1 and $CD8^+$ T cell responses
	cDC2	IRF4, ID2, RBPJ, KLF4, ZBTB46	$CD11c^+$, $MHCII^+$, $CD11b^+$, $CD172a^+$	Involved in humoral and cellular immunity against extracellular pathogens, T follicular helper cell, T_H2 and T_H17 immunity
	MoDCs	MAFB, KLF4, ZBTB46	$CD11c^+$, $MHCII^+$, $CD11b^+$,$Ly6C^+$, $CD64^+$, $CD206^+$, $CD209^+$, $CD14^+$, $CCR2^+$	Induction of $CD8^+$ T cell response, T_H1, T_H2, T_H17 and Treg immunity
pDCs		E2-2, IRF7	$B220^+$, $CD317^+$, SIGLEC-H^+, $CD11c^{low}$, $MHCII^{low}$	Type I interferon secretion upon viral infection, poor antigen presentation, activate $CD8^+$ T cell response

$CD4^+$ T cell-mediated immunity (Schlitzer et al. 2013; Williams et al. 2013). In addition to cDC1s, plasmacytoid DCs (pDCs) develop from both common DC precursors (CDPs) and lymphoid progenitors, which can rapidly produce type I interferons (IFN-I) upon virus infection (Liu 2005). Compared to cDCs, pDCs showed relatively poor ability of priming T cells. Moreover, circulating $Ly6C^{hi}$ monocytes can be differentiate into monocyte-derived DCs (MoDCs) under inflammatory conditions and possess pro-inflammatory functions (Schlitzer et al. 2015; Merad et al. 2013).

To generate effective T cell responses, DCs present antigens on MHC molecules (signal 1), upregulate costimulatory molecules (signal 2) and secret soluble cytokines (signal 3), such as IL-12 and type I interferons (Curtsinger and Mescher 2010). IL-12 is mainly produced by cDC1s and contributes to the priming of $CD8^+$ T and T_H1 cells (Ruffell et al. 2014; Martinez-Lopez et al. 2015). Additionally, DCs can also produce chemokines, including CXC-chemokine ligand 9 (CXCL9), CXCL10 and CXCL11, to attract antigen-specific T cells (Spranger et al. 2017; Chow et al. 2019). Taken together, DCs play an essential role in anti-pathogen and antitumor immunity by conditioning the microenvironment, as well as attracting and priming antigen-specific T cells. To date, it has been demonstrated that ubiquitin signaling plays an essential role in modulating DC's function and DC-mediate immune responses (Fig. 7.1). Herein, we summarized the recent findings of ubiquitination in regulating DC's maturation and function (Table 7.2).

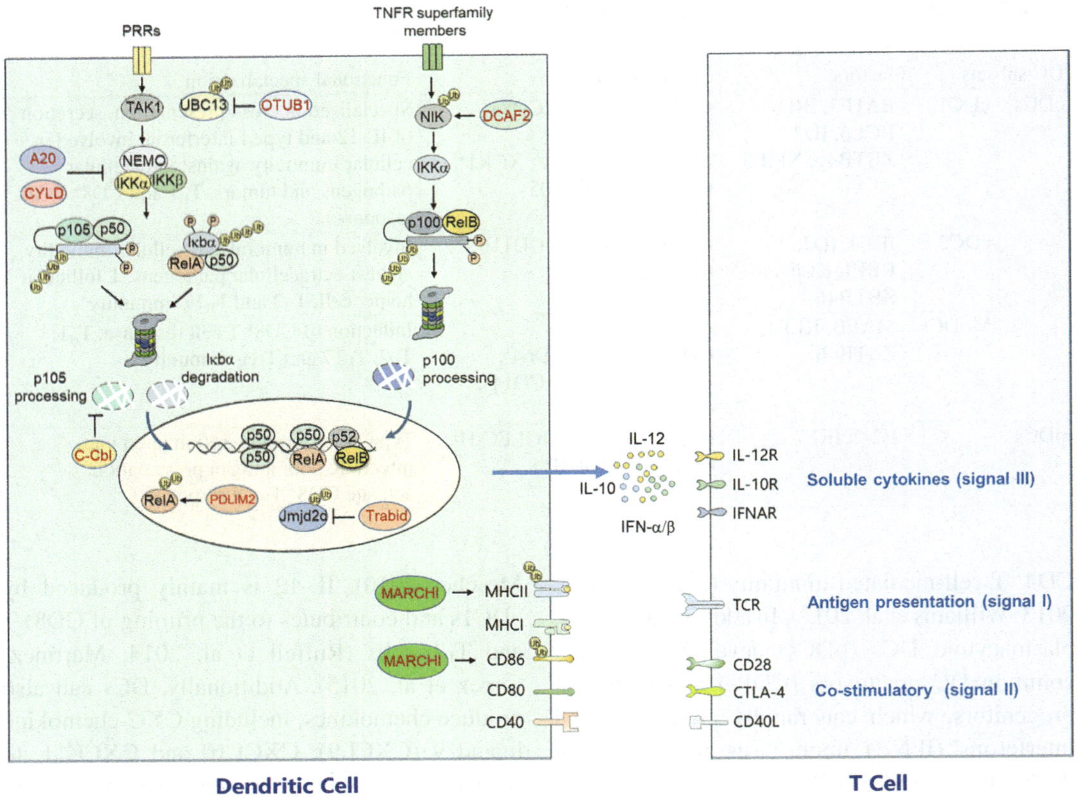

Fig. 7.1 Ubiquitin signaling regulates DC's maturation and function. To prime T cell responses, DCs present antigens on MHC molecules (signal 1), upregulate costimulatory molecules (signal 2) and secret soluble cytokines (signal 3). Ubiquitin signaling involved in DC maturation, antigen presentation and cytokine production. An ovarian tumor (OTU) family deubiquitinase, A20, cleaves the K63-linked ubiquitin chains of NEMO and inhibits the activation of NF-κb signaling. A nuclear E3 ubiquitin ligase, PDLIM2, negatively regulates NF-κb activity by promoting p65 poly-ubiquitination and its degradation. OTUB1 cleaves the K48-linked ubiquitin chains of UBC13, thereby inducing the activation of TAK1 by ubiquitination of IRAK1 and TRAF6. CYLD inhibits NF-κb signaling activation by cleaving K63-linked polyubiquitin chains from NEMO, TRAF2, and RIP1. DCAF2 negatively regulates non-canonical NF-κb activation by controlling NIK stability, and enhances the production of IL-23 by DCs. MARCH1 mediate ubiquitination and degradation of MHC II and CD86 of DCs. Trabid cleaves the ubiquitin chains of Jmjd2d, the histone demethylase, and recruits c-Rel and p50 to the promoters of *Il12* and *Il23a* genes

Table 7.2 Summary of ubiquitin enzymes regulating DCs' activation and function

Ubiquitin enzyme	Type	Substrates	Function in DCs	References
MARCH1	E3 ligase	MHC II, CD86	Regulate DC maturation, antigen presentation, migration, DC-mediated T cell activation	Furuta et al. (2013), Cho et al. (2015), De Gassart et al. (2008), Thibodeau et al. (2008), Corcoran et al. (2011), Tze et al. (2011), Majdoubi et al. (2019)
A20	DUB	NEMO	Inhibit DC maturation and cytokine production, maintain immune homeostasis, suppress DC-mediated T cell activation and differentiation	Kool et al. (2011), Hammer et al. (2011), Song et al. (2008), Breckpot et al. (2009)
TRABID	DUB	JMJD2D	Regulate IL-12 and IL-23 production by DCs, promote DC-mediated Th1 and Th17 generation	Jin et al. (2016)
OTUB1	DUB	UBC13	Induction of cytokine production	Mulas et al. (2021)
PDLIM2	E3 ligase	RelA (p65)	Suppress proinflammatory cytokine production	Tanaka et al. (2007)
$CRL4^{DCAF2}$	DUB	NIK	Regulate the induction of IL-23 and TH17 generation	Huang et al. (2018)
UCH-L1	DUB	MHC I	DC antigen cross-presentation	Reinicke et al. (2019)
c-Cbl	E3 ligase	p50 and p105	Involved in DC activation and proinflammatory cytokine production	Chiou et al. (2011), Li et al. (2019), Xu et al. (2022)
ASB2a	E3 ligase	Actin-binding protein filamins	Regulate DC migration	Lamsoul et al. (2013)
HRD1	E3 ligase	BLIMP	Promote DC MHC II expression	Yang et al. (2014)
USP14	DUB	MHC I	Regulate MHC I antigen presentation	Palmer et al. (2018)
CYLD	DUB		Regulate DC cytokine production and DC-mediated T cell activation	Wurm et al. (2015), Kovalenko et al. (2003)
RNF41	E3 ligase	Clec9A	Regulate DC cross-presentation	Tullett et al. (2020)

7.3 Regulation of DC Maturation and Antigen Presentation by Ubiquitination

7.3.1 Antigen Processing and Presentation

DCs maintain an immature or steady-state to keep immune tolerance under homeostatic conditions (Hammer and Ma 2013). When encountered the pathogens, DCs undergo a transition from an immature to mature status, accompanied with upregulation of MHC II and costimulatory molecules (Ganguly et al. 2013). The antigen presentation is dependent on properly intracellular transport of MHC II, which present a broad arrange of antigens to CD4^{+} T cells. It has been shown that ubiquitination is an important mechanism which controls MHC II intracellular transport (Shin et al. 2006). A member of the membrane-associated RING-CH (MARCH) E3 ubiquitin ligase, MARCH1, has been shown to mediate endosomal trafficking and degradation of MHC II molecule via its cytoplasmic lysine (Furuta et al. 2013; Cho et al. 2015). It has been revealed that MARCH1-mediated MHC II ubiquitination promotes the degradation of the internalized MHC II/peptide complexes and prevents their recycling (Cho et al. 2015). Therefore, MARCH1-deficient DCs significantly increased MHC II/peptide complexes levels on the cell surface. Consistently, LPS-induced DC maturation downregulated MARCH1 expression, accompanied with attenuation of MHC II ubiquitination and increased MHC II surface stabilization (Cho et al. 2015; De Gassart et al. 2008). On the other hand, IL-10 enhances the expression of MARCH1 and decreases MHC II surface expression, thereby contributing to IL-10-mediated immunosuppressive function (Thibodeau et al. 2008). These findings indicate that targeting MARCH1 serves a potential strategy to regulate MHC expression and antigen presentation in vaccine development and tumor immunotherapy.

Antigen cross-presentation by DCs is essential for priming effective $CD8^+$ T cell response in both infectious and cancer models (Kurts et al. 2010). A recent study have shown that, the E3 ubiquitin ligase, RNF41, negatively regulates the cross-presentation of dead cell-derived antigens by DCs (Tullett et al. 2020). Mechanistically, RNF41 binds to Clec9A and ubiquitinates its extracellular domains, thereby promoting the interaction of Clec9A with ER-associated proteins and inducing its degradation (Tullett et al. 2020). Another study has shown that Ubiquitin C-terminal hydrolase-L1 (UCH-L1) regulates the recycling of MHC I as well as antigen cross-presentation (Reinicke et al. 2019). In line with this, UCH-L1-deficient mice showed reduced DC-mediated cross-priming of $CD8^+$ T cell responses post *listeria monocytogenes* infection. UCH-L1 expression levels regulate the co-localization of intracellular MHC I with lysosomal compartments. It has been revealed that ubiquitin-specific protease 14 (USP14) promotes MHC I-mediated antigen presentation, and the antigen is specifically derived from defective ribosomal products (DRiP Ag) (Palmer et al. 2018). Moreover, USP14 inhibition does not affect protein synthesis but partially delayed protein degradation. Taken together, these findings reveal that E3 ubiquitin ligases and DUBs that modulate MHC expression as well as antigen presentation and have implications for vaccine development and cancer immunotherapy.

7.3.2 Regulation of Co-stimulatory Molecules and DC Maturation by Ubiquitination

In addition to MHC II molecule, MARCH1 also ubiquitinates the costimulatory molecule, CD86, in DCs by promoting intracellular degradation of the transmembrane domains (Corcoran et al. 2011). Additionally, CD83 competes with CD86 for interaction with MARCH1, thereby inhibiting CD86 ubiquitination and inducing the expression of CD86 of DCs (Tze et al. 2011). Ubiquitination also affects DC maturation by regulating the transcription level of MHC II and costimulatory molecules. Hrd1, an E3 ubiquitin ligase, promotes MHC II gene transcription by inducing the degradation of BLIMP protein through ubiquitination (Yang et al. 2014). Another study have shown that TBK1 is an important regulator of DC function (Xiao et al. 2017). TBK1 deficiency in DCs resulted in upregulation of co-stimulatory molecules and promotes anti-cancer T cell responses. TBK1 ablation in DCs causes T cell activation and autoimmune symptoms by negatively regulating the function of STAT3, a negative regulator of IFNAR1 signaling. These data suggest to us that TBK1 in DCs serves as a pivotal regulator of maintenance of immune tolerance.

7.4 Regulation of DC Migration by Ubiquitination

The migratory ability of DCs is essential for the initiation of host immunity against pathogens and immune tolerance (Worbs et al. 2017; Lindquist et al. 2004). Ubiquitination is also involved in the regulation of DC migration in various immune contexts (Majdoubi et al. 2019; Lamsoul et al. 2013). In addition to MHC II molecules, MARCH1 is also important for the migration of CD206-positive monocyte-derived DCs to skin draining lymph nodes (Majdoubi et al. 2019). MARCH1 deficiency in CD206-positive MoDCs exhibits MHC II overexpression and downregulation of IRF4 and CCR7 expression, accompanied with reduced migratory ability (Majdoubi et al. 2019). On the other hand, GM-CSF could restore CD206-positive MoDC migration by promoting IRF4 expression. Another study has revealed that, the E3 ubiquitin ligase, Cullin RING ligase CRL5 ASB2α, is highly expressed in immature dendritic cells (DCs) and is decreased post DC maturation (Lamsoul et al. 2013). Using ASB2 conditional knockout mice, it has been shown that ASB2α promotes polyubiquitination and drives proteasome-mediated degradation of actin-binding protein filamins, thereby facilitating cell spreading and the formation of adhesive structures in immature DCs (Lamsoul et al. 2013). While enhanced DC migration during immature stages is essential for triggering rapid immune responses against invading pathogens, the termination of DC trafficking at mature stage of DCs is required to maintain the immune homeostasis.

7.5 Regulation of NF-κB Signaling in DC by Ubiquitination

The canonical NF-κb signaling pathway is an essential mediator of TLR-stimulated DC activation (Kaisho and Tanaka 2008; Harhaj and Dixit 2012). Recent studies demonstrated that ubiquitination plays a crucial role in the regulation of NF-κb signaling in DC maturation. A20, an ovarian tumor (OTU) family deubiquitinase, cleaves the K63-linked ubiquitin chains of NEMO, thereby inhibiting NF-κb signaling activation (Kool et al. 2011; Hammer et al. 2011). It has been shown that A20 negatively regulates the maturation, pro-inflammatory cytokine production, and immune-regulatory function of DCs (Song et al. 2008; Breckpot et al. 2009). A20 knockdown in DCs promotes the induction of costimulatory molecules and proinflammatory cytokines in response to TLR ligands, which leads to hyper-activation of T cell responses. Consistently, the A20-deficient DCs undergo spontaneous maturation at steady conditions, with enhanced expression of the costimulatory molecules. In line with this, DC-conditional A20-deficient mice present an aberrant frequency of effector/memory-like T cells from the peripheral lymphoid organs, indicative of T cell hyper-activation under homeostatic status (Kool et al. 2011; Hammer et al. 2011). Interestingly, MyD88 deletion in DC-conditional A20-knockout mice does not reverse spontaneous DC maturation or aberrant T cell activation, but prevents A20-deficient DCs from hyper-production of proinflammatory cytokines (Hammer et al. 2011). In addition, A20 deficiency in DCs impairs their functions in stimulating regulatory T (Treg) cells and further enhancing effector T cell responses (Song et al. 2008). To treat tumor-bearing mice, the A20-deficient DCs are more potent than wild-type DCs in suppressing the tumor growth, characterized by enhanced infiltration of effector T cells and resistance to Treg-mediated immune-suppression (Song et al. 2008). Another study shows that the rhomoboid family protease, Rhbdd3, facilitates A20 mediated deubiquitination of NEMO via the ubiquitin-binding association (UBA) domain (Liu et al. 2014). Consistent with A20 studies, Rhbdd3 deficiency causes aberrant DC activation and enhanced induction of the inflammatory cytokines by TLR ligands (Liu et al. 2014).

Trabid, an A20-related deubiquitinase, plays an essential role in regulation of TLR-induced IL-12 and IL-23 production in DCs (Jin et al. 2016). Although Trabid is not required for the induction of NF-κb nuclear translocation, it cleaves the ubiquitin chains of Jmjd2d, the histone demethylase, thereby stabilizing and recruiting c-Rel and p50 to the promoters of *Il12* and *Il23a* genes.

Trabid facilitates the production of IL-12 and IL-23 in activated DCs, which in turn are crucial for generation of T_H1 and T_H17 subsets of effector T cells. Therefore, Trabid-mediated DC function plays a pivotal role in autoimmune responses (Jin et al. 2016). Recent study has shown that OTUB1, an ovarian tumor (OTU) family deubiquitinase, plays an essential role in regulating NF-kB signaling in activated DCs (Mulas et al. 2021). OTUB1 deficiency in DCs leads to impaired induction of proinflammatory cytokine production by ligands of TLRs. Mechanistically, OTUB1 cleaves the K48-linked ubiquitin chains of the E2-conjugating enzyme UBC13, thereby inducing the activation of TAK1 by ubiquitination of IRAK1 and TRAF6 (Mulas et al. 2021). These data suggest an essential role of OTUB1 in mediating NF-kb-dependent cytokine production of activated DCs. CYLD, another deubiquitinase, also plays an essential role in regulating NF-κb activity (Wurm et al. 2015). Full-length CYLD

(FL-CYLD) inhibits NF-κb signaling activation via cleaving K63-linked polyubiquitin chains from NEMO, TRAF2, and RIP1, whereas the short splice variant of CYLD (sCYLD) lacking the NEMO and TRAF2-binding sites promotes the activation of NF-κb signaling (Kovalenko et al. 2003). Interestingly, in mice lacking exon 7 and 8 of CYLD, sCYLD is overexpressed whereas FL-CYLD is absent. Conditional deletion of exon 7 and 8 of CYLD in DCs enhanced the activation of CD8α$^+$ DC and the induction of TNF, IL-10, and IL-12 in mice infected with *Listeria Monocytogenes* (Wurm et al. 2015).

PDLIM2, a nuclear E3 ubiquitin ligase, negatively regulates NF-κb activity by promoting p65 poly-ubiquitination and mediating its proteasome-dependent degradation (Tanaka et al. 2007). PDLIM2 deficiency leads to accumulation of nuclear p65, as well as induction of large amounts of proinflammatory cytokines in response to TLR ligands by DCs. This finding demonstrates a pathway by which PDLIM2 mediates NF-κb activation through p65 intranuclear degradation. The E3 ligase, c-Cbl, is involved in modulating the activation of DCs (Chiou et al. 2011; Li et al. 2019). c-Cbl deficiency in DCs leads to enhanced induction of proinflammatory cytokines, which is accompanied with an increased NF-kb activity (Chiou et al. 2011). Mechanistically, c-Cbl deficiency reduces both p50 and p105 levels via its RING-domain, thereby suppressing the pro-inflammatory cytokine production by DCs. This study suggests that c-Cbl negatively regulates DC function, implicating c-Cbl as a potential target for improving the efficacy of DC-based anti-tumor immunotherapy. On the other hand, another evolutionary conserved family of Cbl protein, Cbl-b seems to play limited role in the induction of antigen-specific T cell responses by DCs. Cbl-b deficiency in DCs slightly increased the proinflammatory cytokine production upon TLR stimulation (Wallner et al. 2013). Recent study showed that deletion of Cbl-b and c-Cbl in DCs led to accumulation of conventional type I DCs (cDC1s) with more activated status (Xu et al. 2022). Mechanistically, Cbl-b and c-Cbl target signal transducer and activator of transcription 5 (STAT5), thereby suppressing the pro-apoptotic protein Bim, and promoting ubiquitination-mediated degradation and cell apoptosis in DCs. These studies suggested a functional redundancy of Cbl-b and c-Cbl in DC homeostasis and maturation; however, the underlying molecular mechanisms by which Cbl-b is involved in DC activation need further investigation.

The non-canonical NF-κb signaling pathway is also involved in DC activation and cytokine production (Jie et al. 2018; Hofmann et al. 2011; Katakam et al. 2015). Ubiquitination has been shown to play a crucial role in the regulation of non-canonical NF-κb signaling in DC activation. A recent study demonstrates that an E3 ubiquitin ligase, CRL4^{DCAF2}, is involved in the induction of the inflammatory cytokine, IL-23, by DCs (Huang et al. 2018). Mechanistically, CRL4^{DCAF2} negatively regulates non-canonical NF-kb activation by controlling NF-κb-inducing kinase (NIK) stability, thereby facilitating NIK accumulation and RelB nuclear translocation. Interestingly, CRL4^{DCAF2} control NIK stability independent of TRAF3 degradation. In line with this, DC-conditional CRL4^{DCAF2}-knockout mice displayed more susceptible to autoimmune disease. Together, this study demonstrates that CRL4^{DCAF2} is crucial for the activation of non-canonical NF-κb signaling and provides a potential mechanism that controls inflammatory diseases.

7.6 Regulation of DC-Mediated Immune Homeostasis by Ubiquitination

In addition to inducing immune responses, DCs are also involved in central and peripheral tolerance, which depends on cytokine production, immunomodulatory molecule expression, and signaling pathway-mediated maturation of DCs (Iberg et al. 2017). It has been demonstrated that ubiquitination plays an important role in the regulation of DC-induced immune tolerance. As discussed above, A20 inhibits DC activation by negatively regulate NF-κb signaling (Hammer et al. 2011). A20-deficiency in DCs promotes upregulation of the surface costimulatory molecules, as well as activated T cell expansion. In line with this, the DC-conditional A20-deficient mice exhibit systemic autoimmune symptoms spontaneously, including lymphocyte-dependent colitis and arthritis (Hammer et al. 2011). An ubiquitin-binding protein, ABIN-1 expression in DCs is required to preserve immune tolerance (Callahan et al. 2013). ABIN-1 ablation in DCs exhibits exaggerated NF-κB and MAPK signaling, accompanied by significant accumulation of IL-23 post TLR stimulation. In line with this, challenge of ABIN-1 DC-conditional knockout mice with TLR7 ligand results in accumulation of large numbers of Th17 and γδ T cells, which exacerbated psoriasis symptoms (Callahan et al. 2013).

As we discussed previously, the E3 ubiquitin ligase, $CRL4^{DCAF2}$, is required for maintaining T cell homeostasis under steady-state conditions (Huang et al. 2018). $CRL4^{DCAF2}$ deficiency in DCs perturbs T cell homeostasis and develop spontaneous autoimmunity, characterized by the increased frequency of T_H17 effector T cells as well as elevated anti-nuclear antibodies in the serum. In addition, DCs regulates the differentiation of $CD4^+$ T cells by secreting various cytokines (Walsh and Mills 2013). Ubiquitination modulates DC cytokine production, which guides the differentiation of $CD4^+$ T cells into different T cell subsets, including Th1, Th2, Th17, and Treg. The proinflammatory cytokines IL-12 and IL-23 are involved in regulation of Th1 and Th17 cells in autoimmune and inflammatory diseases (Gaffen et al. 2014; Teng et al. 2015). The deubiquitinase, Trabid, promotes the induction of the expression of IL-12 and IL-23 from DCs, which, in turn, regulates the differentiation of Th1 and Th17 in autoimmune diseases (Jin et al. 2016). Mechanistically, Trabid regulate the histone modification at the promoters of IL-12 and IL23, which involved in deubiquitination and stabilization of the histone demethylase Jmjd2d. These data suggest to us that Trabid mediates the secretion of IL-12 and IL-23 in DCs through the epigenetic regulation, thereby modulating Th1 and Th17 differentiation.

7.7 Concluding Remarks and Future Perspectives

As professional antigen-presenting cells, DCs play key roles in the regulation of innate and adaptive immune responses. As the most potent antigen-presenting cells, DCs are able to activate naïve T cells and induce antigen-specific immune responses in infection or tumor models, better understanding of how DCs are regulated in these scenarios may enhance therapeutic breakthrough in clinical trials. Ubiquitination was identified to play an essential role in DC maturation and function. To date, multiple E3 ubiquitin ligase and DUBs have been identified as key regulators of different physiological aspects of DCs, such as DC activation, maturation, and DC-mediated adaptive immunity. Therefore, the ubiquitin enzymes can be potential therapeutic targets for DC-based therapy of autoimmune diseases, infectious diseases, and cancer. Given the existence of large number of E3s and DUBs in mammalian cells, the ubiquitin-based therapeutic approaches offer potential opportunities in drug development for various diseases.

For future studies, it is important to identify novel E3s and DUBs involved in regulation of DC's function and elucidate the underlying molecular mechanisms. To translate the lab discoveries into clinical trials, a major issue is to identify small molecular inhibitors targeting E3 ligases and DUBs with high affinity and selectivity, which can be a promising immunotherapy approach. Moreover, small molecular inhibitors that selectively target the DC ubiquitination mechanisms in infectious and autoimmune diseases need to be further investigated. In addition, it has been well-documented that different DC subpopulations have unique functions and features. For instance, the cDC1 subset is important for promoting the cross-presentation and pathogen-controlling cytotoxic T cell responses, while the cDC2 subset is crucial for induction of antigen-specific $CD4^+$ T cell responses. The underlying mechanisms of how ubiquitination modulating the function of different DC subsets remain largely unknown and the major task is to illustrate their underlying molecular mechanisms. A better understanding of the diversity and functions of ubiquitin signaling in DC subsets will promote novel approaches for vaccine development and cancer immunotherapy.

Acknowledgements This study was supported by grants from the National Key R&D Program of China (2021YFC2701802 and 2022YFA1304003 to Z. J.), the National Natural Science Foundation of China (32170902 and 32470970 to Z. J.), the Scientific Research Foundation of State Key Laboratory of Vaccines for Infectious Diseases, Xiang An Biomedicine Laboratory (2023XAKJ0101027), and the Fundamental Research Funds for the Central Universities (20720210113 and 20720220003 to Z. J.).

References

Berndsen CE, Wolberger C (2014) New insights into ubiquitin E3 ligase mechanism. Nat Struct Mol Biol 21:301–307

Bottcher JP, Reis Sousa C (2018) The role of type 1 conventional dendritic cells in cancer immunity. Trends Cancer 4:784–792

Breckpot K et al (2009) Attenuated expression of A20 markedly increases the efficacy of double-stranded RNA-activated dendritic cells as an anti-cancer vaccine. J Immunol 182:860–870

Cabeza-Cabrerizo M, Cardoso A, Minutti CM, Pereira da Costa M, Reis Sousa C (2021) Dendritic cells revisited. Annu Rev Immunol 39:131–166

Callahan JA et al (2013) Cutting edge: ABIN-1 protects against psoriasis by restricting MyD88 signals in dendritic cells. J Immunol 191:535–539

Chen ZJ, Sun LJ (2009) Nonproteolytic functions of ubiquitin in cell signaling. Mol Cell 33:275–286

Chiang MC et al (2016) Differential uptake and cross-presentation of soluble and necrotic cell antigen by human DC subsets. Eur J Immunol 46:329–339

Chiou SH et al (2011) The E3 ligase c-Cbl regulates dendritic cell activation. EMBO Rep 12:971–979

Cho KJ, Walseng E, Ishido S, Roche PA (2015) Ubiquitination by March-I prevents MHC class II recycling and promotes MHC class II turnover in antigen-presenting cells. Proc Natl Acad Sci USA 112:10449–10454

Chow MT et al (2019) Intratumoral activity of the CXCR3 chemokine system is required for the efficacy of anti-PD-1 therapy. Immunity 50:1498–1512

Corcoran K et al (2011) Ubiquitin-mediated regulation of CD86 protein expression by the ubiquitin ligase membrane-associated RING-CH-1 (MARCH1). J Biol Chem 286:37168–37180

Curtsinger JM, Mescher MF (2010) Inflammatory cytokines as a third signal for T cell activation. Curr Opin Immunol 22:333–340

De Gassart A et al (2008) MHC class II stabilization at the surface of human dendritic cells is the result of maturation-dependent MARCH I down-regulation. Proc Natl Acad Sci USA 105:3491–3496

Deshaies RJ, Joazeiro CA (2009) RING domain E3 ubiquitin ligases. Annu Rev Biochem 78:399–434

Furuta K, Walseng E, Roche PA (2013) Internalizing MHC class II-peptide complexes are ubiquitinated in early endosomes and targeted for lysosomal degradation. Proc Natl Acad Sci USA 110:20188–20193

Gaffen SL, Jain R, Garg AV, Cua DJ (2014) The IL-23-IL-17 immune axis: from mechanisms to therapeutic testing. Nat Rev Immunol 14:585–600

Ganguly D, Haak S, Sisirak V, Reizis B (2013) The role of dendritic cells in autoimmunity. Nat Rev Immunol 13:566–577

Hammer GE, Ma A (2013) Molecular control of steady-state dendritic cell maturation and immune homeostasis. Annu Rev Immunol 31:743–791

Hammer GE et al (2011) Expression of A20 by dendritic cells preserves immune homeostasis and prevents colitis and spondyloarthritis. Nat Immunol 12:1184–1193

Harhaj EW, Dixit VM (2012) Regulation of NF-kappaB by deubiquitinases. Immunol Rev 246:107–124

Hershko A, Ciechanover A (1998) The ubiquitin system. Annu Rev Biochem 67:425–479

Hildner K et al (2008) Batf3 deficiency reveals a critical role for CD8alpha+ dendritic cells in cytotoxic T cell immunity. Science 322:1097–1100

Hofmann J, Mair F, Greter M, Schmidt-Supprian M, Becher B (2011) NIK signaling in dendritic cells but not in T cells is required for the development of effector T cells and cell-mediated immune responses. J Exp Med 208:1917–1929

Hu H, Sun SC (2016) Ubiquitin signaling in immune responses. Cell Res 26:457–483

Huang T et al (2018) CRL4(DCAF2) negatively regulates IL-23 production in dendritic cells and limits the development of psoriasis. J Exp Med 215:1999–2017

Iberg CA, Jones A, Hawiger D (2017) Dendritic cells as inducers of peripheral tolerance. Trends Immunol 38:793–804

Jie Z et al (2018) NIK signaling axis regulates dendritic cell function in intestinal immunity and homeostasis. Nat Immunol 19:1224–1235

Jin J et al (2016) Epigenetic regulation of the expression of Il12 and Il23 and autoimmune inflammation by the deubiquitinase Trabid. Nat Immunol 17:259–268

Kaisho T, Tanaka T (2008) Turning NF-kappaB and IRFs on and off in DC. Trends Immunol 29:329–336

Katakam AK et al (2015) Dendritic cells require NIK for CD40-dependent cross-priming of CD8+ T cells. Proc Natl Acad Sci USA 112:14664–14669

Kool M et al (2011) The ubiquitin-editing protein A20 prevents dendritic cell activation, recognition of apoptotic cells, and systemic autoimmunity. Immunity 35:82–96

Kovalenko A et al (2003) The tumour suppressor CYLD negatively regulates NF-kappaB signalling by deubiquitination. Nature 424:801–805

Kurts C, Robinson BW, Knolle PA (2010) Cross-priming in health and disease. Nat Rev Immunol 10:403–414

Lamsoul I et al (2013) ASB2alpha regulates migration of immature dendritic cells. Blood 122:533–541

Li X, Gong L, Gu H (2019) Regulation of immune system development and function by Cbl-mediated ubiquitination. Immunol Rev 291:123–133

Lindquist RL et al (2004) Visualizing dendritic cell networks in vivo. Nat Immunol 5:1243–1250

Liu YJ (2005) IPC: professional type 1 interferon-producing cells and plasmacytoid dendritic cell precursors. Annu Rev Immunol 23:275–306

Liu J et al (2014) Rhbdd3 controls autoimmunity by suppressing the production of IL-6 by dendritic cells via K27-linked ubiquitination of the regulator NEMO. Nat Immunol 15:612–622

Majdoubi A et al (2019) Downregulation of MHC class II by ubiquitination is required for the migration of CD206(+) dendritic cells to skin-draining lymph nodes. J Immunol 203:2887–2898

Martinez-Lopez M, Iborra S, Conde-Garrosa R, Sancho D (2015) Batf3-dependent CD103+ dendritic cells are major producers of IL-12 that drive local Th1 immunity against Leishmania major infection in mice. Eur J Immunol 45:119–129

Merad M, Sathe P, Helft J, Miller J, Mortha A (2013) The dendritic cell lineage: ontogeny and function of dendritic cells and their subsets in the steady state and the inflamed setting. Annu Rev Immunol 31:563–604

Mildner A, Jung S (2014) Development and function of dendritic cell subsets. Immunity 40:642–656

Mulas F et al (2021) The deubiquitinase OTUB1 augments NF-kappaB-dependent immune responses in dendritic cells in infection and inflammation by stabilizing UBC13. Cell Mol Immunol 18:1512–1527

Palmer AL et al (2018) Inhibition of the deubiquitinase Usp14 diminishes direct MHC class I antigen presentation. J Immunol 200:928–936

Reinicke AT et al (2019) Deubiquitinating enzyme UCH-L1 promotes dendritic cell antigen cross-presentation by favoring recycling of MHC class I molecules. J Immunol 203:1730–1742

Ruffell B et al (2014) Macrophage IL-10 blocks CD8+ T cell-dependent responses to chemotherapy by suppressing IL-12 expression in intratumoral dendritic cells. Cancer Cell 26:623–637

Schlitzer A et al (2013) IRF4 transcription factor-dependent CD11b+ dendritic cells in human and mouse control mucosal IL-17 cytokine responses. Immunity 38:970–983

Schlitzer A, McGovern N, Ginhoux F (2015) Dendritic cells and monocyte-derived cells: two complementary and integrated functional systems. Semin Cell Dev Biol 41:9–22

Shin JS et al (2006) Surface expression of MHC class II in dendritic cells is controlled by regulated ubiquitination. Nature 444:115–118

Song XT et al (2008) A20 is an antigen presentation attenuator, and its inhibition overcomes regulatory T cell-mediated suppression. Nat Med 14:258–265

Spranger S, Dai D, Horton B, Gajewski TF (2017) Tumor-residing Batf3 dendritic cells are required for effector T cell trafficking and adoptive T cell therapy. Cancer Cell 31:711–723

Steinman RM (2012) Decisions about dendritic cells: past, present, and future. Annu Rev Immunol 30:1–22

Sun SC (2008) Deubiquitylation and regulation of the immune response. Nat Rev Immunol 8:501–511

Swatek KN, Komander D (2016) Ubiquitin modifications. Cell Res 26:399–422

Tanaka T, Grusby MJ, Kaisho T (2007) PDLIM2-mediated termination of transcription factor NF-kappaB activation by intranuclear sequestration and degradation of the p65 subunit. Nat Immunol 8:584–591

Teng MW et al (2015) IL-12 and IL-23 cytokines: from discovery to targeted therapies for immune-mediated inflammatory diseases. Nat Med 21:719–729

Thibodeau J et al (2008) Interleukin-10-induced MARCH1 mediates intracellular sequestration of MHC class II in monocytes. Eur J Immunol 38:1225–1230

Tullett KM et al (2020) RNF41 regulates the damage recognition receptor Clec9A and antigen cross-presentation in mouse dendritic cells. elife 9:e63452

Tze LE et al (2011) CD83 increases MHC II and CD86 on dendritic cells by opposing IL-10-driven MARCH1-mediated ubiquitination and degradation. J Exp Med 208:149–165

Wallner S et al (2013) The role of the e3 ligase cbl-B in murine dendritic cells. PLoS One 8:e65178

Walsh KP, Mills KH (2013) Dendritic cells and other innate determinants of T helper cell polarisation. Trends Immunol 34:521–530

Wang A, Zhu F, Liang R, Li D, Li B (2019) Regulation of T cell differentiation and function by ubiquitin-specific proteases. Cell Immunol 340:103922

Wculek SK et al (2020) Dendritic cells in cancer immunology and immunotherapy. Nat Rev Immunol 20:7–24

Williams JW et al (2013) Transcription factor IRF4 drives dendritic cells to promote Th2 differentiation. Nat Commun 4:2990

Worbs T, Hammerschmidt SI, Forster R (2017) Dendritic cell migration in health and disease. Nat Rev Immunol 17:30–48

Wurm R et al (2015) Protective dendritic cell responses against listeriosis induced by the short form of the deubiquitinating enzyme CYLD are inhibited by full-length CYLD. Eur J Immunol 45:1366–1376

Xiao Y et al (2017) The kinase TBK1 functions in dendritic cells to regulate T cell homeostasis, autoimmunity, and antitumor immunity. J Exp Med 214:1493–1507

Xu F et al (2022) Ablation of Cbl-b and c-Cbl in dendritic cells causes spontaneous liver cirrhosis via altering multiple properties of CD103(+) cDC1s. Cell Death Dis 8:142

Yang H et al (2014) Hrd1-mediated BLIMP-1 ubiquitination promotes dendritic cell MHCII expression for CD4 T cell priming during inflammation. J Exp Med 211:2467–2479

Zhou X, Sun SC (2021) Targeting ubiquitin signaling for cancer immunotherapy. Signal Transduct Target Ther 6:16

Ubiquitin Signaling in the Immune System

8

Liang Jin

Abstract

Protein ubiquitination is a post-translational modification of proteins that is widespread in eukaryotic cells. It was identified initially while investigating the mechanisms of intracellular protein degradation. There is mounting evidence that ubiquitination and its reverse process, deubiquitination, play a critical regulatory role in the process of intrinsic and adaptive immune responses by regulating the function of various immune cell types, thereby influencing the development of a variety of major human diseases such as autoimmune diseases, infectious diseases, and malignancies. This article will discuss current advances in protein ubiquitination-mediated control of several immune cell functions and its significance in multiple sclerosis (MS).

Keywords

Ubiquitination · Autoimmunity · Multiple sclerosis

Ubiquitination is a post-translational modification process that proteins undergo, primarily involving the covalent attachment of ubiquitin to the lysine (K) residues of target proteins. This reaction is sequentially mediated by ubiquitin-activating enzyme (E1), ubiquitin-conjugating enzyme (E2), and ubiquitin ligase (E3) (Zhang et al. 2020). The formation of ubiquitin chains is a result of the carboxy-terminal glycine residue of one ubiquitin forming an isopeptide bond with a lysine residue or the amino-terminal methionine (M1) of another ubiquitin. Given the presence of seven lysine and one M1, various ubiquitin chains emerge, such as those linked through K6-, K11-, K27-, K29-, K33-, K48-, K63-, and M1-connections, in addition to some mixed-type chains. The specificity in the ubiquitination process largely depends on the E3 enzyme. It is estimated that about 634 mammalian proteins can be categorized as ubiquitin ligases, which assist or directly facilitate the transfer of ubiquitin from E2 enzymes to target substrates (Deshaies and Joazeiro 2009; Berndsen and Wolberger 2014). Additionally, ubiquitination is both dynamic and reversible. This means that established ubiquitin chains can be removed by a category of proteases called deubiquitinating enzymes (DUBs) (Sun 2008). While ubiquitination is generally associated with directing proteins for degradation in the 26S proteasome, it has been found that certain ubiquitin chains, such as those linked through K63- and M1-, regulate signal transduction via non-degradative pathways (Iwai 2012). These chains can serve as scaffolds in signaling, fostering interactions between distinct proteins. Furthermore, a growing class of

L. Jin (✉)
School of Life Science and Technology, China Pharmaceutical University, Nanjing, Jiangsu, China
e-mail: ljstemcell@cpu.edu.cn

H. Hu, X. Fu (eds.), *Ubiquitination in Immune System and Immune Response*, Advances in Experimental Medicine and Biology 1466, https://doi.org/10.1007/978-981-97-7288-9_8

proteins with ubiquitin-binding domains (UBDs) can recognize and interact with target proteins when they bind to specific types of ubiquitin chains, thereby facilitating the assembly of signaling complexes (Husnjak and Dikic 2012). Ubiquitination plays a pivotal role in numerous biological processes, especially in various aspects of immune function.

8.1 Regulation of Immune Cell Function by Ubiquitination

8.1.1 Ubiquitination Regulates Macrophage Apoptosis and Autophagy

Macrophages express a variety of pattern recognition receptors (PRRs), such as Toll-like receptors, mannose receptors, and scavenger receptors (Mulder et al. 2019). When pathogens invade, macrophages identify pathogen-associated molecular patterns (PAMPs) to regulate anti-infective responses, combating these invaders by phagocytosing them, releasing cytokines, and producing antimicrobial peptides. When pathogens at the site of infection become difficult to control, macrophages generate bactericidal ROS and undergo apoptosis (Subramanian et al. 2013), indicating that microbial defense is reaching its final stages. In the early stages of infection, the ubiquitin-proteasome system (UPS) delays macrophage apoptosis by regulating the degradation of Bcl-2 family members, such as Bax, through K48-linked ubiquitination. The apoptosis-sensitive gene (SAG) is a crucial component of the UPS. When stimulated by PAMPs, macrophages express both pro- and anti-apoptotic factors to maintain their functionality. However, when SAG expression decreases, it leads to the accumulation of apoptosis-related proteins like Bax and SARM, resulting in an imbalance between Bcl-2 and Bax in the mitochondria. This imbalance then triggers the release of cytochrome C, activating caspase-9 and caspase-3, which promotes macrophage apoptosis (Chang and Ding 2014). Therefore, SAGUPS promotes the ubiquitin-mediated degradation of apoptosis-related proteins, aiding the survival of infected macrophages and swiftly initiating early infection immune responses.

Furthermore, autophagy is a crucial mechanism for macrophages to resist invasive pathogens, with ubiquitination playing a significant role. Parkin, an E3 ubiquitin ligase, facilitates the ubiquitination of phagosomes containing Mycobacterium tuberculosis, enhancing autophagy and pathogen clearance. Mice and fruit flies lacking Parkin are sensitive to infections from certain intracellular bacteria, highlighting Parkin's essential role in immunity (Manzanillo et al. 2013). Additionally, TRAF6 enhances macrophage autophagy mediated by TLR4 (Toll-like receptor 4) through the K63-linked ubiquitination of Beclin-1. In contrast, A20 can inhibit LPS-induced autophagy. Mice lacking A20 show severe organ damage, are highly sensitive to LPS and tumor necrosis factor (TNF), and risk neonatal death. Apart from phagocytosing pathogens, macrophages also activate inflammatory signaling pathways such as NF-κB and MAPK in response to microbial PAMP activation, contributing to the body's inflammatory response (Shi and Kehrl 2010).

8.1.2 Ubiquitination Regulates MHC II and NF-κB Signaling Pathway in Dendritic Cells

Dendritic cells (DCs) are the primary antigen-presenting cells (APCs) specialized in recognizing, capturing, and processing antigens. They present these antigens to naive T cells to initiate and expand their activation. They act as intermediaries between innate and adaptive immunity, playing a crucial role in initiating adaptive immune responses, especially in antigen presentation and immune activation (Waisman et al. 2017). After processing antigens, DCs present these on their cell membrane as pMHC and interact with $CD4^+$ or $CD8^+$ T cells, providing the initial signal for T cell activation (Overgaard et al. 2015). Fully matured DCs abundantly display co-stimulatory molecules like CD80, CD86, and CD40, providing additional signals for further T cell activation, making these mature DCs the essential APCs for activating naive T cells (Zhu et al. 2020).

Research has shown that the transcription factor BLIMP1 can inhibit the production of the MHC II activation factor CIITA. In the absence of BLIMP1, DCs exhibit a higher level of MHC II on $CD4^+$ T cells, thereby enhancing immune responses (Kim et al. 2013). Additionally, the activation of TLR boosts the presence of the E3 ubiquitin ligase Hrd1, accelerating the ubiquitination of BLIMP1 in DCs, thus enhancing MHC II expression. By reducing MHC II expression, specific Hrd1 mutations in DCs display a reduced phenotype in MOG-specific $CD4^+$ T cell presentation and offer protection against EAE (Yang et al. 2014).

The process of ubiquitination plays a pivotal role in the expression and transport of MHC II on the cell surface, especially through the ubiquitin-dependent protein degradation in lysosomes (Shin et al. 2006; van Niel et al. 2006). The E3 ubiquitin ligase MARCH1, a member of the MARCH family, ubiquitinates the β-subunit of MHC II through its cytoplasmic lysine (Oh and Shin 2015). MARCH1 not only ubiquitinates MHC II but also affects the intracellular degradation of CD86 in DCs (Baravalle et al. 2011; Corcoran et al. 2011). The interaction between MARCH1 and CD83 does not lead to its ubiquitination. CD83 competes with MARCH1 for interaction with CD86, inhibiting the ubiquitination of CD86 and promoting its expression on the surface of DCs (Tze et al. 2011). In essence, MARCH1 negatively regulates DC maturation by controlling the ubiquitination of MHC II and CD86, with this process diminishing during DC maturation. Recent research has shown that during Salmonella infection, the NEDD4 family's WWP2 E3 ubiquitin ligase is essential for SteD-dependent ubiquitination of MHC II. The tumor suppressor protein TMEM127 acts as an adaptor, binding to WWP2, enabling WWP2 to ubiquitinate MHC II, aiding in its degradation (Kool et al. 2011).

The typical NF-κB signaling pathway is a critical regulator for the activation and functional maturation of DCs post-TLR stimulation (Kaisho and Tanaka 2008). Ubiquitination appears pivotal in controlling NF-κB signaling during DC tolerance and activation. Several E3 ubiquitin ligases and DUBs present in DCs modulate NF-κB activation by ubiquitinating or deubiquitinating key regulators of NF-κB, namely NEMO (also known as IKKγ) and NF-κB subunits (p65 and c-Rel). TLR activation typically results in the recruitment of adapter protein MyD88 and kinases IRAK1 and IRAK4. These IRAKs subsequently disassociate from the receptor complex and bind to E3 ubiquitin ligase TRAF6, leading in collaboration with E2 ubiquitin-conjugating enzyme Ubc13 to the polyubiquitination of IRAK1/4 and TRAF6 (Griewahn et al. 2019). Similar to TNFR stimulation, the LUBAC (linear ubiquitin chain assembly complex) is essential for the full activation of the MyD88-dependent pathway for NF-κB. The LUBAC complex extends K63-linked polyubiquitin chains to M1-linked chains, recruiting and activating the IKK complex through its adapter NEMO, thereby releasing NF-κB transcription factors.

Ubiquitination also plays a vital role in the non-canonical NF-κB pathway in DC cytokine production (Zhu et al. 2020). Research has unveiled the participation of the E3 ubiquitin ligase $CRL4^{DCAF2}$ in the expression of the inflammatory cytokine IL-23 induced by GM-CSF. $CRL4^{DCAF2}$ negatively regulates IL-23 production in DCs by controlling the stability and activation of the NF-κB inducing kinase (NIK) in the non-canonical NF-κB pathway. Early studies suggested that activation of cellular inhibitor of apoptosis proteins 1 and 2 (cIAP1/2) proceeds through its K63 ubiquitination mediated by TRAF2, promoting NIK degradation (Vallabhapurapu et al. 2008). Intriguingly, $CRL4^{DCAF2}$ directly fosters the polyubiquitination and subsequent degradation of NIK, independent of cIAP activity. Conditional deficiency of DCAF2 in DCs amplifies IL-23 production, fostering psoriasis development (Iberg and Hawiger 2020).

8.1.3 Ubiquitination Regulates TCR Signaling and the Activation of T Cells

CD4 T cells play a pivotal role in adaptive immunity against pathogens and are also involved in autoimmunity, asthma, allergies, and anti-tumor immunity (Zhu et al. 2010). After antigen stimulation, naive CD4 T cells get activated through the binding of T cell receptors (TCRs) and co-stimulatory receptors (Murphy et al. 2000; Smith-Garvin et al. 2009). Within a specific cytokine milieu, CD4 T cells differentiate into five subtypes of CD4 helper T cells (Th), including Th1, Th2, Th17, T follicular helper cells (Tfh), and Treg (Baravalle et al. 2011; Zhu et al. 2010). These CD4 T cell subsets secrete various cytokines and possess unique functions. Extracellular stimulatory signals can activate downstream intracellular signaling pathways by activating tyrosine kinases adjacent to TCRs, promoting the assembly of signaling transduction complexes, activating pathways such as NF-κB, MAPK, and calcium ions, and ultimately initiating transcription factors like NF-κB, AP-1, and nuclear factor of activated T cells (NFAT). As a result, effector protein molecules are modulated, leading to cell proliferation and aiding in the completion of T cell activation (Smith-Garvin et al. 2009; Malynn and Ma 2010). In the absence of co-stimulatory signals, TCR activation promotes T cell inactivity. Several E3 ubiquitin ligases, including GRAIL (gene related to anergy in lymphocytes), Cbl-b, and Itch, regulate the induction of T cell immune tolerance.

Ubiquitin-mediated T cell activation is a complex mechanism. Among the E3 ubiquitin ligases, Casitas B-lineage lymphoma (Cbl-b) is the most recognized and acts as the primary gatekeeper for T cell activation. It gets recruited to the TCR at the immunological synapse during T cell activation (Wiedemann et al. 2005) and inhibits TCR downstream through various mechanisms there. Cbl-b interacts with key TCR signaling molecules like LCK, SLP76, ZAP70, Vav1, PKCθ, and PI3K through its multiple protein-interacting domains (Bachmaier et al. 2000). In collaboration with the E3 ligase Itch, Cbl-b can mediate Lys33-linked polyubiquitination of TCR-ζ, which doesn't target the TCR receptor for degradation or endocytosis but rather prevents its phosphorylation and subsequent binding with downstream ZAP70 kinase (Huang et al. 2010). Through these interactions, and by ubiquitinating parts of these signaling molecules (like the p85 regulatory subunit of PI3K and PLCγ1), Cbl-b substantially suppresses T cell activation (Zhang et al. 2008). Thus, mice lacking Cbl-b have hyperactive T cells that don't require CD28 for activation (Chiang et al. 2000), significantly enhancing T cell anti-tumor immunity in vivo (Chiang et al. 2007). GRAIL is an E3 ligase located in the endoplasmic reticulum and acts as a negative regulator that prevents T cell activation. TCR-mediated T cell activation upregulates GRAIL expression, which in turn impedes T cell activation. While overexpression of GRAIL reduces IL-2 and IL-4 secretion by T cells (Anandasabapathy et al. 2003), T cells lacking GRAIL exhibit enhanced proliferation and cytokine secretion, unrelated to CD28 co-stimulation (Nurieva et al. 2010).

The strength and persistence of TCR signaling hinge on a dynamic equilibrium encompassing the internalization, recycling, and breakdown of cell membrane receptors. While it's well acknowledged that upon antigen challenge, TCR/CD3 structures undergo ubiquitination (Hou et al. 1994), and this ubiquitination is pivotal for pulling surface receptors inward (Haglund and Dikic 2012), we are just scratching the surface in understanding the ubiquitin-centric mechanisms that dictate TCR's inward journey and its anchoring stability on the cell's facade. Notably, Cbl-b is reputed to champion TCR's inward move, either solo or pairing up with the E3 ligase known as c-Cbl. Additionally, Cbl-b meddles in the realm of activation and clustering, thereby compromising the integrity of the immune synapse and subsequently putting a damper on TCR signaling (Krawczyk et al. 2000). In a similar vein, the E3 ligase named GRAIL takes it upon itself to dial down the TCR/CD3 combo's presence on the cell surface, orchestrating this by polyubiquitinating and dispatching CD3ζ units to the proteasomal shredder (Nurieva et al. 2010).

Diving deeper, as highlighted earlier, the E3 ligase Cbl-b emerges as a pivotal dampener of signals channeled through CD28. This claim gets its empirical backing from the fact that in the absence of Cbl-b, T cells rise to the activation occasion even when CD28 is missing from the scene. This hints that a void in Cbl-b might very well be the antidote to most T-cell oddities spotted in the CD28-absent mice in a live setting (Chiang et al. 2000). Delving into the mechanics of it, Cbl-b throws a spanner in the CD28-reliant PI3K machinery by cozying up to and tagging the PI3K p85 unit with ubiquitin (Alcazar et al. 2009). These ubiquitin ties, while not dispatching their linked entities to degradation, play gatekeeper preventing p85β from mingling with CD28. As a downstream ripple effect, T cells devoid of Cbl-b exhibit an upsurge in Akt's activation zeal and a revved-up NF-κB signaling relay, a phenomenon once theorized to stem from Cbl-b's restraining hand on p85 (Qiao et al. 2008).

8.1.4 Ubiquitination Regulates the Development and Activation of B Cells

B cells originate from hematopoietic stem cells located in the bone marrow. Through the progression stages of early pre-B cells, intermediate pre-B cells, and late pre-B cells, these cells mature into immature B cells (Clurman et al. 1996). In these early developmental stages, NF-κB signaling plays a pivotal role in fostering the formation of B cells by counteracting the TNF-α signaling mechanism. The B-cell-activating factor, commonly known as BAFF, triggers the non-canonical NF-κB signaling pathway. This factor predominantly influences the maturation of advanced B cells in the spleen. It has been proposed that DYRK1a acts as a kinase in response to BAFF stimulation, bolstering the survival of B cells that is brought about by BAFF (Li et al. 2021). TRAF3 acts as a deterrent in the non-canonical NF-κB signaling pathway. It facilitates the ubiquitination of NIK's (NF-κB-inducing kinase) N-terminal domain using a K48-linked ubiquitin ligase. Within this mechanism, NIK partners with TRAF2-cIAP1/2 and TRAF3 to form a complex. Here, TRAF3 interacts with NIK, bringing in TRAF2-cIAP1/2. The ubiquitin ligase, cIAP1/2, then propels the ubiquitin-mediated degradation of NIK by connecting to K48-linked polyubiquitin chains. This process ensures NIK levels are maintained at a relatively lower range (Varfolomeev et al. 2006).

Another line of research has brought to light the role of OTUD7B, a significant regulatory molecule in the non-canonical NF-κB signaling pathway. OTUD7B can also function as a K48-specific deubiquitinating enzyme. By binding with TRAF3 and mediating its deubiquitination, it impedes TRAF3's proteasome degradation, ultimately activating the non-canonical NF-κB signaling pathway. MYSM is identified as a deubiquitination enzyme that impacts transcription factors' function by interacting with chromatin, mediating H2A's deubiquitination. Reports suggest that MYSM1 stimulates the Ebf1 promoter to engage with transcriptional regulators, enhancing Ebf1 expression through the deubiquitination of histone H2AK119 found in CLPs and early pre-B cells. This interaction affects the transition from intermediate pre-B cells to late pre-B cells, leading to a reduction in B cell numbers (Jiang et al. 2011; Forster et al. 2017).

Both DCs and B cells hold paramount importance as antigen-presenting cells (APCs). They have the capability to identify, process, and bind antigenic peptides to molecules similar to MHC II, making them recognizable by $CD4^+$ T lymphocytes. However, these APCs deploy entirely distinct strategies for ubiquitinating and regulating MHC II. Distinctions were observed in the length of the ubiquitin chain bound to MHC II in DCs and B cells: immature DCs exhibit 4–6 ubiquitins, while B cells display only 2–3. Experimental elongation of the ubiquitin chain leads to MHC II's efficient transport to lysosomes, whereas MHC II without two ubiquitins doesn't achieve this. Additional research points to shared mechanisms governing MHC II abundance across DCs, germinal center (GC) B cells, and follicular B cells. Within GC B cells and follicular B cells, MARCH-1 promotes MHC II degradation. It achieves this by appending ubiquitin entities to a conserved lysine in the MHC II β chain, obstructing its recycling to the surface for further presentation and directing it for degradation, thus refreshing the peptide-MHC II (pMHC II) on the cell surface (Bannard et al. 2016). Impairments in MHC II ubiquitination lead to a buildup of specific pMHC II on GC B cells, implying that MHC II turnover facilitates the replacement of older complexes.

8.1.5 Ubiquitination Regulates Type I Interferon and Proinflammatory Factors in Microglia Secretion

Microglial cells are naturally occurring macrophages in the central nervous system (CNS), playing a pivotal role in maintaining CNS's normal functions and in the immunological response to external infections (Borst et al. 2021). A study indicated that, compared with WT normal mice, the microglial cells from mice with a knocked-out E3 ubiquitin ligase Peli1 gene, when stimulated by LPS or viral infection, exhibited a pronounced increase in the expression of type I interferons (IFN-Is). Concurrently, there was a substantial enhancement in the phosphorylation of kinases TBK1 and IKKε. This suggests that Peli1 acts to suppress the activity of kinases TBK1 and IKKε, thereby reducing IFN-Is expression in microglial cells. It further underscores that Peli1 inhibits the signaling pathways associated with the activation of TBK1 and IKK in microglia, ultimately curtailing TLR-mediated IFN-I production (Xiao et al. 2015).

Additionally, Peli1 is involved in another vital signaling pathway in microglia, which regulates inflammation and autoimmune reactions within the CNS. Evidence has shown that various ligands for MyD88-dependent TLRs can stimulate microglia to produce chemokines and pro-inflammatory cytokines (Huang et al. 2017). The absence of Peli1 significantly affects the expression of these genes in microglia (Xiao et al. 2013). Peli1 triggers the K63 ubiquitination of c-IAP2, thereby activating the ligase activity of c-ubiquitin IAP2, leading to the K48 ubiquitination and degradation of TRAF3 mediated by c-IAP2. As previously mentioned, TRAF3 is distinct from other members of the TRAF protein family not only because it is involved in multiple signaling pathways but also due to its diverse functions in each pathway, exhibiting both positive and negative regulatory effects. Specifically, TRAF3 inhibits the MyD88-driven activation of MAPK. Hence, Peli1 indirectly modulates the TLR signaling pathway by prompting microglia to release pro-inflammatory factors, enhancing T cell migration from peripheral immune organs to the CNS, thereby exacerbating the progression of autoimmune inflammation in the CNS (Hacker et al. 2011).

8.1.6 Ubiquitin Regulates Immune Cells in MS/EAE

Autoimmune diseases arise when the immune system abnormally targets and attacks self-antigens, leading to chronic inflammation and tissue damage. Multiple sclerosis (MS) stands out as a chronic inflammatory demyelinating disorder and is the most prevalent autoimmune disease of the central nervous system. This condition is characterized by the invasion of macrophages, T cells, and B cells into the central nervous system, resulting in immunologically-triggered demyelination, neural injury, and the formation of plaques (Sospedra and Martin 2005). The rodent EAE (experimental autoimmune encephalomyelitis) model can emulate some pivotal characteristics of MS, making it a widely utilized tool for investigating the pathogenesis and potential treatments of MS (Constantinescu et al. 2011).

The infiltration of T cells and other immune cells into the central nervous system is a pathological hallmark of MS, with T cells recognized as vital mediators in both MS and EAE. Interestingly, in active MS lesions, the count of $CD8^+$ T cells surpasses that of $CD4^+$ T cells. However, in the EAE context, $CD4^+$ T cells predominantly dominate the demyelinating lesions (van Nierop et al. 2017; Machado-Santos et al. 2018). In EAE, T cell subgroups, specifically Th1 cells producing IFN-γ, Th17 cells secreting IL-17, and T cells expressing GM-CSF, play a significant role. Each of these T cell subsets can independently instigate EAE (Stromnes et al. 2008; Codarri et al. 2011). T cell responses are rigorously modulated by TCR signaling, and it's evident that DUBs, including USP9X,

USP15, and CYLD, can influence this signal (Ruan et al. 2020). Within autoimmune scenarios of the central nervous system, the encephalitogenic nature of T cells relies on their differentiation, activation, and survival, all of which are intricately governed by ubiquitination and DUBs. Th17 cells are pivotal in the context of MS/EAE and represent therapeutic targets for MS (Dos Passos et al. 2016). The differentiation of Th17 cells is orchestrated by the transcription factor RORγt, which also drives the production of GM-CSF, a determining cytokine in EAE development (Codarri et al. 2011). The protein levels of RORγt are partly regulated by UPS, with both USP4 (Yang et al. 2015) and USP17 (Han et al. 2014) being crucial for maintaining RORγt stability. On the other hand, DUBA (OTUD5) facilitates the degradation of RORγt by stabilizing the E3 ligase UBR5, which leads to the ubiquitination of RORγt and its subsequent proteasomal degradation (Rutz et al. 2015). Beyond protein abundance, the activity of RORγt is also subject to ubiquitin-mediated regulation. Ubiquitination of RORγt at K446 impairs its interaction with the co-activator SRC1, subsequently inhibiting Th17 differentiation. This specific ubiquitination at the K446 site can be reversed by USP15, thus underscoring the positive regulatory role of USP15 in Th17 differentiation (He et al. 2016). The development, differentiation, and functionality of Th1 cells are overseen by the transcription factor T-bet (Szabo et al. 2000). It has been demonstrated that T-bet's stability is positively regulated by USP10, as USP10 directly interacts with T-bet to mediate its deubiquitination (Pan et al. 2014). IFN-γ, a signature cytokine of Th1 cells, induces various immune responses, including Th1 differentiation, via activating the JAK-STAT1 signaling pathway. The signaling induced by IFN-γ is profoundly amplified by USP13, which achieves this by deubiquitinating and stabilizing STAT1 (Yeh et al. 2013). In the absence of GRAIL, there's a marked increase in IFN-γ secretion by Th1 cells. GRAIL emerges as a crucial E3 ligase inducing an unresponsive state in $CD4^+$ T cells. Earlier, oral administration of myelin basic protein (MBP) had shown promise in ameliorating MBP-induced EAE. However, this oral tolerance is disrupted due to the loss of T cell unresponsiveness in the absence of GRAIL (Kriegel et al. 2009). The unresponsive function induced by GRAIL is safeguarded by USP8, which is indispensable for the deubiquitination and stability of GRAIL (Soares et al. 2004). Consequently, USP8 might be a robust modulator of T cell functions and EAE.

$Peli1^{-/-}$ mice were found to spontaneously develop an autoimmune disease characterized by multi-organ inflammation and autoantibody production. Transfer of $Peli1^{-/-}$ T cells to recombination-activating gene 1-deficient ($Rag1^{-/-}$) mice induced much higher EAE disease scores than wild-type This suggests a critical role for Peli1 in suppressing T cell activation. Further studies revealed that Peli1 negatively regulates T cell activation and suppresses the development of autoimmunity by targeting K48 ubiquitination and degradation of c-Rel (Huang et al. 2021). Therefore, Peli1 has a vital role in the maintenance of T cell peripheral tolerance. Peli1 can also contribute to EAE via positively regulating microglial cell activation in the CNS, in addition to inhibiting autoimmune disorders by negatively regulating T cell activation. Peli1 promotes TLR-mediated signaling in microglia, as previously established. In vivo experiments have demonstrated that $Peli1^{-/-}$ can activate the MAPK signaling pathway by inducing K63 ubiquitination of c-IAP2, triggering ubiquitin ligase activity of c-IAP2 and catalyzing K48 ubiquitination and degradation of TRAF3, thereby inducing pro-inflammatory cytokine production and recruitment of autoimmune T cells from peripheral lymphoid organs to the CNS (Hacker et al. 2011).

8.2 Prospectives

This review summarizes some of ubiquitination's major functions in the control of innate and adaptive immune responses. A large amount of E3 ubiquitin ligases and DUBs have been identified as critical immune regulators, with roles ranging from innate immunity and inflammation to DC maturation and T cell activation and differentiation. Additionally, ubiquitination is critical for maintaining central and peripheral immunological tolerance and preventing the onset of autoimmunity (e.g., multiple sclerosis). Studies using mutant mouse models have been instrumental in elucidating the physiological functions of E3 ubiquitin ligases and DUBs, and this area of research will further add important information to our understanding of the role of ubiquitination in the immune system.

References

Alcazar I et al (2009) p85beta phosphoinositide 3-kinase regulates CD28 coreceptor function. Blood 113(14):3198–3208

Anandasabapathy N et al (2003) GRAIL: an E3 ubiquitin ligase that inhibits cytokine gene transcription is expressed in anergic CD4+ T cells. Immunity 18(4):535–547

Bachmaier K et al (2000) Negative regulation of lymphocyte activation and autoimmunity by the molecular adaptor Cbl-b. Nature 403(6766):211–216

Bannard O et al (2016) Ubiquitin-mediated fluctuations in MHC class II facilitate efficient germinal center B cell responses. J Exp Med 213(6):993–1009

Baravalle G et al (2011) Ubiquitination of CD86 is a key mechanism in regulating antigen presentation by dendritic cells. J Immunol 187(6):2966–2973

Berndsen CE, Wolberger C (2014) New insights into ubiquitin E3 ligase mechanism. Nat Struct Mol Biol 21(4):301–307

Borst K, Dumas AA, Prinz M (2021) Microglia: immune and non-immune functions. Immunity 54(10):2194–2208

Chang SC, Ding JL (2014) Ubiquitination by SAG regulates macrophage survival/death and immune response during infection. Cell Death Differ 21(9):1388–1398

Chiang YJ et al (2000) Cbl-b regulates the CD28 dependence of T-cell activation. Nature 403(6766):216–220

Chiang JY et al (2007) Ablation of Cbl-b provides protection against transplanted and spontaneous tumors. J Clin Invest 117(4):1029–1036

Clurman BE et al (1996) Turnover of cyclin E by the ubiquitin-proteasome pathway is regulated by cdk2 binding and cyclin phosphorylation. Genes Dev 10(16):1979–1990

Codarri L et al (2011) RORgammat drives production of the cytokine GM-CSF in helper T cells, which is essential for the effector phase of autoimmune neuroinflammation. Nat Immunol 12(6):560–567

Constantinescu CS et al (2011) Experimental autoimmune encephalomyelitis (EAE) as a model for multiple sclerosis (MS). Br J Pharmacol 164(4):1079–1106

Corcoran K et al (2011) Ubiquitin-mediated regulation of CD86 protein expression by the ubiquitin ligase membrane-associated RING-CH-1 (MARCH1). J Biol Chem 286(43):37168–37180

Deshaies RJ, Joazeiro CA (2009) RING domain E3 ubiquitin ligases. Annu Rev Biochem 78:399–434

Dos Passos GR et al (2016) Th17 cells pathways in multiple sclerosis and neuromyelitis optica spectrum disorders: pathophysiological and therapeutic implications. Mediat Inflamm 2016:5314541

Forster M et al (2017) MYSM1-dependent checkpoints in B cell lineage differentiation and B cell-mediated immune response. J Leukoc Biol 101(3):643–654

Griewahn L, Koser A, Maurer U (2019) Keeping cell death in check: ubiquitylation-dependent control of TNFR1 and TLR signaling. Front Cell Dev Biol 7:117

Hacker H, Tseng PH, Karin M (2011) Expanding TRAF function: TRAF3 as a tri-faced immune regulator. Nat Rev Immunol 11(7):457–468

Haglund K, Dikic I (2012) The role of ubiquitylation in receptor endocytosis and endosomal sorting. J Cell Sci 125(Pt 2):265–275

Han L et al (2014) The E3 deubiquitinase USP17 is a positive regulator of retinoic acid-related orphan nuclear receptor gammat (RORgammat) in Th17 cells. J Biol Chem 289(37):25546–25555

He Z et al (2016) Ubiquitination of RORgammat at lysine 446 limits Th17 differentiation by controlling coactivator recruitment. J Immunol 197(4):1148–1158

Hou D et al (1994) Activation-dependent ubiquitination of a T cell antigen receptor subunit on multiple intracellular lysines. J Biol Chem 269(19):14244–14247

Huang H et al (2010) K33-linked polyubiquitination of T cell receptor-zeta regulates proteolysis-independent T cell signaling. Immunity 33(1):60–70

Huang XP et al (2017) Peli1 contributions in microglial activation, neuroinflammatory responses and neurological deficits following experimental subarachnoid hemorrhage. Front Mol Neurosci 10:398

Huang X et al (2021) The ubiquitin ligase Peli1 inhibits ICOS and thereby Tfh-mediated immunity. Cell Mol Immunol 18(4):969–978

Husnjak K, Dikic I (2012) Ubiquitin-binding proteins: decoders of ubiquitin-mediated cellular functions. Annu Rev Biochem 81:291–322

Iberg CA, Hawiger D (2020) Natural and induced tolerogenic dendritic cells. J Immunol 204(4):733–744

Iwai K (2012) Diverse ubiquitin signaling in NF-kappaB activation. Trends Cell Biol 22(7):355–364

Jiang XX et al (2011) Control of B cell development by the histone H2A deubiquitinase MYSM1. Immunity 35(6):883–896

Kaisho T, Tanaka T (2008) Turning NF-kappaB and IRFs on and off in DC. Trends Immunol 29(7):329–336

Kim SJ, Gregersen PK, Diamond B (2013) Regulation of dendritic cell activation by microRNA let-7c and BLIMP1. J Clin Invest 123(2):823–833

Kool M et al (2011) The ubiquitin-editing protein A20 prevents dendritic cell activation, recognition of apoptotic cells, and systemic autoimmunity. Immunity 35(1):82–96

Krawczyk C et al (2000) Cbl-b is a negative regulator of receptor clustering and raft aggregation in T cells. Immunity 13(4):463–473

Kriegel MA, Rathinam C, Flavell RA (2009) E3 ubiquitin ligase GRAIL controls primary T cell activation and oral tolerance. Proc Natl Acad Sci USA 106(39):16770–16775

Li Y et al (2021) DYRK1a mediates BAFF-induced noncanonical NF-kappaB activation to promote autoimmunity and B-cell leukemogenesis. Blood 138(23):2360–2371

Machado-Santos J et al (2018) The compartmentalized inflammatory response in the multiple sclerosis brain is composed of tissue-resident CD8+ T lymphocytes and B cells. Brain 141(7):2066–2082

Malynn BA, Ma A (2010) Ubiquitin makes its mark on immune regulation. Immunity 33(6):843–852

Manzanillo PS et al (2013) The ubiquitin ligase parkin mediates resistance to intracellular pathogens. Nature 501(7468):512–516

Mulder WJM et al (2019) Therapeutic targeting of trained immunity. Nat Rev Drug Discov 18(7):553–566

Murphy KM et al (2000) Signaling and transcription in T helper development. Annu Rev Immunol 18:451–494

Nurieva RI et al (2010) The E3 ubiquitin ligase GRAIL regulates T cell tolerance and regulatory T cell function by mediating T cell receptor-CD3 degradation. Immunity 32(5):670–680

Oh J, Shin JS (2015) Molecular mechanism and cellular function of MHCII ubiquitination. Immunol Rev 266(1):134–144

Overgaard NH et al (2015) $CD4^+/CD8^+$ double-positive T cells: more than just a developmental stage? J Leukoc Biol 97(1):31–38

Pan L et al (2014) Deubiquitination and stabilization of T-bet by USP10. Biochem Biophys Res Commun 449(3):289–294

Qiao G et al (2008) T-cell receptor-induced NF-kappaB activation is negatively regulated by E3 ubiquitin ligase Cbl-b. Mol Cell Biol 28(7):2470–2480

Ruan J, Schluter D, Wang X (2020) Deubiquitinating enzymes (DUBs): DoUBle-edged swords in CNS autoimmunity. J Neuroinflammation 17(1):102

Rutz S et al (2015) Deubiquitinase DUBA is a post-translational brake on interleukin-17 production in T cells. Nature 518(7539):417–421

Shi CS, Kehrl JH (2010) TRAF6 and A20 regulate lysine 63-linked ubiquitination of Beclin-1 to control TLR4-induced autophagy. Sci Signal 3(123):ra42

Shin JS et al (2006) Surface expression of MHC class II in dendritic cells is controlled by regulated ubiquitination. Nature 444(7115):115–118

Smith-Garvin JE, Koretzky GA, Jordan MS (2009) T cell activation. Annu Rev Immunol 27:591–619

Soares L et al (2004) Two isoforms of otubain 1 regulate T cell anergy via GRAIL. Nat Immunol 5(1):45–54

Sospedra M, Martin R (2005) Immunology of multiple sclerosis. Annu Rev Immunol 23:683–747

Stromnes IM et al (2008) Differential regulation of central nervous system autoimmunity by T(H)1 and T(H)17 cells. Nat Med 14(3):337–342

Subramanian K et al (2013) CD163 and IgG codefend against cytotoxic hemoglobin via autocrine and paracrine mechanisms. J Immunol 190(10):5267–5278

Sun SC (2008) Deubiquitylation and regulation of the immune response. Nat Rev Immunol 8(7):501–511

Szabo SJ et al (2000) A novel transcription factor, T-bet, directs Th1 lineage commitment. Cell 100(6):655–669

Tze LE et al (2011) CD83 increases MHC II and CD86 on dendritic cells by opposing IL-10-driven MARCH1-mediated ubiquitination and degradation. J Exp Med 208(1):149–165

Vallabhapurapu S et al (2008) Nonredundant and complementary functions of TRAF2 and TRAF3 in a ubiquitination cascade that activates NIK-dependent alternative NF-kappaB signaling. Nat Immunol 9(12):1364–1370

van Niel G et al (2006) Dendritic cells regulate exposure of MHC class II at their plasma membrane by oligoubiquitination. Immunity 25(6):885–894

van Nierop GP et al (2017) Phenotypic and functional characterization of T cells in white matter lesions of multiple sclerosis patients. Acta Neuropathol 134(3):383–401

Varfolomeev E et al (2006) The inhibitor of apoptosis protein fusion c-IAP2.MALT1 stimulates NF-kappaB activation independently of TRAF1 AND TRAF2. J Biol Chem 281(39):29022–29029

Waisman A et al (2017) Dendritic cells as gatekeepers of tolerance. Semin Immunopathol 39(2):153–163

Wiedemann A et al (2005) T-cell activation is accompanied by an ubiquitination process occurring at the immunological synapse. Immunol Lett 98(1):57–61

Xiao Y et al (2013) Peli1 promotes microglia-mediated CNS inflammation by regulating Traf3 degradation. Nat Med 19(5):595–602

Xiao Y et al (2015) Peli1 negatively regulates type I interferon induction and antiviral immunity in the CNS. Cell Biosci 5:34

Yang H et al (2014) Hrd1-mediated BLIMP-1 ubiquitination promotes dendritic cell MHCII expression for CD4 T cell priming during inflammation. J Exp Med 211(12):2467–2479

Yang J et al (2015) Cutting edge: ubiquitin-specific protease 4 promotes Th17 cell function under inflammation by deubiquitinating and stabilizing RORgammat. J Immunol 194(9):4094–4097

Yeh HM et al (2013) Ubiquitin-specific protease 13 regulates IFN signaling by stabilizing STAT1. J Immunol 191(6):3328–3336
Zhang R, Zhang N, Mueller DL (2008) Casitas B-lineage lymphoma b inhibits antigen recognition and slows cell cycle progression at late times during CD4+ T cell clonal expansion. J Immunol 181(8):5331–5339
Zhang Z et al (2020) GRM4 inhibits the proliferation, migration, and invasion of human osteosarcoma cells through interaction with CBX4. Biosci Biotechnol Biochem 84(2):279–289
Zhu J, Yamane H, Paul WE (2010) Differentiation of effector CD4 T cell populations (*). Annu Rev Immunol 28:445–489
Zhu B et al (2020) Roles of ubiquitination and deubiquitination in regulating dendritic cell maturation and function. Front Immunol 11:586613

9 Regulation of the Inflammasome Activation by Ubiquitination Machinery

Feng Liu and Chengjiang Gao

Abstract

Inflammasomes are multiprotein complexes that assemble in response to the detection of stress- or infection-associated stimuli and lead to the activation of caspase-1 and consequent maturation of caspase-1 target molecules such as interleukin (IL)-1β and IL-18. Although inflammasome is the essential component of the innate immunity system to defense against insults, inappropriate or prolonged activation of inflammasome may be harmful and is associated with various diseases, e.g., gout, atherosclerosis, diabetes, and Alzheimer's disease. Therefore, regulating inflammasome activation is crucial for maintaining immune homeostasis. Studies have found that post-translational modifications (PTMs), e.g., ubiquitination and phosphorylation, are critical for inflammasome activation. Ubiquitination is an important form of post-translational modification of proteins that plays a pivotal role in various cellular functions. In recent years, its function in regulating inflammasome assembly has been a hot topic of interest. This study discussed the function and mechanism of the ubiquitin system controlling inflammasome activation and highlighted the challenges of this research area.

Keywords

Inflammasome · Ubiquitination · NLRP3 · Caspase

F. Liu · C. Gao (✉)
Key Laboratory of Infection and Immunity of Shandong Province, Shandong University, Jinan, Shandong, P.R. China

Department of Immunology, School of Biomedical Sciences, Shandong University, Jinan, Shandong, P.R. China
e-mail: cgao@sdu.edu.cn

9.1 Introduction

Inflammation is an essential body response to infection or injury, regulated by several physiological and pathological processes (Medzhitov 2008). The initiation of inflammation is the trigger of the innate immune system. The pattern recognition receptors (PRRs) expressed by the innate immune cells recognize particular damage-associated molecular patterns (DAMPs) or microbe-associated molecular patterns (MAMPs), subsequently mobilizing specific signaling cascades to modulate phagocytosis and degradation of the microbe and regulate gene expression, which in turn promotes the production of inflammatory cytokines (Medzhitov 2008; Yeung et al. 2018). Although inflammation is the essential component of the innate immunity system against diverse organismal injuries, the deregulation of inflammation has a detrimental

H. Hu, X. Fu (eds.), *Ubiquitination in Immune System and Immune Response*, Advances in Experimental Medicine and Biology 1466, https://doi.org/10.1007/978-981-97-7288-9_9

effect on human health. This deregulation process has been associated with diseases, e.g., gout (Zhao et al. 2022), atherosclerosis (Slocum et al. 2016), diabetes (Tsalamandris et al. 2019), and Alzheimer's disease (Kinney et al. 2018). Therefore, regulating the initiation and progression of the inflammatory response at different levels is critical in keeping homeostasis.

In 2002, a novel PRR called inflammasome was identified by Martinon et al., who reported it to participate in the response against PAMPs or DAMPs (Martinon et al. 2002). The assembly and activation of inflammasome is essential in initiating inflammation. Inflammasomes are activated in response to the detection of stress- or infection-associated stimuli, such as extracellular ATP or bacterial toxins (Strowig et al. 2012; Broz and Dixit 2016). This activation then leads to an auto-activation of caspase-1 and consequent maturation of caspase-1 target molecules such as IL-1β and IL-18 and initiates a process of cell death known as pyroptosis (Strowig et al. 2012; Broz and Dixit 2016). The mechanisms of regulating inflammasome activation have been a hot topic of interest. Studies have found that post-translational modifications (PTMs) of the inflammasome, e.g., ubiquitination and phosphorylation, are critical for inflammasome activation (Man and Kanneganti 2015).

Since the discovery of ubiquitin-mediated protein degradation, our understanding of how the ubiquitin system fundamentally controls cell function and homeostasis has exponentially increased. It has been found that besides regulating protein homeostasis through protein degradation, the ubiquitin system is crucial for regulating other cellular outcomes, such as signal transduction, protein-protein interaction, and alteration of protein subcellular localization (Chaugule and Walden 2016; Fulda et al. 2012).

A clear understanding of the ubiquitin system in regulating inflammasome activation can contribute to developing therapeutic interventions for inflammasome-mediated diseases. This study summarized the current knowledge on the regulation of inflammasome activation by ubiquitin machinery and proposed the challenges and future directions that this area of research.

9.2 Inflammasome and Ubiquitination

9.2.1 Inflammasome Activation

Innate immune cells, such as monocytes, macrophages, and neutrophils, recognize danger signals, DAMPs, and MAMPs through their PRRs and initiate the inflammatory response cascade. The activation of inflammasomes has a crucial role in the initial steps of the inflammatory response (Latz et al. 2013). Inflammasomes are multimeric protein complexes responsible for the cleavage of caspase-1 into its active form and the processing of pro-IL-1β and pro-IL-18 into their mature forms (Strowig et al. 2012). Inflammasomes typically comprise a sensor, an adaptor, and an effector protein (Franchi et al. 2012). The sensor part, mainly NOD-like receptors (NLRs), recognizes the danger signal and initiates inflammasome assembly. The adaptor protein, apoptosis-associated speck-like protein containing a caspase-recruitment domain CARD (ASC), forms prion-like aggregation to mediate the linkage of upstream activating signal from the sensor part and the downstream effector. The effector protein, the cysteine protease pro-caspase-1, executes the function of cleaving the pro-IL-1β and pro-IL-18 into biologically active IL-1β and IL-18; IL-1β participates in the immune response by recruiting the innate immune cells to the site of infection and regulating the activity and function of adaptive immune cells (Dinarello 2009), while IL-18 regulates the activation of cytotoxic natural killer (NK) cells and T cells (Dinarello 2009). Caspase-1 mediates the maturation of IL-1β and IL-18 and induces a proinflammatory form of cell death called pyroptosis (Miao et al. 2011).

Five PRRs, including NLRP1, NLRP3, NLRC4, pyrin, and absent-in-melanoma 2 (AIM2), have been reported to constitute inflammasomes. NLRP3 inflammasome is best characterized and widely studied since it can respond to various stimuli (Man and Kanneganti 2015; Lamkanfi and Dixit 2014). The activation of the canonical NLRP3 inflammasome includes two steps: the priming step and the activation step

(Broz and Dixit 2016). In the priming step, the NF-κB pathway is activated with TLR engagement, leading to the upregulation of some inflammasome components, e.g., NLRP3 and pro-IL-1β. The basal quantity of NLRP3 in resting macrophages is insufficient for inflammasome activation. Therefore, the priming step is essential for increasing the NLRP3 transcription and licensing NLRP3 for the activation step (Bauernfeind et al. 2009; Cornut et al. 2020). In resting macrophages, pro-IL-1β is not expressed consistently and needs the trigger of TLR stimulation (Cornut et al. 2020). The activation step is induced by the activating signals such as mitochondrial reactive oxygen species (mtROS), adenine triphosphate (ATP), K^+ efflux, Ca^{2+} flux, lysosomal disruption, different misfolded proteins, pore-forming toxins, crystalline substances, and viral RNA (Martinon et al. 2006; Allen et al. 2009; Mariathasan et al. 2006; Hornung et al. 2008). After sensing the activating signal, the NLRP3 proteins are oligomerized through the homotypic interactions of the NACHT domain (Latz et al. 2013). The centrosomal NIMA-related kinase7 (NEK7) engages to form the active NLRP3 inflammasome and exposes the NLRP3-PYD domain to recruit ASC through a PYD-PYD interaction (Andreeva et al. 2021). Next, ASC forms prion-like aggregates and constitutes a platform to activate caspase-1 through proximity-induced self-cleavage between p20 and p10 of caspase-1 (Schmidt et al. 2016; Cai et al. 2014). Following the cleavage and activation of caspase-1, caspase-1 proteolytically cleaves the pro-IL-1β and IL-18 to yield mature IL-1β and IL-18 (Schmidt et al. 2016). Additionally, activated caspase-1 also cleaves and activates gasdermin D (GSDMD), which then translocates to the plasma membrane and forms pores that release IL-1β and IL-18 into the extracellular space. Meanwhile, the pore formation of GSDMD induces proinflammatory form of cell death known as pyroptosis (Shi et al. 2015). If NLRP3 is adequately present in cells, short-term priming with TLR stimulus is enough to render an active inflammasome upon activating signal independent of transcription (Juliana et al. 2012).

Except for the canonical NLRP3 inflammasome activation pathway, so-called noncanonical pathway is also triggered by distinct inflammatory caspases, murine caspase-11, and its human orthologs caspase-4 and caspase-5 (Kayagaki et al. 2011, 2015). Intracellular LPS directly activates caspase-11, which cleaves gasdermin D, forming a membrane pore that leads to cell death and potassium efflux. This ion efflux then functions as the activating signal, leading to NLRP3 inflammasome assembly.

9.2.2 Ubiquitination Mechanism

Ubiquitination is an important form of post-translational modification of proteins that has a pivotal role in various cellular functions. Ubiquitin (Ub) is a 76 amino acid protein that is usually conjugated to the lysine residue of substrate proteins by forming an isopeptide bond (Komander and Rape 2012). The addition of ubiquitin to a substrate is mediated by sequential catalyzation of three enzymes, namely, ubiquitin-activating enzyme (E1), ubiquitin-conjugating enzyme (E2), and ubiquitin-ligating enzyme (E3) (Wilkinson 1987; Ardley and Robinson 2005). The sequential addition of Ub is initiated with the recognition of the C-terminus of Ub by the E1 enzyme, which then forms a thioester bond. Through trans-thioesterification, Ub loaded on E1 is then transferred to the active cysteine site of the E2 enzyme. Finally, the Ub is linked to the substrate by the catalyzation of E3, forming an isopeptide bond between the C-terminal glycine of the Ub and the lysine residue of the substrate protein. E1 and E2 enzymes have 2 and approximately 40 members, respectively, whereas E3 has around 600 members. E3 ligases can be classified into three major categories: really interesting new gene (RING), homology to E6AP C terminus (HECT), and RING between RING (RBR). The RING family can simultaneously bind to both the E2-Ub thioester and the substrate and directly transfer Ub to the substrate. HECT and RBR function in a two-step reaction: first, Ub is transferred from the E2-Ub complex to the active-site cysteine in the E3, and then E3 mediates the transfer of Ub to the substrate and forms an isopeptide bond.

The C-terminal G75 and G76 residues of ubiquitin mediate the conjugation to a substrate's lysine (K) residue. Different types of ubiquitination have been identified depending on the number of ubiquitins and the conjugation manner. Ubiquitination can occur just with one ubiquitin molecule (monoubiquitination) but also repeatedly in the same protein at different sites (multi-monoubiquitination) (Komander and Rape 2012). Ubiquitination can also occur with multiple ubiquitin molecules at the same site, namely polyubiquitination. In this case, the first ubiquitin serves as the acceptor of additional ubiquitins that conjugate to each other at any one of their lysine residues (K6, K11, K27, K29, K33, K48, K63 or methionine 1 (M1, linear chains)) (Komander and Rape 2012). Hybrid polyubiquitin chains also exist, in which additional ubiquitins conjugate to each other through different K residues (Komander and Rape 2012). Interestingly, ubiquitination can exist on residues other than lysine, such as serine or threonine, which needs further elucidation (McDowell and Philpott 2013). The type of ubiquitin chains attached to the substrate is important for the target protein's fate, function, and regulation (Akutsu et al. 2016). K48-linked polyubiquitin chains have been well-studied for protein degradation by the proteasome, and K63-linked polyubiquitin chains can act as a scaffold for the recruitment of other proteins and mediate cellular signaling. Linear M1 polyubiquitin chains are also involved in signalling pathways. Other polyubiquitin chains have also been studied with different fucntions.

Ubiquitination can be removed by deubiquitinases (DUBs), which are essential for keeping the equilibrium of the ubiquitin system (Clague et al. 2013). Deubiquitinases are a family with ~100 members that can be divided into seven subfamilies. Six thiol-protease families are the ubiquitin-specific proteases (USP), ubiquitin C-terminal hydrolases (UCH), ovarian tumor domain-containing proteases (OTU), Machado–Joseph disease (MJD)/Josephin domain DUBs, MIU-containing novel DUB family (MINDY), and zinc finger-containing ubiquitin peptidase 1 (ZUFSP). In addition, there is one zinc-metalloprotease group, i.e., the JAB1/MPN/Mov34 metalloenzyme family (Clague et al. 2013; Kwasna et al. 2018; Abdul Rehman et al. 2016). DUBs can remove free ubiquitin from Ub precursors (de novo Ub synthesis) or edit ubiquitin chains from their substrates. DUBs cooperate with the E1-E2-E3 system to keep the balance of the ubiquitin system.

9.2.3 Regulation of NLRP3 Inflammasome by Ubiquitination

9.2.3.1 Ubiquitination of NLRP3

NLRP3 comprises three typical domains, PYD, NACHT, and LRR, and is the most widely investigated inflammasome sensor (Fig. 9.1). The ubiquitination of NLRP3 has been thoroughly studied, and various E3 ligases and DUBs have been reported to mediate this modification (Fig. 9.1). The dopamine (DA) neurotransmitter blocks NLRP3 inflammasome activation through the dopamine D1 receptor (DRD1). DRD1 signaling then inhibits NLRP3 inflammasome via a second messenger cyclic adenosine monophosphate (cAMP), which binds to NLRP3 and pro-

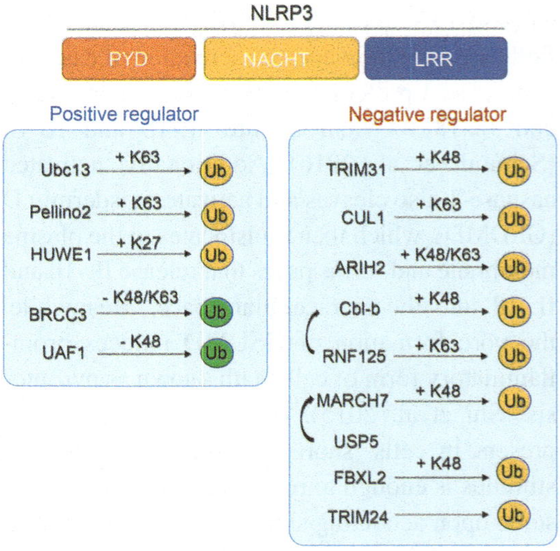

Fig. 9.1 The regulation of NLRP3 ubiquitination

motes ubiquitination and degradation via the E3 ligase MARCH7 (Yan et al. 2015), which interacts with the NBD or LRR domain of NLRP3 and catalyzes the K48-linked ubiquitination of NLRP3 for autophagy-dependent degradation (Yan et al. 2015). ARIH2 (Ariadne homolog 2), an E3 ubiquitin ligase, interacts with the NACHT domain of NLRP3 and mediates the negative regulation of NLRP3 inflammasome by inducing mixed K48- and K63-linked ubiquitination (Kawashima et al. 2017). The E3 ligase TRIM31 directly binds to NLRP3 and promotes K48-linked polyubiquitination and proteasomal degradation of NLRP3 (Song et al. 2016). The study proved that TRIM31 deficiency promotes NLRP3 inflammasome activation and affects dextran sodium sulfate (DSS)-induced colitis and alum-induced peritonitis in vivo (Song et al. 2016). In addition, a study found that SCF^{FBXL2} E3 ligase recognizes W73 on NLRP3 for interaction and targets K689 on NLRP3 for ubiquitination and degradation (Han et al. 2015).

The FBXL2 is regulated by FBXO2, an E3 ligase component induced by LPS treatment (Han et al. 2015). Han et al. utilized a unique small molecule inhibitor of FBXO3 to restore FBXL2 levels, decreasing NLRP3 protein levels and reducing the release of cellular IL-1β and IL-18 (Han et al. 2015). Moreover, cullin1 (CUL1), a key component of the Skp1-cullin1-F-box E3 ligase, catalyzes NLRP3 ubiquitination to inhibit its activation under normal physiological conditions, whereas it disassociates from NLRP3 to liberate the repressed NLRP3 inflammasome in response to inflammatory stimuli (Wan et al. 2019). The C-terminus of CUL1 interacts with NLRP3 and catalyzes the K63-linked ubiquitination, which regulates NLRP3 inflammasome activation without the effect of degradation (Wan et al. 2019). Interestingly, a sequential ubiquitination process for NLRP3 has been reported. E3 ubiquitin ligase RNF125 catalyzes K63-linked ubiquitination on the LRR domain of NLRP3, and then Cbl-b binds to the K63-linked ubiquitin chain and targets NLRP3 at K496 for K48-linked ubiquitination and proteasome-mediated degradation (Tang et al. 2020). The sequential action of RNF125 and Cbl-b restricts NLRP3 inflammasome activation, thus keeping the NLRP3 inflammasomes in check and restraining endotoxemia (Tang et al. 2020).

The E3 ligase Pellino2 has a dual role in NLRP3 regulation. On the one hand, Pellino2 promotes the ubiquitination of NLRP3 during the priming phase of activation, a step required for consequent inflammasome activation. Humphries et al. found that Pellino2 ubiquitinates IRAK1, blocks its interaction with NLRP3, and the function of IRAK1 to inhibit NLRP3 (Humphries et al. 2018).

HUWE1, an HECT E3 ubiquitin ligase, interacts with the NACHT domain of NLRP3 and catalyzes K27-linked polyubiquitination of NLRP3, which facilitates NLRP3 assembly, ASC speck formation, and caspase-1 activation (Guo et al. 2020). HUWE1 is also found to interact with AIM2 and NLRC4 (Guo et al. 2020). The E3 ligase TRIM24 is reported as a transcriptional factor negatively regulating inflammatory responses (Perez-Lloret et al. 2016). A recent study revealed that TRIM24 negatively regulates NLRP3/caspase-1/IL-1β-mediated pyroptosis and cell migration of human endometrial stromal cells (hESC). TRIM24 interacts with NLRP3, and the upregulation of TRIM24 facilitates the ubiquitination of NLRP3 in ectopic hESC (Hang et al. 2021). Hang et al. suggested that TRIM24 may participate in the progression of endometriosis by regulating the ubiquitination of NLRP3 and subsequent NLRP3/caspase-1/IL-1β-mediated pyroptotic pathway (Hang et al. 2021).

There are also some indirect E3 ligases regulating NLRP3 inflammasome activation. TRAF6 has been proven to be associated with the non-transcriptional priming of the NLRP3 inflammasome via its E3 ligase activity (Xing et al. 2017). TRAF6 is well-known for its role in NF-κB-mediated transcription of NLRP3. Xing and colleagues discovered that TRAF6 facilitates NLRP3 oligomerization and an ASC-NLRP3 interaction (Xing et al. 2017). However, NLRP3 is not the direct substrate of TRAF6 (Xing et al. 2017). The TRIM33 has also been reported as an indirect E3 ligase regulating NLRP3 inflammasome activation. The interaction between the cytosolic dsRNA sensor DHX33 and NLRP3 triggers inflammasome activation with the help of TRIM33 (Weng et al. 2014). TRIM33 can initiate the K63-linked ubiquitination of DHX33 at K218, which leads to the formation of the DHX33-NLRP3 inflammasome complex and promotes the activation of the NLRP3 inflammasome (Weng et al. 2014). Moreover, the RBR E3 ligase Parkin has been reported to regulate NLRP3 inflammasome activation. Parkin induces the ubiquitin-modifying enzyme anti-apoptotic signaling protein 20 (A20), which suppresses NF-κB activation and blocks the priming process of the NLRP3 inflammasome (Mouton-Liger et al. 2018). Parkin negatively regulates NLRP3 inflammasome activation since Parkin-deficient cells have increased NLRP3 activation (Mouton-Liger et al. 2018).

Several DUBs have been reported as important regulators for NLRP3. Py et al. (2013) reported that G5, a small molecule inhibitor of DUBs, inhibits the deubiquitination and activation of NLRP3, providing potent evidence that NLRP3 activity is repressed by its ubiquitination. Next, they screened an expression library for DUBs, identifying the DUB BRCC3 as a critical regulator of NLRP3 activity by promoting its deubiquitination and characterized NLRP3 as a substrate for the cytosolic BRCC3-containing BRISC complex (Py et al. 2013). Furthermore, ABRO1, another BRISC deubiquitinase complex subunit, has been reported as an essential factor for optimal NLRP3 activation (Ren et al. 2019). Ren et al. showed that upon LPS stimulation, S194 phosphorylation-dependent ABRO1 binding to NLRP3 recruits BRISC to remove the K63-linked polyubiquitin chain (Ren et al. 2019). DUBs USP7 and USP47 can positively regulate inflammasome activation independently of transcription, whereas their deubiquiting targets have not been revealed (Palazon-Riquelme et al. 2018). Ubiquitin specific peptidase 1 (USP1)-associated factor 1 (UAF1) forms three DUB enzyme complexes, namely, UAF1/USP1, UAF1/USP12, and USP1/USP46 (Murai et al. 2011). Song et al. reported that UAF1 deubiquitinase complexes function as a positive regulator for NLRP3 inflammasome activation in vitro and in vivo (Song et al. 2020). The UAF1/USP1 deubiquitinase complex removes K48-linked ubiquitination of NLRP3, inhibiting proteasomal degradation of NLRP3 and promoting NLRP3 inflammasome activation (Song et al. 2020). The same study also proved that the UAF1/USP12 and UAF1/USP46 complexes promote NF-κB activation and increase NLRP3 and IL-1β levels by regulating p65 expression (Song et al. 2020). Recently, Cai et al. reported that the DUB USP5 restrains the activation of the NLRP3 inflammasome independent of its catalytic activity (Cai et al. 2021). USP5 promotes the K48-linked polyubiquitination of NLRP3 and facilitates its degradation through the autophagy-lysosomal pathway by recruiting the E3 ligase MARCH7 (Cai et al. 2021). USP5 negatively regulates the activation of NLRP3 inflammasome in cells and alum-induced peritonitis in vivo (Cai et al. 2021).

Only one E2 ubiquitin-conjugating enzyme, Ubc13, has been reported to mediate NLRP3 inflammasome activation. Ubc13 associates with NLRP3, catalyzes its K63-linked polyubiquitination at K565 and K687 and promotes NLRP3 inflammasome activation (Ni et al. 2021).

9.2.3.2 Ubiquitination of ASC

ASC is the central adaptor for inflammasomes composed of PYD and CARD domains and is also tightly regulated by ubiquitination (Fig. 9.2). Shi et al. reported that K63-linked polyubiquitination of ASC recruits the autophagic adaptor p62 and leads to the degradation by autophagy (Shi et al. 2012). This mechanism keeps both AIM2 and NLRP3 inflammasomes in check (Shi et al. 2012). Similarly, as a treatment to amelio-

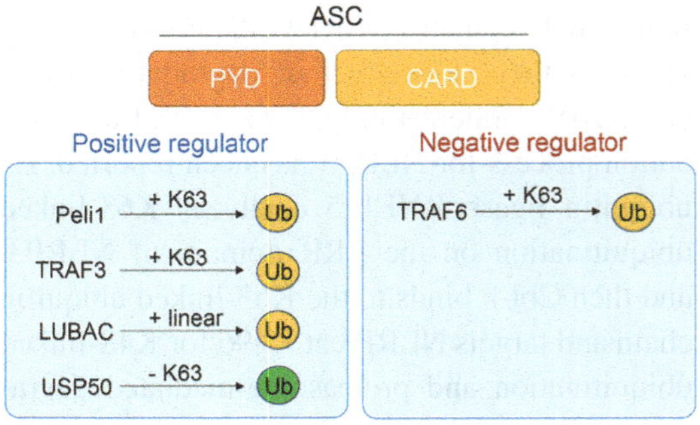

Fig. 9.2 The regulation of ASC ubiquitination

rate burn wound progression, far infrared can induce the TRAF6-mediated ASC K63-linked polyubiquitination and lead to autophagy degradation (Chiu et al. 2016). The ubiquitination of ASC also serves as a response to pathogen infection (Lopez-Castejon 2020). Severe acute respiratory syndrome coronavirus ORF3a protein promotes the NLRP3 inflammasome by regulating TRAF3-dependent ubiquitination of ASC (Siu et al. 2019); ORF3a interacts with TRAF3 and ASC and facilitates ASC speck formation (Siu et al. 2019). During SARS-CoV infection or ORF3a overexpression, TRAF3-dependent K63-linked ubiquitination of ASC is more significant (Siu et al. 2019). Meanwhile, vesicular stomatitis virus (VSV) infection induces MAVS-mediated recruitment of TRAF3 to ASC and enhances the K63-linked ubiquitination of ASC at K174, which is crucial for ASC speck formation and inflammasome activation (Guan et al. 2015). A deficiency of MAVS or TRAF3 reduces ASC ubiquitination and cytosolic aggregation, leading to reduced inflammasome response upon VSV infection (Guan et al. 2015). The K63-linked ubiquitination reported by Shi et al. (2012), Chiu et al. (2016), Siu et al. (2019), and Guan et al. (2015) show the opposite effect for inflammasome activation, suggesting that the same type of ubiquitin chains on ASC may exhibit different functions, which may be due to the different ubiquitin-conjugation sites. Linear polyubiquitination of ASC by the linear ubiquitin assembly complex (LUBAC) has also been reported. LUBAC is formed by HOIP, HOIL, and SHARPIN (SHANK-associated RH domain interactor) and regulates the addition of linear chains to the methionine 1 (M1) of ASC (Rodgers et al. 2014). Macrophages deficient in HOIL-1 show reduced IL-1β secretion after inflammasome activation induced by NLRP3 activators, suggesting that the linear polyubiquitination of ASC promotes inflammasome activation (Rodgers et al. 2014). Recently, Zhang et al. reported that E3 ligase Peli1 conjugates K63-linked ubiquitin chains to K55 of ASC and strengthens ASC-NLRP3 interaction and ASC oligomerization, thereby contributing to inflammasome activation (Zhang et al. 2021). Peli1 deficiency impairs the ubiquitination of ASC and inhibits inflammasome activation in vitro and in vivo (Zhang et al. 2021).

To date, only one DUB has been reported to regulate ASC. Lee et al. found that the DUB, USP50, deubiquitinates K63-linked polyubiquitin chains of ASC and promotes NLRP3 inflammasome activation. USP50-deficient cells show impaired inflammasome activation (Lee et al. 2017). However, USP50 is considered catalytically inactive because it lacks one of the canonical residues required for catalysis (Quesada et al. 2004). Therefore, the underlying mechanism of USP50 regulating ASC still needs to be further explored.

9.2.3.3 Ubiquitination of Caspase-1

The E3 ligase IAPs (inhibitors of apoptosis proteins) regulate procaspase-1 activation with both positive and negative regulatory functions. Labbe et al. found that IAPs induce K63-linked polyubiquitination of caspase-1; in vivo experiments with cIAP1-, cIAP2-, or XIAP-deficient mice show attenuated caspase-1 activation (Labbe et al. 2011). In contrast, Vince et al. reported that genetic depletion of cIAP1, cIAP2, and XIAP activates caspase-1 and pro-IL-1β cleavage (Vince et al. 2012). The deficiency of IAPs produces ROS to activate both the NLRP3 inflammasome and caspase-8 (Vince et al. 2012). Given the positive and negative regulatory functions of IAPs in caspase-1, the detailed mechanism still needs further thorough validation. Except for IAPs, the LUBAC member SHARPIN has also been found to regulate caspase-1. Douglas et al. reported that SHARPIN-deficient mice present an over-activation of the inflammasome in the skin and proposed that this is mediated by caspase-1 (Douglas et al. 2015). However, SHARPIN-deficient bone marrow-derived macrophages show a reduction in inflammasome activation compared with wild-type cells, possibly due to the difference between cell types (Douglas et al. 2015). For sepsis in the lungs, SHARPIN interacts with caspase-1 to prevent its activation and the following cleavage and maturation of IL-1β and IL-18 in a LUBAC-independent manner (Nastase et al. 2016). The above reports highlight that cellular environments may obviously affect the functions of E3 ligase. The ZIKA virus infection can also activate inflammasome by manipulating the caspase-1 ubiquitination (Zheng et al. 2018). NS1, a nonstructural protein of the ZIKA virus, recruits USP8 to remove K11-linked ubiquitin chains at K134 of caspase-1 and strengthens the stability of caspase-1 (Zheng et al. 2018).

9.2.3.4 Ubiquitination of IL-1β

Ubiquitination is also a key factor for regulating the function of IL-1β. Pro-IL-1β can be ubiquitinated in BMDC and BMDM (Ainscough et al. 2014). The ubiquitination of IL-1β can lead to proteasomal degradation, while proteasome inhibitors prevent this process (Ainscough et al. 2014). The E2 UBE2L3 regulates pro-IL-1β ubiquitination and leads to its degradation by the proteasome (Eldridge et al. 2017). The activation of inflammasome causes UBE2L3 depletion, which further leads to the reduction of pro-IL-1β degradation and promotes the activation of inflammasome (Eldridge et al. 2017). The regulatory loop between UBE2L3 and caspase-1 keeps the inflammasome activation in check.

Duong et al. employed mass spectrometric and mutational analyses and identified that K133 of pro-IL-1β is a physiological ubiquitination site that supports its processing (Duong et al. 2015). Another study found that A20 deficiency enhances the ubiquitination of pro-IL-1β in a RIPK3-dependent manner (Duong et al. 2015). Consistently, Zhang et al. found that DUB POH1 interacts with pro-IL-1β, removes its K63-linked polyubiquitin chains, and reduces the efficacy of pro-IL-1β cleavage (Zhang et al. 2018).

9.2.4 Control of Other Inflammasomes by the Ubiquitin System

In addition to NLRP3 inflammasome, ubiquitination also affects the activation of other inflammasomes. NLRP1 inflammasome is the first NLR-mediated inflammasome discovered, and its best-characterized stimuli are anthrax lethal toxins (Le-Tx) (Mitchell et al. 2019). Le-Tx can directly cleave NLRP1 to degrade the NLRP1 N-terminus polypeptide and free the NLRP1 C-terminus polypeptide containing the CARD domain. The free NLRP1 C-terminus polypeptide can further process and activate caspase-1. Xu et al. found that E3 ligase UBR2 cooperates with an E2 ubiquitin-conjugating enzyme UBE2O and ubiquitinates NLRP1 to target it for proteasomal degradation in response to Le-Tx treatment (Xu et al. 2019). UBR2-mediated degradation of Le-Tx-cleaved NLRP1 induces the release of the CARD domain for subsequent caspase-1 activation (Xu et al. 2019). In addition, some other effectors that mediate NLRP1 proteasomal degradation of the N-terminus, e.g., the Shigella E3 ligase IpaH7.8 or inhibition of serine dipeptidase DPP8 and DPP9, may also trigger NLRP1 inflammasome activation (Chui et al. 2019; Sandstrom et al. 2019).

AIM2 has also been reported to be regulated by ubiquitination. Upon DNA virus infection, E3 ligase TRIM11 binds with AIM2 via its PS domain and undergoes auto-polyubiquitination at K458 to promote an association between TRIM11 and the autophagic cargo receptor p62 to mediate AIM2 degradation via autophagy (Liu et al. 2016). TRIM11, as a secondary receptor, can deliver AIM2 to the autophagosomes for degradation in a p62-dependent manner and is a negative regulator of AIM2 inflammasome (Liu et al. 2016). A recent study found that USP21 regulates AIM2 inflammasome; upon DNA stimulation, USP21 increases AIM2 protein stability through its deubiquitination activity (Hong et al. 2021). Also, USP21-mediated deubiquitination of AIM2 is required to assemble the AIM2 inflammasome (Hong et al. 2021). Depletion of USP21 does not affect the DNA-binding capacity of AIM2 but represses the formation of the AIM2-ASC complex (Hong et al. 2021). HUWE1 is also reported for regulating AIM2 ubiquitination, whereas it affects the NLRP3 and NLRC4 at the same time (Guo et al. 2020).

Ubiquitination of NLRP7 in resting conditions mediates its recruitment to the endolysosome for degradation. Bednash et al. showed that the DUB STAMPBP (or AMSH) impedes NALP7 trafficking to lysosomes to increase NALP7 abundance through deubiquitinating NLRP7 and makes it available for the formation of active inflammasome (Bednash et al. 2017).

The activation of NLRC4 inflammasome can also be regulated by ubiquitination. SUG-1 (protease regulatory subunit 8 homolog) ubiquitinates NLRC4 and allows it to interact with caspase-8, which triggers its activation (Kumar et al. 2010). One of the point mutations in NLRC4 that causes autoinflammatory syndromes (H443P) enhances the interaction between SUG-1 and caspase-8 (Raghawan et al. 2017). This mechanism elucidates how this mutation leads to enhanced inflammation.

9.3 Conclusion and Perspective

Over the past years, many studies have focused on regulating inflammasome by ubiquitination. In vitro and in vivo studies taking advantage of the use of cellular and mouse models deficient in particular components of the ubiquitin system, e.g., E3 and DUB, as well as of the tools such as selective ubiquitin chain antibodies or specific ubiquitin probes revealed the underlying mechanism of ubiquitination regulating inflammasome. However, there are still existing issues that need to be resolved. For example, the TRAF3-mediated ASC K63-linked ubiquitination promotes inflammasome activation, while K63-linked ubiquitination is also reported to mediate the autophagy of ASC. The same type of ubiquitin chains exhibits the opposite effect on ASC. Thus, the exact sites or types of ubiquitin chains involved in the specific inflammasome component ubiquitination need to be further elucidated. Also, the ubiquitination regulation of the activation of the noncanonical or alternative NLRP3 activation pathways is still not fully understood.

It is of great interest to target ubiquitination for therapeutic purposes for inflammasome-related disease, especially after the success of the proteasome inhibitor bortezomib in myeloma treatment (Field-Smith et al. 2006). However, further studies should elucidate how E1-E2-E3 and DUBs are regulated in different cell types or in response to different inflammatory stimulations are required. This area of research is still growing, and extensive efforts are needed for breakthrough discoveries regarding the ubiquitin system regulating inflammasome.

References

Abdul Rehman SA, Kristariyanto YA, Choi SY et al (2016) MINDY-1 is a member of an evolutionarily conserved and structurally distinct new family of deubiquitinating enzymes. Mol Cell 63(1):146–155. https://doi.org/10.1016/j.molcel.2016.05.009

Ainscough JS, Frank Gerberick G, Zahedi-Nejad M et al (2014) Dendritic cell IL-1alpha and IL-1beta are polyubiquitinated and degraded by the proteasome. J Biol Chem 289(51):35582–35592. https://doi.org/10.1074/jbc.M114.595686

Akutsu M, Dikic I, Bremm A (2016) Ubiquitin chain diversity at a glance. J Cell Sci 129(5):875–880. https://doi.org/10.1242/jcs.183954

Allen IC, Scull MA, Moore CB et al (2009) The NLRP3 inflammasome mediates in vivo innate immunity to influenza A virus through recognition of viral RNA. Immunity 30(4):556–565. https://doi.org/10.1016/j.immuni.2009.02.005

Andreeva L, David L, Rawson S et al (2021) NLRP3 cages revealed by full-length mouse NLRP3 structure control pathway activation. Cell 184(26):6299–6312 e22. https://doi.org/10.1016/j.cell.2021.11.011

Ardley HC, Robinson PA (2005) E3 ubiquitin ligases. Essays Biochem 41:15–30. https://doi.org/10.1042/EB0410015

Bauernfeind FG, Horvath G, Stutz A et al (2009) Cutting edge: NF-kappaB activating pattern recognition and cytokine receptors license NLRP3 inflammasome activation by regulating NLRP3 expression. J Immunol 183(2):787–791. https://doi.org/10.4049/jimmunol.0901363

Bednash JS, Weathington N, Londino J et al (2017) Targeting the deubiquitinase STAMBP inhibits NALP7 inflammasome activity. Nat Commun 8:15203. https://doi.org/10.1038/ncomms15203

Broz P, Dixit VM (2016) Inflammasomes: mechanism of assembly, regulation and signalling. Nat Rev Immunol 16(7):407–420. https://doi.org/10.1038/nri.2016.58

Cai X, Chen J, Xu H et al (2014) Prion-like polymerization underlies signal transduction in antiviral immune defense and inflammasome activation. Cell 156(6):1207–1222. https://doi.org/10.1016/j.cell.2014.01.063

Cai B, Zhao J, Zhang Y et al (2021) USP5 attenuates NLRP3 inflammasome activation by promoting autophagic degradation of NLRP3. Autophagy 18(5):990–1004. https://doi.org/10.1080/15548627.2021.1965426

Chaugule VK, Walden H (2016) Specificity and disease in the ubiquitin system. Biochem Soc Trans 44(1):212–227. https://doi.org/10.1042/BST20150209
Chiu HW, Chen CH, Chang JN, Chen CH, Hsu YH (2016) Far-infrared promotes burn wound healing by suppressing NLRP3 inflammasome caused by enhanced autophagy. J Mol Med 94(7):809–819. https://doi.org/10.1007/s00109-016-1389-0
Chui AJ, Okondo MC, Rao SD et al (2019) N-terminal degradation activates the NLRP1B inflammasome. Science 364(6435):82–85. https://doi.org/10.1126/science.aau1208
Clague MJ, Barsukov I, Coulson JM, Liu H, Rigden DJ, Urbe S (2013) Deubiquitylases from genes to organism. Physiol Rev 93(3):1289–1315. https://doi.org/10.1152/physrev.00002.2013
Cornut M, Bourdonnay E, Henry T (2020) Transcriptional regulation of inflammasomes. Int J Mol Sci 21(21):8087. https://doi.org/10.3390/ijms21218087
Dinarello CA (2009) Immunological and inflammatory functions of the interleukin-1 family. Annu Rev Immunol 27:519–550. https://doi.org/10.1146/annurev.immunol.021908.132612
Douglas T, Champagne C, Morizot A, Lapointe JM, Saleh M (2015) The inflammatory caspases-1 and -11 mediate the pathogenesis of dermatitis in sharpin-deficient mice. J Immunol 195(5):2365–2373. https://doi.org/10.4049/jimmunol.1500542
Duong BH, Onizawa M, Oses-Prieto JA et al (2015) A20 restricts ubiquitination of pro-interleukin-1beta protein complexes and suppresses NLRP3 inflammasome activity. Immunity 42(1):55–67. https://doi.org/10.1016/j.immuni.2014.12.031
Eldridge MJG, Sanchez-Garrido J, Hoben GF, Goddard PJ, Shenoy AR (2017) The atypical ubiquitin E2 conjugase UBE2L3 is an indirect caspase-1 target and controls IL-1beta secretion by inflammasomes. Cell Rep 18(5):1285–1297. https://doi.org/10.1016/j.celrep.2017.01.015
Field-Smith A, Morgan GJ, Davies FE (2006) Bortezomib (velcadetrade mark) in the treatment of multiple myeloma. Ther Clin Risk Manag 2(3):271–279. https://doi.org/10.2147/tcrm.2006.2.3.271
Franchi L, Munoz-Planillo R, Nunez G (2012) Sensing and reacting to microbes through the inflammasomes. Nat Immunol 13(4):325–332. https://doi.org/10.1038/ni.2231
Fulda S, Rajalingam K, Dikic I (2012) Ubiquitylation in immune disorders and cancer: from molecular mechanisms to therapeutic implications. EMBO Mol Med 4(7):545–556. https://doi.org/10.1002/emmm.201100707
Guan K, Wei C, Zheng Z et al (2015) MAVS promotes inflammasome activation by targeting ASC for K63-linked ubiquitination via the E3 ligase TRAF3. J Immunol 194(10):4880–4890. https://doi.org/10.4049/jimmunol.1402851
Guo Y, Li L, Xu T et al (2020) HUWE1 mediates inflammasome activation and promotes host defense against bacterial infection. J Clin Invest 130(12):6301–6316. https://doi.org/10.1172/JCI138234
Han S, Lear TB, Jerome JA et al (2015) Lipopolysaccharide primes the NALP3 inflammasome by inhibiting its ubiquitination and degradation mediated by the SCFFBXL2 E3 ligase. J Biol Chem 290(29):18124–18133. https://doi.org/10.1074/jbc.M115.645549
Hang Y, Tan L, Chen Q, Liu Q, Jin Y (2021) E3 ubiquitin ligase TRIM24 deficiency promotes NLRP3/caspase-1/IL-1beta-mediated pyroptosis in endometriosis. Cell Biol Int 45(7):1561–1570. https://doi.org/10.1002/cbin.11592
Hong Y, Lee SO, Oh C et al (2021) USP21 deubiquitinase regulates AIM2 inflammasome activation. J Immunol 207(7):1926–1936. https://doi.org/10.4049/jimmunol.2100449
Hornung V, Bauernfeind F, Halle A et al (2008) Silica crystals and aluminum salts activate the NALP3 inflammasome through phagosomal destabilization. Nat Immunol 9(8):847–856. https://doi.org/10.1038/ni.1631
Humphries F, Bergin R, Jackson R et al (2018) The E3 ubiquitin ligase Pellino2 mediates priming of the NLRP3 inflammasome. Nat Commun 9(1):1560. https://doi.org/10.1038/s41467-018-03669-z
Juliana C, Fernandes-Alnemri T, Kang S, Farias A, Qin F, Alnemri ES (2012) Non-transcriptional priming and deubiquitination regulate NLRP3 inflammasome activation. J Biol Chem 287(43):36617–36622. https://doi.org/10.1074/jbc.M112.407130
Kawashima A, Karasawa T, Tago K et al (2017) ARIH2 ubiquitinates NLRP3 and negatively regulates NLRP3 inflammasome activation in macrophages. J Immunol 199(10):3614–3622. https://doi.org/10.4049/jimmunol.1700184
Kayagaki N, Warming S, Lamkanfi M et al (2011) Non-canonical inflammasome activation targets caspase-11. Nature 479(7371):117–121. https://doi.org/10.1038/nature10558
Kayagaki N, Stowe IB, Lee BL et al (2015) Caspase-11 cleaves gasdermin D for non-canonical inflammasome signalling. Nature 526(7575):666–671. https://doi.org/10.1038/nature15541
Kinney JW, Bemiller SM, Murtishaw AS, Leisgang AM, Salazar AM, Lamb BT (2018) Inflammation as a central mechanism in Alzheimer's disease. Alzheimers Dement 4:575–590. https://doi.org/10.1016/j.trci.2018.06.014
Komander D, Rape M (2012) The ubiquitin code. Annu Rev Biochem 81:203–229. https://doi.org/10.1146/annurev-biochem-060310-170328
Kumar Y, Radha V, Swarup G (2010) Interaction with Sug1 enables Ipaf ubiquitination leading to caspase 8 activation and cell death. Biochem J 427(1):91–104. https://doi.org/10.1042/BJ20091349
Kwasna D, Abdul Rehman SA, Natarajan J et al (2018) Discovery and characterization of ZUFSP/ZUP1, a distinct deubiquitinase class important for genome

stability. Mol Cell 70(1):150–164 e6. https://doi.org/10.1016/j.molcel.2018.02.023

Labbe K, McIntire CR, Doiron K, Leblanc PM, Saleh M (2011) Cellular inhibitors of apoptosis proteins cIAP1 and cIAP2 are required for efficient caspase-1 activation by the inflammasome. Immunity 35(6):897–907. https://doi.org/10.1016/j.immuni.2011.10.016

Lamkanfi M, Dixit VM (2014) Mechanisms and functions of inflammasomes. Cell 157(5):1013–1022. https://doi.org/10.1016/j.cell.2014.04.007

Latz E, Xiao TS, Stutz A (2013) Activation and regulation of the inflammasomes. Nat Rev Immunol 13(6):397–411. https://doi.org/10.1038/nri3452

Lee JY, Seo D, You J et al (2017) The deubiquitinating enzyme, ubiquitin-specific peptidase 50, regulates inflammasome activation by targeting the ASC adaptor protein. FEBS Lett 591(3):479–490. https://doi.org/10.1002/1873-3468.12558

Liu T, Tang Q, Liu K et al (2016) TRIM11 suppresses AIM2 inflammasome by degrading AIM2 via p62-dependent selective autophagy. Cell Rep 16(7):1988–2002. https://doi.org/10.1016/j.celrep.2016.07.019

Lopez-Castejon G (2020) Control of the inflammasome by the ubiquitin system. FEBS J 287(1):11–26. https://doi.org/10.1111/febs.15118

Man SM, Kanneganti TD (2015) Regulation of inflammasome activation. Immunol Rev 265(1):6–21. https://doi.org/10.1111/imr.12296

Mariathasan S, Weiss DS, Newton K et al (2006) Cryopyrin activates the inflammasome in response to toxins and ATP. Nature 440(7081):228–232. https://doi.org/10.1038/nature04515

Martinon F, Burns K, Tschopp J (2002) The inflammasome: a molecular platform triggering activation of inflammatory caspases and processing of proIL-beta. Mol Cell 10(2):417–426. https://doi.org/10.1016/s1097-2765(02)00599-3

Martinon F, Petrilli V, Mayor A, Tardivel A, Tschopp J (2006) Gout-associated uric acid crystals activate the NALP3 inflammasome. Nature 440(7081):237–241. https://doi.org/10.1038/nature04516

McDowell GS, Philpott A (2013) Non-canonical ubiquitylation: mechanisms and consequences. Int J Biochem Cell Biol 45(8):1833–1842. https://doi.org/10.1016/j.biocel.2013.05.026

Medzhitov R (2008) Origin and physiological roles of inflammation. Nature 454(7203):428–435. https://doi.org/10.1038/nature07201

Miao EA, Rajan JV, Aderem A (2011) Caspase-1-induced pyroptotic cell death. Immunol Rev 243(1):206–214. https://doi.org/10.1111/j.1600-065X.2011.01044.x

Mitchell PS, Sandstrom A, Vance RE (2019) The NLRP1 inflammasome: new mechanistic insights and unresolved mysteries. Curr Opin Immunol 60:37–45. https://doi.org/10.1016/j.coi.2019.04.015

Mouton-Liger F, Rosazza T, Sepulveda-Diaz J et al (2018) Parkin deficiency modulates NLRP3 inflammasome activation by attenuating an A20-dependent negative feedback loop. Glia 66(8):1736–1751. https://doi.org/10.1002/glia.23337

Murai J, Yang K, Dejsuphong D, Hirota K, Takeda S, D'Andrea AD (2011) The USP1/UAF1 complex promotes double-strand break repair through homologous recombination. Mol Cell Biol 31(12):2462–2469. https://doi.org/10.1128/MCB.05058-11

Nastase MV, Zeng-Brouwers J, Frey H et al (2016) An essential role for SHARPIN in the regulation of caspase 1 activity in sepsis. Am J Pathol 186(5):1206–1220. https://doi.org/10.1016/j.ajpath.2015.12.026

Ni J, Guan C, Liu H et al (2021) Ubc13 promotes K63-linked polyubiquitination of NLRP3 to activate inflammasome. J Immunol 206(10):2376–2385. https://doi.org/10.4049/jimmunol.2001178

Palazon-Riquelme P, Worboys JD, Green J et al (2018) USP7 and USP47 deubiquitinases regulate NLRP3 inflammasome activation. EMBO Rep 19(10):44766. https://doi.org/10.15252/embr.201744766

Perez-Lloret J, Okoye IS, Guidi R et al (2016) T-cell-intrinsic Tif1alpha/Trim24 regulates IL-1R expression on TH2 cells and TH2 cell-mediated airway allergy. Proc Natl Acad Sci USA 113(5):568–576. https://doi.org/10.1073/pnas.1522287113

Py BF, Kim MS, Vakifahmetoglu-Norberg H, Yuan J (2013) Deubiquitination of NLRP3 by BRCC3 critically regulates inflammasome activity. Mol Cell 49(2):331–338. https://doi.org/10.1016/j.molcel.2012.11.009

Quesada V, Diaz-Perales A, Gutierrez-Fernandez A, Garabaya C, Cal S, Lopez-Otin C (2004) Cloning and enzymatic analysis of 22 novel human ubiquitin-specific proteases. Biochem Biophys Res Commun 314(1):54–62. https://doi.org/10.1016/j.bbrc.2003.12.050

Raghawan AK, Sripada A, Gopinath G et al (2017) A disease-associated mutant of NLRC4 shows enhanced interaction with SUG1 leading to constitutive FADD-dependent caspase-8 activation and cell death. J Biol Chem 292(4):1218–1230. https://doi.org/10.1074/jbc.M116.763979

Ren G, Zhang X, Xiao Y et al (2019) ABRO1 promotes NLRP3 inflammasome activation through regulation of NLRP3 deubiquitination. EMBO J 38(6):100376. https://doi.org/10.15252/embj.2018100376

Rodgers MA, Bowman JW, Fujita H et al (2014) The linear ubiquitin assembly complex (LUBAC) is essential for NLRP3 inflammasome activation. J Exp Med 211(7):1333–1347. https://doi.org/10.1084/jem.20132486

Sandstrom A, Mitchell PS, Goers L, Mu EW, Lesser CF, Vance RE (2019) Functional degradation: a mechanism of NLRP1 inflammasome activation by diverse pathogen enzymes. Science 364(6435):1330. https://doi.org/10.1126/science.aau1330

Schmidt FI, Lu A, Chen JW et al (2016) A single domain antibody fragment that recognizes the adaptor ASC defines the role of ASC domains in inflammasome assembly. J Exp Med 213(5):771–790. https://doi.org/10.1084/jem.20151790

Shi CS, Shenderov K, Huang NN et al (2012) Activation of autophagy by inflammatory signals limits IL-1beta production by targeting ubiquitinated inflammasomes for destruction. Nat Immunol 13(3):255–263. https://doi.org/10.1038/ni.2215

Shi J, Zhao Y, Wang K et al (2015) Cleavage of GSDMD by inflammatory caspases determines pyroptotic cell death. Nature 526(7575):660–665. https://doi.org/10.1038/nature15514

Siu KL, Yuen KS, Castano-Rodriguez C et al (2019) Severe acute respiratory syndrome coronavirus ORF3a protein activates the NLRP3 inflammasome by promoting TRAF3-dependent ubiquitination of ASC. FASEB J 33(8):8865–8877. https://doi.org/10.1096/fj.201802418R

Slocum C, Kramer C, Genco CA (2016) Immune dysregulation mediated by the oral microbiome: potential link to chronic inflammation and atherosclerosis. J Intern Med 280(1):114–128. https://doi.org/10.1111/joim.12476

Song H, Liu B, Huai W et al (2016) The E3 ubiquitin ligase TRIM31 attenuates NLRP3 inflammasome activation by promoting proteasomal degradation of NLRP3. Nat Commun 7:13727. https://doi.org/10.1038/ncomms13727

Song H, Zhao C, Yu Z et al (2020) UAF1 deubiquitinase complexes facilitate NLRP3 inflammasome activation by promoting NLRP3 expression. Nat Commun 11(1):6042. https://doi.org/10.1038/s41467-020-19939-8

Strowig T, Henao-Mejia J, Elinav E, Flavell R (2012) Inflammasomes in health and disease. Nature 481(7381):278–286. https://doi.org/10.1038/nature10759

Tang J, Tu S, Lin G et al (2020) Sequential ubiquitination of NLRP3 by RNF125 and Cbl-b limits inflammasome activation and endotoxemia. J Exp Med 217(4):2091. https://doi.org/10.1084/jem.20182091

Tsalamandris S, Antonopoulos AS, Oikonomou E et al (2019) The role of inflammation in diabetes: current concepts and future perspectives. Eur Cardiol 14(1):50–59. https://doi.org/10.15420/ecr.2018.33.1

Vince JE, Wong WW, Gentle I et al (2012) Inhibitor of apoptosis proteins limit RIP3 kinase-dependent interleukin-1 activation. Immunity 36(2):215–227. https://doi.org/10.1016/j.immuni.2012.01.012

Wan P, Zhang Q, Liu W et al (2019) Cullin1 binds and promotes NLRP3 ubiquitination to repress systematic inflammasome activation. FASEB J 33(4):5793–5807. https://doi.org/10.1096/fj.201801681R

Weng L, Mitoma H, Trichot C et al (2014) The E3 ubiquitin ligase tripartite motif 33 is essential for cytosolic RNA-induced NLRP3 inflammasome activation. J Immunol 193(7):3676–3682. https://doi.org/10.4049/jimmunol.1401448

Wilkinson KD (1987) Protein ubiquitination: a regulatory post-translational modification. Anticancer Drug Des 2(2):211–229

Xing Y, Yao X, Li H et al (2017) Cutting edge: TRAF6 mediates TLR/IL-1R signaling-induced nontranscriptional priming of the NLRP3 inflammasome. J Immunol 199(5):1561–1566. https://doi.org/10.4049/jimmunol.1700175

Xu H, Shi J, Gao H et al (2019) The N-end rule ubiquitin ligase UBR2 mediates NLRP1B inflammasome activation by anthrax lethal toxin. EMBO J 38(13):e101996. https://doi.org/10.15252/embj.2019101996

Yan Y, Jiang W, Liu L et al (2015) Dopamine controls systemic inflammation through inhibition of NLRP3 inflammasome. Cell 160(1-2):62–73. https://doi.org/10.1016/j.cell.2014.11.047

Yeung YT, Aziz F, Guerrero-Castilla A, Arguelles S (2018) Signaling pathways in inflammation and anti-inflammatory therapies. Curr Pharm Des 24(14):1449–1484. https://doi.org/10.2174/1381612824666180327165604

Zhang L, Liu Y, Wang B et al (2018) POH1 deubiquitinates pro-interleukin-1beta and restricts inflammasome activity. Nat Commun 9(1):4225. https://doi.org/10.1038/s41467-018-06455-z

Zhang L, Ko CJ, Li Y et al (2021) Peli1 facilitates NLRP3 inflammasome activation by mediating ASC ubiquitination. Cell Rep 37(4):109904. https://doi.org/10.1016/j.celrep.2021.109904

Zhao J, Wei K, Jiang P et al (2022) Inflammatory response to regulated cell death in gout and its functional implications. Front Immunol 13:888306. https://doi.org/10.3389/fimmu.2022.888306

Zheng Y, Liu Q, Wu Y et al (2018) Zika virus elicits inflammation to evade antiviral response by cleaving cGAS via NS1-caspase-1 axis. EMBO J 37(18):9947. https://doi.org/10.15252/embj.201899347

The Function of Ubiquitination in T-Cell Development

10

Zhengcan Peng, Huiyuan Zhang, and Hongbo Hu

Abstract

Thymus is an important primary lymphoid organ for T cell development. After T-lineage commitment, the early thymic progenitors (ETPs) develop into CD4−CD8− (DN), CD4+CD8+ (DP) and further CD4+ SP or CD8+ SP T cells. Under the help of thymic epithelial cells (TEC), dendritic cell (DC), macrophage, and B cells, ETPs undergo proliferation, T cell receptor (TCR) rearrangement, β-selection, positive selection, and negative selection, and thus leading to the generation of T cells that are diverse repertoire immunocompetent but not self-reactive. Additionally, some self-reactive thymocytes give rise to Treg under the help of TEC and DC. The regulation of T cell development is complicated. As a post-translational modification, ubiquitination regulates signal transduction in diverse biological processes. Ubiquitination functions in T cell development through regulating key signal pathway or maturation and function of related cells. In this review, the regulation of T cell development by ubiquitination is summarized and discussed.

Keywords

T cell development · Ubiquitination · Treg development · Negative and positive selection · Thymic epithelial cell (TEC)

10.1 T-Cell Development

Intrathymic T-cell development starts from the seeding of progenitors (thymus seeding progenitors, TSP) derived from bone marrow, which enter the thymus as ETPs (Awong and Zúñiga-Pflücker 2014; Kumar et al. 2018; Six et al. 2007). ETPs proliferate vastly and gradually develop towards T cell lineage orientation, driven by the thymic microenvironment. The process is divided into three stages, $CD4^{-}CD8^{-}$ double negative (DN), $CD4^{+}CD8^{+}$ double positive (DP), and $CD4^{+}$ single positive (SP) or $CD8^{+}$ single positive (SP) stages. Thymocytes develop into mature T cells with the ability to recognize self-major histocompatibility complex MHC-antigens and

Z. Peng · H. Zhang
Center for Immunology and Hematology, State Key Laboratory of Biotherapy and Cancer Center, West China Hospital, Sichuan University, Chengdu, China

H. Hu (✉)
Center for Immunology and Hematology, State Key Laboratory of Biotherapy and Cancer Center, West China Hospital, Sichuan University, Chengdu, China

Chongqing International Institute for Immunology, Chongqing, China
e-mail: hongbohu@scu.edu.cn

H. Hu, X. Fu (eds.), *Ubiquitination in Immune System and Immune Response*, Advances in Experimental Medicine and Biology 1466, https://doi.org/10.1007/978-981-97-7288-9_10

non-self-reactivity (Hosokawa and Rothenberg 2021; Res and Spits 1999; Yui and Rothenberg 2014). NOTCH1 signaling is a key signal for T cell lineage commitment, and the activation of NOTCH1 signaling requires NOTCH ligands such as DLL4 and DLL1 expressed by thymic epithelial cells (Radtke et al. 2004; Tanigaki and Honjo 2007). CD34, a surface marker for human hematopoietic stem and precursor cells, marks the early progenitor cells in thymus (Galy et al. 1993). In CD34$^+$ cells, CD1a expression marks the complete commitment of the T-lineage. Human thymic CD4$^-$CD8$^-$ double negative (DN) cells can be divided into three groups by the expression of CD34, CD7, and CD1a surface molecules, as CD34$^+$CD7$^-$CD1a$^-$ (Thy1), CD34$^+$CD7$^+$CD1a$^-$ (Thy2), and CD34$^+$CD7$^+$CD1a$^+$ (Thy3) (Hao et al. 2008; Casero et al. 2015; Plum et al. 2008; Weerkamp et al. 2006; Staal et al. 2016). Of these, Thy1 cells are the earliest T precursor cells that have been shown to have multi-lineage potentials in in vitro differentiation experiments, including T cells, B cells, NK cells, and myeloid cells (Hao et al. 2008; Weerkamp et al. 2006; Haddad et al. 2006). Recent single-cell sequencing analysis revealed that Thy2 cells are also a heterogeneous population of cells that can be classified into three groups of cells by CD44 and CD2 expression, CD44$^+$CD2$^-$ (Thy2a), CD44$^+$CD2$^+$ (Thy2b) and CD44$^-$CD2$^+$ (Thy2c) (Le et al. 2020). Among them, downregulation of CD44 expression implies stronger T-lineage potentials, so Thy2c and Thy3 cells represent precursor cells are T-cell restricted and no longer have alternative lineage potential (Le et al. 2020). CD2 is a marker for T-lineage specification. Thy2a, which does not express CD2, is similar with to Thy1 regarding multiple lineage differentiation, and Thy2b cells, expressing CD2, show potential of T, NK, and myeloid cells and no longer has the ability to develop into B cells (Le et al. 2020). CD34$^+$CD38$^-$ cells have also been considered as the earliest precursor cells. The expression of CD38 is earlier than CD7, indicated by that half of the cells in CD7 are CD38 positive (Weerkamp et al. 2006; Dik et al. 2005).

In mice, DN cells are divided into four populations, DN1, DN2, DN3, and DN4, according to the expression profiles of CD44 and CD25$^-$, CD44$^+$ CD25$^-$, CD44$^+$CD25$^+$, CD44$^-$CD25$^+$, and CD44$^-$CD25$^-$ (Germain 2002; Krueger et al. 2017; Seo and Taniuchi 2016; Godfrey et al. 1993). DN1 contains mouse ETP (CD44$^+$KIT^{hi}CD25$^-$) with multiple lineage potential (Godfrey et al. 1993; Ceredig and Rolink 2002). DN2 is divided into two phases, DN2a and DN2b. The commitment of T lineage potential occurs at the transition from DN2a to DN2b, when Tcrb, TCRG, and TCRD genes begin to be rearrange (Petrie and Zúñiga-Pflücker 2007; Godfrey et al. 1994). DN2a cells are also multiple-lineage potent and can differentiate to T, NK, and innate lymphoid cells (ILCs). DN2b develops towards T cells and loses alternative lineage potential (Lu et al. 2005; Shen et al. 2003; Masuda et al. 2005). From DN2 to DN3, DN cells develop into either γδ T cells or αβ T cells (Smeets et al. 2014; Joachims et al. 2006).

10.2 Selections in T Cell Development

During T-cell development, thymocytes undergo several runs of TCR rearrangement and selection to eventually become mature self-tolerant CD4$^+$ or CD8$^+$ T cells (Kumar et al. 2018). TCRs are divided into αβTCR and γδTCR, and during T-cell development, fragments of the variable (V-region), diverse (D-region), and linker (J-region) regions of the TCRα, TCRβ, TCRγ, and TCRδ chains undergo the recombination arrangement, forming a diverse TCR library, is called V(D)J rearrangement (Awong and Zúñiga-Pflücker 2014; Kumar et al. 2018; Liu et al. 2021). TCR rearrangement is mediated by the recombination activation genes RAG1 and RAG2 (Jones and Simkus 2009). TCR gene rearrangement in the human thymus follows the sequential order of *TCRD*, *TCRG*, *TCRB* and *TCRA* (Dik et al. 2005; Ktorza et al. 1996; Yassai and Gorski 2000). As early as in the CD34$^+$CD38$^-$CD1a$^-$ phase, *TCRD* gene rearrangement begins (Dik et al. 2005). *TCRB* gene rearrangement begins in

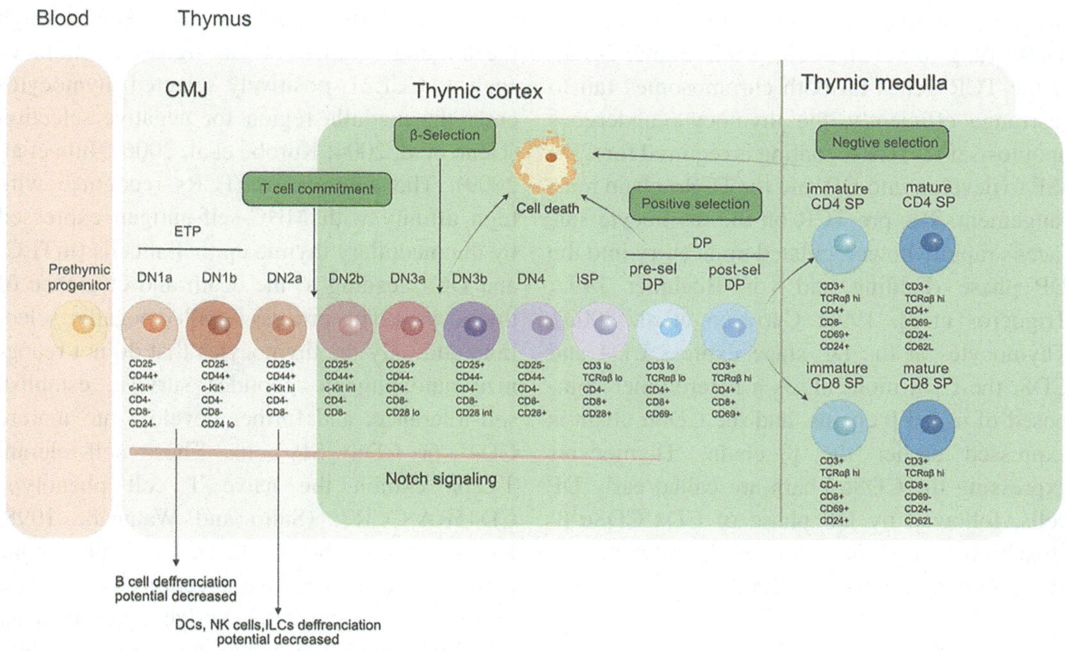

Fig. 10.1 T cell developmental stages in thymus. T cell development and maturation will go through three checkpoints including—β-selection, positive selection, and negative selection in DN, ISP, DP, and SP stages which is defined based on specific major cell-surface phenotype markers as shown. Prethymic progenitors migrate into the thymus and become ETPs (include DN1a and DN1b), which still retain the ability to generate other cells except B cell, but with the function of thymic environment and notch signaling, the potential to generate non-T cells is decreased gradually and commit to T cell lineage as green shading shown. β-selection happens between DN3a and DN3b, followed by DN4 and go through ISP and DP stage with the expression of *Cd8* and *Cd4* genes. DP cells then undergo positive selection and negative selection and become CD4 SP or CD8 SP with the ability to recognize and incorporate with MHC II or MHC I but with low affinity to self-antigen. The downregulation of CD69 and CD24 expression marks the maturation of T cells. *CMJ* cortico-medullary junction, *DC* dendritic cell, *NK* natural killer, *ILC* innate lymphoid cell, *ETPs* early T cell progenitors, *DN* double negative, *ISP* immature single-positive cell, *DP* double positive, *SP* single positive

the $CD34^+CD38^+CD1a^-$ phase, followed by DN cells expressing CD4 molecules and becoming immature CD4 single positive T cells (CD4 ISP) (Kumar et al. 2018; Krueger et al. 2017). CD4 ISP have the potential to develop to γδT or αβT cells (Kumar et al. 2018; Krueger et al. 2017) (Fig. 10.1).

10.2.1 β-Selection

In humans, $CD34^+CD38^+CD1a^+$ cells undergo β-selection (β-selection); in mice, it occurs at the DN3 stage (Ceredig and Rolink 2002; Joachims et al. 2006). β-selection is an important node in αβT cell development, where precursor T cells with a producible rearranged TCR β-chain are selected for the next step of development, when the successfully rearranged β chain binds to the pre-TCRα chain and CD3 to form a complex called pre-TCR (Joachims et al. 2006; Michie and Zúñiga-Pflücker 2002). Pre-TCR does not require ligand activation and continuously provides signals for survival, proliferation, and allelic rejection to developing T cells (Voll et al. 2000; Aifantis et al. 2001; Yamasaki et al. 2006; Aifantis et al. 1997; Falk et al. 1996). Allelic rejection means that when a TCR gene on one chromosome is rearranged, the TCR gene on the other chromosome cannot be rearranged, and only when the first rearrangement fails, the TCR gene on the second chromosome begins to be rearrange, aiming to ensure the specificity of TCR (Yang-Iott et al. 2010; von Boehmer 2005). If the TCR genes on both chromosomes fail to rearrange efficiently, the thymocyte undergoes apoptosis. Pre-TCR signaling is required for CD4 ISP to develop into DP, and for TCRα chain rearrangement. The pre-TCR on the thymocyte surface is rapidly downregulated upon entry into the DP phase (Fehling and von Boehmer 1997; Trigueros et al. 1998; Carrasco et al. 2003). Thymocytes at the DP stage express CD4 and CD8, the CD8 molecule is a heterodimer composed of α and β chains, and the CD8α chain is expressed earlier the β chain. Thymocytes expressing the CD8α chain are called early DP cells, followed by DP phase of $CD4^{+}CD8\alpha^{+}\beta^{+}$ (Joachims et al. 2006; Blom et al. 1999; Taghon et al. 2009; Carrasco et al. 2002).

10.2.2 Positive and Negative Selection

Following β-selection, the TCRα chain begins to be rearranged, and the TCRβ chain forms TCRαβ with the rearranged TCRα chain (Dik et al. 2005; Levelt et al. 1995). Thymic epithelial cells (cTEC) in the thymic cortical region express MHCI or MHCII molecules, and thymocytes with successfully rearranged TCRα chain recognize self-antigens bound to MHCI or MHCII molecules with low or moderate affinity, thereby activates the TCR and provide survival signaling for DP cell. DP cells which fail to recognize autoantigens bound to MHCI or MHCII molecules undergo apoptosis. This process is known as positive selection (Saito and Watanabe 1998; Klein et al. 2014).

After positive selection, DP cells develop into $CD4^{+}$ cells that recognize MHCII molecules or $CD8^{+}$ SP cells that recognize MHCI molecules. In these thymocytes successfully positively selected, the expression of early activating molecules CD69, CCR7 and subsequently CD27 molecules is upregulated, while *RAG1* and *RAG2* gene expression is downregulated (Yamashita et al. 1993; Ueno et al. 2004; Kurobe et al. 2006; Nitta et al. 2009; Brandle et al. 1994). Through the binding of CCR7 to its ligand chemokines such as CCL21, positively selected thymocytes enter the medulla region for negative selection (Ueno et al. 2004; Kurobe et al. 2006; Nitta et al. 2009). The autoreactive TCRs recognize with high affinity with MHC-self-antigen expressed by the medullary thymic epithelial cells (mTEC) and DCs, leading to the death and clearance of thymocytes. This process is called negative selection, and only the thymocytes that do not recognize autoantigenic peptides survive, establish self-tolerance, and further develop into mature $CD4^{+}$ or $CD8^{+}$ SP cells. These self-tolerant T-cells exhibit the naive T cell phenotype $CD45RA^{+}CCR7^{+}$ (Saito and Watanabe 1998; Kyewski and Derbinski 2004). T cells undergone both positive and negative selections have diverse TCRs that recognize the antigen presented by self MHC-molecule while not showing autoreactivity, thus egress into the periphery and execute immune functions.

10.3 The Role of Ubiquitination in T Cell Development

10.3.1 Introduction of Ubiquitination

Ubiquitination is a post-translational modification of proteins which plays an important role in many aspects of cell function (Hu and Sun 2016; Hershko and Ciechanover 1998). Ubiquitin is a small protein of 76 amino acids and covalently bound to a lysine residue (K) of the target protein through the catalysis of ubiquitin activating enzyme E1, ubiquitin binding enzyme E2, and ubiquitin ligase E3, thereby causing target protein degradation or endorsing target protein for new functions (Park et al. 2014; Pickart 2001). The glycine residue at the end of a ubiquitin molecule is attached to the K residue or the amino-terminal methionine (M1) of another ubiquitin molecule via an isopeptide bond, thus forming a ubiquitin chain (Berndsen and Wolberger 2014). Depending on the amino acid residues attached to the ubiquitin molecule, ubiquitin chains are classified into several types, including K6, K11, K27, K29, K33, K48, K63, and M1, and mixed ubiquitin chains (Pickart 2001). The most studied types of ubiquitination, such as K48 and K11, mainly mediate proteasome-dependent protein degradation, while K63 and M1 ubiquitination mainly mediate signal transduction and DNA repair (Husnjak and Dikic 2012). Ubiquitination is a reversible modification, and deubiquitinating enzymes (DUB) mediate the deubiquitination process by removing the ubiquitin chains attached to target proteins (Sun 2008).

The substrate specificity of ubiquitin is mainly determined by E3 ligases which mediate the binding of ubiquitin molecules to target proteins directly, or transfer of ubiquitin molecules from E2-binding enzymes to target proteins (Berndsen and Wolberger 2014; Sluimer and Distel 2018). There are more than 700 E3 ubiquitin ligase coding genes in the human genome (Park et al. 2014). Two of the most studied E3 families include the RING and HECT E3 family (Pao et al. 2018). E3 ligases containing the RING domain mediate the transfer of ubiquitin from E2-binding enzymes to substrates (Deshaies and Joazeiro 2009), whereas E3 ligases containing HECT form thioester bonds with ubiquitin via cysteine residues on the HECT structural domain, thereby transferring ubiquitin

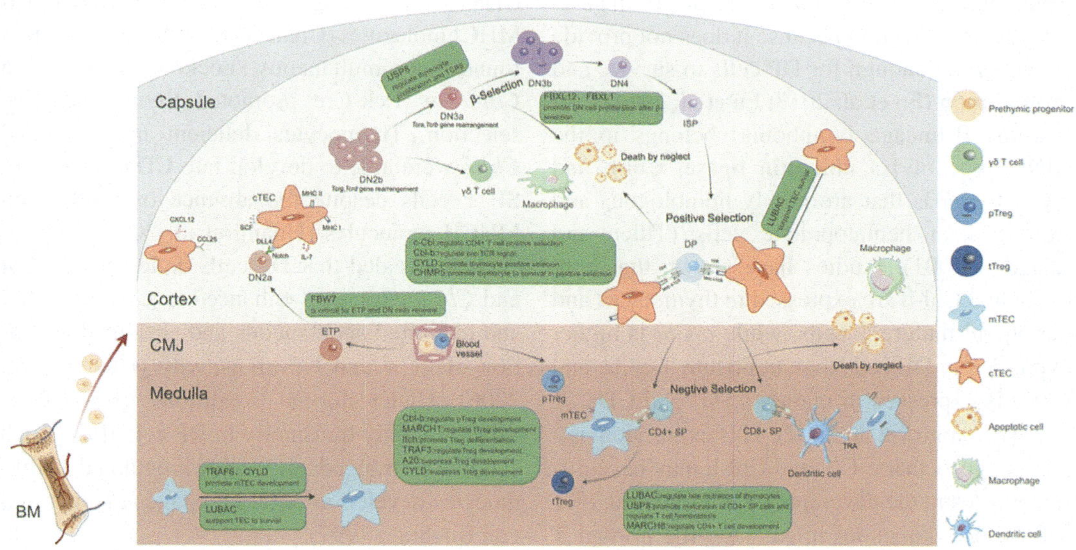

Fig. 10.2 Ubiquitylation is essential for thymocyte, TEC, and Treg. In the postnatal thymus, bone marrow derived prethymic progenitor cells migrate into the thymic parenchyma through the blood vessel in the presence of CXCL12 and CCL25 chemokines. Ubiquitylation plays significant roles in different T cell development stages as shown. *Tcrb* gene rearrangement mostly occurs at the DN3a stage, which indicates αβT cell commitment rather than γδ T cell. Subsequently, Tcra genes rearrange and form a pre-TCR, which triggers activation to DN3b, strong proliferation, progression to DN4, induces the expression of Cd8 and Cd4 genes and go through ISP and DP stage. In the cortex or CMJ, DP thymocytes interact with stromal cells for positive selection which many ubiquitylation associated proteins participate in. In the medulla, further negative selection of SP thymocytes includes the deletion of tissue-specific-antigen-reactive T cells and generation of regulatory T cells. Ubiquitylation is vital for Treg development with promotion or suppression. Meanwhile, ubiquitylation also regulates the development and survival of both cTEC and mTEC. *BM* bone marrow, *ETP* early T cell progenitor, *DLL4* delta like canonical notch ligand 4, *SCF* stem cell factor, *IL-7* Interleukin-7, *cTEC* cortical thymic epithelial cell, *MHC* major histocompatibility complex, *mTEC* medullary thymic epithelial cell, *ETP* early T cell progenitor, *CMJ* cortico-medullary junction, *DN* double negative, *ISP* immature single-positive cell, *DP* double positive, *SP* single positive, *Treg* regulatory T cell, *TRA* tissue restriction antigen, *AIRE* autoimmune regulator

to target proteins (Pickart 2001; Dye and Schulman 2007). Ubiquitination plays an important role in a variety of biological processes, and we will focus on the role of ubiquitination in T-cell development (Fig. 10.2).

10.3.2 Ubiquitination Regulates αβT Cell Development

10.3.2.1 Cbl Regulates Pre-TCR Signaling in Positive and Negative Selection

Positive and negative selection in the thymus is controlled by the strength of TCR signal (Gascoigne et al. 2016). Strong TCR signal leads to the death of T cells in negative selection, and weak TCR signal prevents thymocytes from passing positive selection because it does not provide signal strong enough for DP cells to survive and proliferation (Fu et al. 2013; Fu et al. 2014). Cbl (Casitas B-lineage lymphoma) belongs to the RING type of E3 ubiquitin ligase. Cbl-b and c-Cbl two E3s that are highly homologous and expressed in hematopoietic cells (Thien and Langdon 2001). Studies have shown that both c-Cbl and Cbl-b are expressed in thymocytes and peripheral mature T cells, while *c-Cbl* is higher expressed in DP cells than in mature T cells, and *Cbl-b* is expressed in mature $CD4^+$ and $CD8^+$ T cells (Naramura et al. 2002). This expression pattern suggests that c-Cbl and Cbl-b may have different roles in the development and function of T cells. *Cbl-b* knockout mice show the normal T cell development in thymus, whereas deficiency of *c-Cbl* results in a significant increasement in $CD4^+$ SP thymocytes (Naramura et al. 1998). With TCR transgenic mice, investigators found that deletion of *c-Cbl* promotes $CD4^+$ T cell to pass positive selection of $CD4^+$ T cells, but does not affect positive selection of CD8 T cells; meanwhile, more T cells that were hyporesponsive to TCR stimulation are found in peripheral $CD4^+$ T cells in *c-Cbl* knockout mice than WT mice (Naramura et al. 1998). These results suggest that c-Cbl is a key regulator in positive selection of $CD4^+$ T cells and TCR activation. Molecular mechanism studies revealed that *c-Cbl* knockout enhanced TCR signaling in thymocytes, with enhanced phosphorylation of ZAP70 and downstream ERK1/2 activation (Naramura et al. 1998); *c-Cbl* defect DP cells showed the enhanced expression of CD5, CD69, and CD3 (Naramura et al. 1998), suggesting enhanced positive selection of DP cells. On the one hand, c-Cbl regulates TCR signaling by modulating the activation of ZAP70 and ERK, thus promoting positive selection, and on the other hand knockout of *c-Cbl* resulted in enhanced CD3 and CD4 expression, thus promoting TCR and MHC molecule binding and positive selection of $CD4^+$ T cells (Naramura et al. 1998).

The development of DP cells into $CD4^+$ or $CD8^+$ SP cells is highly dependent on MHCII and MHCI molecules (Chan et al. 1998). Investigators found that simultaneous knockout of *c-Cbl* and *Cbl-b* by Lck-Cre promoted thymic negative selection. Thymocytes deficient in *c-Cbl* and *Cbl-b* were able to develop into $CD4^+$ and $CD8^+$ SP T cells despite the absence of MHCI and MHCII molecules (Huang et al. 2006). Further studies revealed that DP cells deficient in *c-Cbl* and *Cbl-b* expressed enhanced level of pre-TCR and exhibited spontaneous and sustained activation of LCK and NF-κB activity (Huang et al. 2006). Under normal conditions, pre-TCR is expressed only transiently in DP cell after TCRβ chain is rearranged, and is downregulated as thymocytes develop into the DP stage (Fehling and von Boehmer 1997).

Pre-TCR complex formed by TCRβ, pre-TCRα, and CD3 (Groettrup et al. 1993; Berger et al. 1997) persistently activates LCK and NF-kB signaling (Voll et al. 2000; Aifantis et al. 2001) and thus provide DN cells with proliferation signaling as well as allelic rejection signals (Yamasaki et al. 2006; Aifantis et al. 1997). Studies have reported that c-Cbl mediates the ubiquitination and degradation of pre-TCR (Panigada et al. 2002) and therefore regulates pre-TCR signaling in DP cells. In the c-Cbl and Cbl-b double KO mice, pre-TCR signaling is not downregulated, and TCR activation is persistent and continuously providing the survival and proliferation signals to thymocytes, so these thymocytes can still produce $CD4^+$ or $CD8^+$ SP T cells in the presence of Cbl-b and c-Cbl as well as MHCI and MHCII (Huang et al. 2006; Huang and Gu 2008).

10.3.2.2 CYLD and Thymocyte Development

CYLD is a deubiquitinating enzyme that reportedly remove ubiquitin from TRAF2, TRAF6, and NEMO, thereby negatively regulate NF-κB signaling pathway (Brummelkamp et al. 2003; Kovalenko et al. 2003; Trompouki et al. 2003). Mice with deficiency of CYLD (knockout exon 2 of *Cyld*) had impaired thymocyte development, exhibiting an increased proportion of DP cells and a decrease in SP cells and a significant decrease in peripheral T cells, suggesting that CYLD regulates thymocyte development from DP to SP (Reiley et al. 2006). Mechanistic studies revealed that CYLD regulates TCR signaling in DP cells, and that CYLD binds activated LCK and reduce the K63 and K48 ubiquitination LCK, thereby promoting LCK recruitment to ZAP70 and subsequent TCR activation (Reiley et al. 2006). However, at the same time, another study using mice knockout of exons 2 and 3 of *Cyld* gene demonstrated that the defect of CYLD did not affect T-cell development in the thymus; the percentages of DP and SP cells in CYLD KO thymus were similar to those of wild-type mice (Zhang et al. 2006). Interestingly, however, mice defect of exons 2 and 3 of the Cyld gene were more susceptible to colonic inflammation and had an increased incidence of tumors (Zhang et al. 2006). Molecular mechanism studies showed that knockout of *Cyld* (exons 2 and 3) resulted in enhanced NF-κB signaling (Zhang et al. 2006) and T cells lacking CYLD (exons 2 and 3) showed enhanced NEMO ubiquitination and NF-κB signaling activated by TCR. Besides T-cells, CYLD KO macrophages exhibited enhanced TRAF2 ubiquitination and NF-κB activation, suggesting that CYLD regulates the inflammatory response by deubiquitinating NEMO and TRAF2 and negatively regulating the NF-κB signaling pathway (Zhang et al. 2006).

Studies suggested that different CYLD mutants might have different effects on T-cell development. To investigate the precise effects of CYLD on T development, study used Lck-cre to conditionally knock out exon 9 ($Cyld^{\Delta 9}$) of the Cyld gene in T cells, which encodes a catalytic domain required for CYLD to function as a deubiquitinating enzyme (Tsagaratou et al. 2010). The results showed that SP thymocytes and peripheral T cells were significantly reduced and positive selection of thymocytes was significantly blocked in $Cyld^{\Delta 9}$ mice (Tsagaratou et al. 2010). Moreover, the basal levels of NF-κB and JNK activity were enhanced, and knockdown of NEMO in $Cyld^{\Delta 9}$ mice could rescue T cell development and survival, suggesting that CYLD affect NF-κB activation in thymocyte by regulating NEMO, thereby establishing the appropriate TCR signaling threshold to promote positive selection (Tsagaratou et al. 2010).

10.3.2.3 Function of MARCH8 in CD4+ T Cell Selection

TECs and DCs express MHCII molecules, presenting self-antigens to thymocytes and mediating positive and negative selection of $CD4^+$ T cells. The E3 ubiquitin ligase MARCH8 targets and regulates MHCII molecule stability of TECs, and knockout of MARCH8 in mice resulted in increased MHCII expression levels on the surface of cTEC and AIRE-mTEC, while MHCII molecules on the surface of $AIRE^+$ mTEC were unaffected, suggesting that MARCH8 affects MHCII protein stability (Liu et al. 2016; von Rohrscheidt et al. 2016). However, in MARCH8 KO mice, the cell number of thymic and splenic $CD4^+$ T cells were similar to those of wild-type mice, suggesting that $CD4^+$ T cell development was not affected in general (Liu et al. 2016; von Rohrscheidt et al. 2016). Further studies revealed that CD83 and MARCH8 antagonize each other to regulate the stability of MHCII molecule. CD83 is expressed by TEC, activated T cells, B cells, and DCs. CD83 binds to MHCII and maintains MHCII protein stability in DCs, by competing with E3 ubiquitin ligase MARCH1 for MHCII, thereby antagonizing MARCH1-mediated ubiquitination of MHCII (Fujimoto et al. 2002; Tze et al. 2011). Furthermore, deficiency of CD83 in TEC resulted in impaired CD4+ T cell development and a significant reduction in $CD4^+$ SP in the thymus, with unaffected DP and DN cells (von Rohrscheidt et al. 2016). Consistently, MHCIIK225R mutation (which cannot be ubiquitinated) or knockout of MARCH8 recued $CD4^+$ T cell development in *CD83*$^{-/-}$ mice. Taken together, studies suggest that in TEC and DCs, MARCH8 and CD83 antagonistically regulate the stability of MHCII, thereby affecting $CD4^+$ T cell development (Liu et al. 2016; von Rohrscheidt et al. 2016)

10.3.2.4 FBW7 Is Involved in Early Thymocyte Development

FBW7 (also known as FBXW7) is a RING-type E3 ubiquitin ligase, and knockdown of FBW7 leads to embryonic lethality (Tetzlaff et al. 2004; Tsunematsu et al. 2004). FBW7 mediates the ubiquitination and degradation of various proteins critical for hematopoietic cell development such as NOTCH1 and c-MYC (Malyukova et al. 2007; O'Neil et al. 2007; Thompson et al. 2007). Defect on FBW7 in hematopoietic cells leads to a reduction of hematopoietic stem cells (HSC) and impaired HSC self-renewal. The maintenance of cell cycle in *Fbw7*$^{-/-}$ HSC is distributed and *Fbw7*$^{-/-}$ HSC could not enter a resting state, thereby affecting HSC self-renewal (Thompson et al. 2008). Using *Fbw7*$^{-/-}$ Myc$^{-/-}$ mice, the investigators found that the dysfunctional HSC caused by FBW7 deletion was rescued by deficiency of MYC, suggesting that the function of *Fbw7*$^{-/-}$ in HSC was mediated by accumulated c-MYC protein (Reavie et al. 2010). Meanwhile, hematopoietic cell-specific depletion of FBW7 resulted in impaired T-cell development (Thompson et al. 2008). Further analysis of thymocyte composition revealed that hematopoietic cell-specific KO of FBW7 resulted in significantly reduced thymic ETP and DN cells, with almost complete abolition of DN2 and DN3 cells (Thompson et al. 2008). These results suggest a critical role of FBW7 in HSC self-renewal and early T-cell development.

10.3.2.5 LUBAC Regulates NF-κB and T-Cell Development

The linear ubiquitin chain assembly complex (LUBAC) consists of at least three proteins, RNF31 (also known as HOIP), RBCK1 (also known as HOIL-1), and SHARPIN (also known as SIPL1). LUBAC regulates multiple signaling pathways by catalysing the conjugation of linear ubiquitin chain to substrates (Emmerich et al. 2011). Although the optimized catalytic activity of LUBAC requires these three proteins, the contribution of these proteins is not equal. HOIP deletion in mouse leads to the embryonic lethality and completely abrogates the E3 catalytic capacity of LUBAC (Peltzer et al. 2014), whereas cells without SHARPIN still have substantial linear ubiquitination, as the complex of HOIP and HOIL is sufficient for the basic function of LUBAC (Gerlach et al. 2011). Loss-of-function mutations of HOIL and HOIP result in reduced T cell numbers, suggesting that HOIL and HOIP might be involved in T cell development or homeostasis (Boisson et al. 2015). Studies have found that *Cd4*-cre mediated specific knockdown of *Rnf31* (encoding HOIP) or *Rbckl* (encoding HPIL-1) in T cells (*Rnf31*$^{\Delta CD4}$ or *Rbckl*$^{\Delta CD4}$ mice) resulted in the almost complete disappearance of $CD4^+$ or $CD8^+$ αβ T cells in the peripheral organs; meanwhile, the Treg cells exhibit a normal proportion but a significant reduction in absolute numbers (Teh et al. 2016). The remaining $CD4^+$ or $CD8^+$ αβ T cells were almost all $CD44^{hi}CD62L^{lo}$, exhibiting an effector/memory phenotype (Teh et al. 2016). While systemic Sharpin knockout mice (*Sharpin*^{cpdm}) had normal numbers and proportions of $CD4^+$ and $CD8^+$ T cells, the absolute number and proportions of Treg in $CD4^+$αβ T cells were reduced (Teh et al. 2016). Rnf31$^{\Delta CD4}$ and Rbck1$^{\Delta CD4}$ mice had reduced CD4+ and CD8+ SP cells but normal DP cells in the thymus. HOIP and HOIL-1 do not regulate the positive and negative selection in the T-cell development, but regulate the later maturation of thymocytes. *Sharpin*^{cpdm} mice showed normal ratios and absolute numbers of DN, DP, and SP cells in the thymus (Teh et al. 2016). Moreover, $Rnf31^{\Delta CD4}$, $Rbck1^{\Delta CD4}$, and $Sharpin^{cpdm}$ mice showed reduced $CD25^{+}FOXP3^{+}$ Treg as well as $CD25^{-}FOXP3^{+}$ Treg precursor cells in the thymus, suggesting that HOIP, HOIL-1, and SHARPIN promote Treg development in the thymus, while HOIP and HOIL-1 also regulate CD4+ and CD8+ T cell development in the thymus (Teh et al. 2016). Further experiments found that HOIP and HOIL-1 regulated the activation of NF-κB in thymocytes. HOIP deficient thymocytes showed the delayed degradation of IκBα after anti-CD3 and anti-CD28 antibody stimulation, additionally, knockdown of HOIP or HOIL-1 KO thymocytes showed the delayed and attenuated phosphorylation of p65 and p38 MAPK, as well as degradation of IκBα after TNF stimulation (Teh et al. 2016). The investigators used $Rnf31^{\Delta CD4}$ mice and *IKKca* mice (with consistent activated IKKβ) to generate $Rnf31^{\Delta CD4}IKKca$ mice and found that induction of consistent activation of IKKβ rescued the ratio of mature $CD4^{+}$ SP T cells ($CD4^{+}CD24^{low}CD62L^{high}$) and $FOXP3^{+}$ Treg in the thymus caused by RNF31 deficiency, but T cells in periphery were not rescued to normal levels (Teh et al. 2016), suggesting that LUBAC affects thymocyte development by mechanisms other than regulation of NF-κB.

10.3.2.6 USP8 Controls FOXO1 and Bcl-2 in Thymocyte Development

The ubiquitin-specific protease USP8 is a member of the USP deubiquitinating enzyme family, and studies have shown that USP8 regulates T cell development and promotes thymic SP cell maturation (Dufner et al. 2015). T-cell specific knockdown of *Usp8* in mouse using *Cd4-cre* led to the significant reduction in the proportion and cell number of thymic $CD4^{+}$ and $CD8^{+}$ SP T cells, with DP and SP cells unaffected (Dufner et al. 2015). Meanwhile, the percentage of $CD25^{+}FOXP3^{+}$ Treg was significantly reduced in USP8 KO mice (Dufner et al. 2015). Interestingly, TCRβ was barely expressed in thymic $CD4^{+}$ and $CD8^{+}$ SP cells in USP8 KO mice (Dufner et al. 2015). Further studies revealed that USP8 regulates thymocyte proliferation without affecting the negative selection, and Usp8 KO T cell showed impaired proliferation upon TCR stimulation in vitro (Dufner et al. 2015). The transcription factor FOXO1 regulates the expression of *Il7ra*, *Ccr7* and *Sell* (Kerdiles et al. 2009; Ouyang et al. 2009). In Usp8 KO DP cells, the mRNA levels of *Il7ra* and *Ccr7* were reduced, consistently, FOXO1, which binds the *Il7ra* promoter, was significantly reduced in Usp8 KO thymocytes (Dufner et al. 2015). Similar to T-cell specific IL-7R KO mice, USP8 KO mice had a reduced CD8 SP T cell in the TCR^{hi} cells, suggesting that USP8 regulates T cell development directly or indirectly through FOXO1-IL7Rα axis. Furthermore, knockout of Usp8 also resulted in the reduction of $CD4^{+}$ SP cells in the thymus, suggesting that USP8 also regulates T-cell development through other mechanisms (Dufner et al. 2015).

USP8 also regulates T-cell development by correlating with CHMP5, which is an endonuclear somatic sorting complex required for transport (ESCRT) protein and plays a critical role in multivesicular vesicle formation (Shim et al. 2006; Rusten et al. 2011). It has also been shown that CHMP5 inhibits the ubiquitination of IκBα, suggesting function of CHMP5 besides of post-translational modifications (Greenblatt et al. 2015). Indeed, CHMP5 regulates positive selection by a post-translational modification mechanism that promotes thymocyte survival. Positively selected thymocytes start an anti-apoptotic program, expressing the pro-survival protein Bcl-2 etc (Linette et al. 1994; Strasser et al. 1994; Starr et al. 2003), which is inactivated and degraded by oxidation (Azad et al. 2010). It was found that positive selection promoted the binding of CHMP5 and USP8, thereby maintaining CHMP5 protein stability, and CHMP5 further inhibits the oxidation and degradation of Bcl-2, to ensure the survival of thymocytes after positive selection (Adoro et al. 2017). At all stages of thymocyte development, DP cells have the highest expression level of Chmp5 mRNA, and knockout of CHMP5 in T cells mediated by *Cd4-cre* resulted in disturbed thymic structure, with a significant reduction of $CD4^+$ and $CD8^+$ SP cells, as well as $CD24^{lo}TCR\beta^{hi}$ mature T cells in the thymus. In the peripheral and lymphoid organs, the substantial reduction of αβ T cells was observed in CHMP5 KO mice, and these remaining T cells exhibit a $CD44^{hi}$ phenotype, suggesting that T cell development and function are severely impaired in these KO mice (Adoro et al. 2017). Knockout of the pro-apoptotic gene *Bim* or transgenic overexpression of Bcl-2 in mice with T cell-specific KO of CHMP5 reversed the defect of T cell development caused by CHMP5 deficiency. The interaction of CHMP5 and USP8 was detected in $CD69^+TCR\beta^{hi}$ DP cells, which inhibits the oxidation and degradation of Bcl-2, suggesting that CHMP5 promotes thymocyte survival by maintaining Bcl-2 protein stability; however, this interaction is not observed in $CD69^-TCR\beta^{neg\text{-}low}$ DP cells which do not undergo positive selection (Adoro et al. 2017). These studies suggest the multiple function of USP8 in regulating T-cell development.

10.3.2.7 Role of FBX Family (FBXL1 and FBXL12) in Thymocyte Development

Thymocytes undergo two runs of proliferations during the DN phase, the first run occurs before TCR β-chain rearrangement, mediated by signals such as stem cell factor (SCF), IL-7, and NOTCH (Shortman et al. 1990; Penit et al. 1995; Rodewald et al. 1997). The second proliferative phase occurs during β-selection. DN3 cells that successfully rearranged β-chains express pre-TCR, prompting massive cell proliferation (Penit et al. 1995; Kreslavsky et al. 2012; Hoffman et al. 1996; Aifantis et al. 2006). During DN cell proliferation, the cell cycle switch from resting G0 to G1 and the S/G1/M phase, a process controlled by the cyclin and cell cycle-dependent kinase (CDK) complexes (Rowell and Wells 2006), while the activity of the cyclin-CDK complex and cell cycle progression is inhibited by the Cip/Kip family, a family whose members include Cdkn1a, Cdkn1b, and Cdkn1c. Cdkn1b plays a key function in β-selection (Nakayama et al. 1996; Kiyokawa et al. 1996; Fero et al. 1996), by inhibiting the cyclin A-CDK and cyclin E-CDK complexes. Mice knocking out Cdkn1b KO mice have enlarged thymus (Nakayama et al. 1996; Kiyokawa et al. 1996) while overexpression of Cdkn1b results in reduced thymus size, and T-cell development is blocked at the DN3 stage (Tsukiyama et al. 2001).

FBXL1 and FBXL12, the key components of the SCF (Skp-cullin-F-box) ubiquitin ligase complex, are involved in the regulation of thymic DN cell proliferation (Xu et al. 2007; Hao et al. 2005; Zhao et al. 2019). In thymocytes, NOTCH signaling induces FBXL1 expression and pre-TCR induces FBXL12 expression (Sarmento et al. 2005). The FBXL1-SCF and FBXL12-SCF complexes synergistically mediate K48 ubiquitination and proteasome-dependent degradation of Cdkn1b, thereby promote the proliferation of β-selected DN cells (Zhao et al. 2019; Kossatz et al. 2004; Tsvetkov et al. 1999). The size of thymus from Fbxl1 KO mice is significantly reduced. In Fbxl1 KO thymus, DN1 and DN2 cells are normally, while DN3 to DN4 development is severely impaired, the cell proliferation of DN4, ISP, and DP thymus was all abolished (Zhao et al. 2019). The deficiency of Cdkn1b could rescue the shrinking thymus caused by Fbxl1 KO (Nakayama et al. 2004). Systemic KO of *Fbxl12* leads to embryonic lethality (Nishiyama et al. 2015), T-cell specific deficiency of Fbxl12 (*Lck-cre Fbxl12$^{fl/fl}$*), causes the phenotype similar with *Cdkn1b$^{-/-}$* mice (Zhao et al. 2019). In DN cells and total thymocytes of isolated from *Lck-cre Fbxl12$^{fl/fl}$* mice, the Cdkn1b protein levels were increased compared to control cells. KO of Cdkn1b in *Lck-cre Fbxl12$^{fl/fl}$* mice reversed the DN3-DN4 developmental defect caused by FBXL12 deficiency (Zhao et al. 2019). Additionally, FBXL12 and FBXL1double KO resulted in a much more severe reduction of DN4, ISP, and DP cells, suggesting a synergistic effect of FBXL12 and FBXL1 (Zhao et al. 2019). In summary, FBXL1 and FBXL12 ubiquitinated Cdkn1b for degradation, thereby promoting cell proliferation after β-selection.

10.3.3 Treg Development

Treg is a subset of T-cells expressing the signature transcription factor FOXP3 with immunosuppression function (Caramalho et al. 2015). Based on the origin and function, Treg cells are divided into tTreg (thymic-derived Treg, as well as know nautral Treg, nTreg) and pTreg (peripheral-derived Treg, or induced Treg, iTreg) (Caramalho et al. 2015). tTreg develops in the thymus and mainly distributes in the medullary region of the thymus and migrates to the periphery (Aschenbrenner et al. 2007), where it plays a critical role in immune tolerance. FOXP3 and IL-2 are required for the development, maintenance and function of tTreg (Fontenot et al. 2003; Fontenot et al. 2005; Tai et al. 2005). pTreg is differentiated from peripheral naïve CD4 T cells induced by TGF-β. pTreg mainly secretes IL-10 and TGF-β to exert immunosuppressive effects at sites of inflammation (Richards et al. 2015; Lee et al. 2011).

Although most of the self-reactive CD4 T cells are eliminated in the thymic medulla, a subpopulation of self-reactive CD4 T cells develop into Treg cell (Kumar et al. 2018; Itoh et al. 1999; Burchill et al. 2008; Lio and Hsieh 2008). The affinity of TCR to MHC-self peptides is a key factor for thymocyte fate. Weak TCR-self-antigen interaction is necessary for positive selection, and strong interactions lead to negative selection and apoptosis of thymocytes. In contrast, some of the T cells with high affinity TCRs to self-antigens are not diminished, but instead undergo development pathway into Treg cells (Aschenbrenner et al. 2007; Pacholczyk et al. 2006; Picca et al. 2006; Cabarrocas et al. 2006). TCR affinity to self-antigens of Treg is in a certain range, within which the possibility of generating Treg cells increases with higher affinity. More importantly, the affinity of TCR-MHC/self-antigen is generally considered to be between the criteria for positive and negative selection (Klein et al. 2014; Jordan et al. 2001; Caton et al. 2014; Larkin et al. 2008; Weissler and Caton 2014). Treatment of humanized mice with anti-CD28 antibody leads to an increase in the number of $CD25^+FOXP3^+$ Treg in the thymus (Guo et al. 2008; Legrand et al. 2006); consistently, tTreg were substantially reduced in the thymus and spleen of mice deficient of CD28. In addition, the cell number and proportion of tTreg are significantly reduced in patients with mutated ZAP70. Given the critical rules of CD28 and ZAP70 in TCR activation, these results suggest TCR signaling is required for Treg development in thymus (Poliani et al. 2013).

10.3.3.1 Cbl-b in Treg Development

Cbl-b plays a key role in regulating threshold of the TCR signaling. $cblb^{-/-}$ mice demonstrate enhanced IL-2 production and T cell proliferation upon TCR activation (Bachmaier et al. 2000; Chiang et al. 2000). Studies have shown that Cbl-b is involved in the regulation of CD28–related thymic Treg and peripheral Treg development (Zhao et al. 2015). CD28 is important for thymic tTreg development, and the knockout of CD28 resulted in a substantial reduction in thymic and peripheral tTreg, and CD28 co-stimulation induced expression of FOXP3 in thymic DP cells (Tai et al. 2005). As Cbl-b KO resulted in an increase tTreg in thymus, the deficiency of Cbl-b in $CD28^{-/-}$ mice partially rescued tTreg development in the thymus. Mechanism studies showed that upon TCR activation, Cbl-b binds FOXP3, and mediates the ubiquitination and subsequent proteasome-dependent degradation of FOXP3 with the association STUB1. The impaired tTreg development of $CD28^{-/-}$ mice is recovered by proteasome inhibitor treatment, similar to the effect of knocking out Cbl-b in $CD28^{-/-}$ mice, suggesting that Cbl-b regulates tTreg development by ubiquitinating and degrading FOXP3 (Zhao et al. 2015).

Cbl-b also regulates TGF-β-induced pTreg development in periphery. cbl-b regulates pTreg development by regulating FOXP3 expression rather than degradation, exhibiting a different mechanism from that in thymus. The transcription factors FOXO3a and FOXO1 directly bind the FOXP3 promoter to promote FOXP3 transcriptional expression, and the phosphorylation of FOXO3a and FOXO1 abolished their binding to FOXP3 promoter (Harada et al. 2010). Studies have proved that Cbl-b KO leads to a reduction of

TGF-β-induced pTreg development (Harada et al. 2010; Wohlfert et al. 2006). Molecular mechanism studies have found that Cbl-b deficient CD4 T cells showed the enhanced phosphorylation of FOXO3a and FOXO1, but reduced FOXP3 expression after TGF-β stimulation (Harada et al. 2010). Over-expression of FOXO3a in Cbl-b KO mice restored FOXP3 expression, suggesting that Cbl-b promotes FOXP3 expression and tTreg development though regulating FOXO3a (Harada et al. 2010). However, there is unclear how Cbl-b inhibits the phosphorylation of FOXO3a and FOXO1 as an E3 ubiquitin ligase. Another study showed that AKT2 mediates the phosphorylation of FOXO3a and FOXO1 (Qiao et al. 2013). AKT phosphorylation was enhanced in Cbl-b KO T cells, and AKT2 KO neutralized the enhanced phosphorylation of FOXO3a and FOXO1 caused by Cbl-b deletion (Qiao et al. 2013). In the in vitro TGF-β-induced Treg differentiation assay, treatment of *Cbl-b*$^{-/-}$ T cells treated by AKT2 inhibitors produced more Treg production than controls (Qiao et al. 2013), suggesting Cbl-b affects FOXO phosphorylation by regulating AKT phosphorylation, thereby regulating pTreg development (Qiao et al. 2013). As an E3 ubiquitin ligase, how Cbl-b regulates AKT phosphorylation remains to be further investigated. In addition, it has also been shown that stimulation of *Cbl-b*$^{-/-}$ T cell stimulated by TGF-β had reduced phosphorylation of SMAD2 and reduced FOXP3^{+} Treg production than WT cells, suggesting that Cbl-b might Treg differentiation through modulation of the TGF-β/SMAD2 signaling pathway and FOXP3 expression (Wohlfert et al. 2006).

10.3.3.2 MARCH1 Regulates DC Function and Treg Development

Treg development in the thymus requires antigen delivery by DCs and TECs (Hsieh et al. 2012; Laufer et al. 1996). DCs expressing MHC-II mediate negative selection of thymocytes and Treg production (Baravalle et al. 2011; Proietto et al. 2008). MARCH1 is an E3 ubiquitin ligase containing the RING domain, which mediates ubiquitination of MHCII and CD86 in DC cells, thereby promoting endocytosis, lysosomal transport, and degradation of MHCII and CD86 molecules (Baravalle et al. 2011; Matsuki et al. 2007). After DC cells are activated by microbial and inflammatory stimuli, MARCH1 expression is downregulated, therefore the protein levels of MHCII and CD86 are increased to enhance antigen presentation by DCs (De Gassart et al. 2008; Walseng et al. 2010). MARCH1 expression is upregulated in DC to respond to immunosuppressive cytokine such as IL-10, resulting in reduced protein levels of MHCII and CD86 of DCs, and antigen presentation (Tze et al. 2011; Baravalle et al. 2011). In in thymic DC cells, MARCH1 mediates the ubiquitination of MHCII molecules and regulates tTreg development (Oh et al. 2013). It was found that MARCH1 knockdown resulted in a significant reduction of tTreg, and MARCH KO DC cells were unable to support antigen-specific Treg production in vivo and in vitro (Oh et al. 2013). Furthermore, the mice bearing mutated MHCII (which could not be to ubiquitinated) exhibit reduced thymic Treg numbers (Oh et al. 2013; Liu et al. 2020), suggesting the importance of MARCH1-mediated ubiquitination of MHCII for tTreg development.

10.3.3.3 The Role of Itch in Treg Development

Itch is a HECT-type E3 ubiquitin ligase that primarily regulates T cell activation (Liu 2007). Itch also promotes TGF-β-induced Treg production (Venuprasad et al. 2008). Itch$^{-/-}$ T cells are less sensitive to TGF-β stimulation than control cells, resulting in impaired FOXP3 expression (Venuprasad et al. 2008) and reduced Treg differentiation. Mechanism studies revealed that Itch interacts with the transcription factor TIEG1 and mediates its ubiquitination (Venuprasad et al. 2008). Ubiquitinated TIEG1 directly binds the promoter of FOXP3 and promotes the transcription of FOXP3, and double KO of TIEG1 and Itch resulted in more substantially reduction in FOXP3 expression and Treg differentiation (Venuprasad et al. 2008), suggesting that Itch synergizes with TIEG1 in promoting FOXP3 expression and Treg differentiation (Venuprasad et al. 2008).

In addition, it has also been shown that Itch also regulates TGF-β-induced Treg production via Ndfip1 (Beal et al. 2011). Compared to WT cells, $Itch^{-/-}$ T cells were more proliferative and tend to differentiate toward Th2 cells, producing more Th2-cytokines such as IL-4 and IL-5 (Fang et al. 2002; Venuprasad et al. 2006). Itch binds the transcription factor JunB and mediates its ubiquitin and degradation, which requires Ndfip1. Upon T-cell activation, Ndfip1 binds Itch and promotes Itch-mediated JunB degradation (Oliver et al. 2006; Ramon et al. 2011). Knockout of either *Ndfip1* or *Itch* prolongs the half-life of JunB, resulting in enhanced Th2 response and increased IL-4 production (Oliver et al. 2006); meanwhile, TGF-β-induced $FOXP3^+$ Treg differentiation was inhibited (Beal et al. 2011). Deficiency of *Ndfip1* does not affect the interaction of TIEG1 and FOXP3 promoter, suggesting that the function of Ndfip1 and Itch in $FOXP3^+$ Treg differentiation was independent of TIEG1. Further studies have shown that $Ndfip1^{-/-}$ T-cells produced more IL-4 which inhibits FOXP3 expression than WT T-cells (Beal et al. 2011; Dardalhon et al. 2008). The authors proposed that Ndfip1 promotes Itch-mediated ubiquitination and degradation of JunB, thereby inhibiting IL-4 production and promoting $FOXP3^+$ Treg differentiation (Beal et al. 2011).

10.3.3.4 TRAF Protein Regulates Treg Development

TRAF3 is a member of the TRAF (TNFR-associated factor) family, which is mainly involved in downstream signaling of tumor necrosis family receptors (TNFR) and Toll-like receptors (TLR), through the TRAF structural domain and RING structural domain (Hacker et al. 2011). Studies have reported that TRAF3 negatively regulates thymic Treg development (Yi et al. 2014; Xie et al. 2011). TRAF3 regulates the development of thymic Treg precursor cells to tTreg, requiring IL-2 (Lio and Hsieh 2008; Yi et al. 2014). IL-2 binds to IL2R to activate downstream signaling pathways. JAK1 and JAK 3 bind CD122 (IL-2Rβ) and CD132 (common γ-chain), respectively, and mediate self-phosphorylation as well as IL-2Rβ and the transcription factor STAT5 (Johnston et al. 1996; Minami and Taniguchi 1995). Phosphorylated STAT5 is essential for FOXP3 expression (Burchill et al. 2007; Yao et al. 2007). Studies have shown that TRAF3 promotes FOXP3 expression by recruiting the tyrosine TCPTP to the IL-2R complex (Yi et al. 2014), dephosphorylates JAK 1 and JAK 3. The inhibited JAK activation and negatively regulated the transcriptional activity of STAT5 attenuates IL-2 signaling and thereby inhibits thymic Treg development. Whether TRAF3 acts as an E3 ubiquitin ligase or as an adaptor protein in this process is unclear (Yi et al. 2014). In addition, Treg-specific KO of TRAF3 inhibited Treg function and dysregulated CD4 T cell homeostasis, as demonstrated by an increase in Th1-like effector or memory T cells than WT mice (Chang et al. 2014). The activation of follicular helper T cells (Tfh cells) was enhanced in Treg-specific TRAF3 KO mice, resulting in increased germinal center formation and IgG antibody production (Chang et al. 2014). Another molecule in the TRAF family, TRAF6, has also been reported to regulate thymic Treg development and Treg function. TRAF6 facilitates the location of FOXP3 in the nucleus and transcriptional functions by mediating the K63 ubiquitination of FOXP3. In TRAF6 KO Treg cells, TRAF6 FOXP3 is located in the perinuclear region and the immunosuppressive function of Treg is compromised, resulting in enhanced antitumor immune responses in mice (Ni et al. 2019).

10.3.3.5 A20 Inhibits Thymic Treg Development

A20 is a deubiquitinating enzyme which primarily mediates the negative feedback of NF-κB signaling and regulates TNF-induced apoptosis (Vereecke et al. 2009). A20 has important roles in many aspects of T cell activation, survival, and differentiation upon TCR activation. A20 removes ubiquitin chain from MALT1, thereby inhibiting NF-κB activation (Coornaert et al. 2008; Duwel et al. 2009); in addition, A20 inhibits apoptosisi of CD4 T-cells mediated by RIPK3 kinase (Onizawa et al. 2015). A20 has also been reported to inhibit Treg development in the thymus (Fischer et al. 2017). Conditional knockout of A20 in T cells resulted in a significant increased thymic and peripheral Treg (Fischer et al. 2017). A20 deficient Treg cells was less IL-2 dependent, but their proliferation and apoptosis levels were not affected compared with WT cells (Fischer et al. 2017). In T cell-specific A20 knockout mice, thymic Treg precursor cells ($CD4^+CD25^+GITR^+FOXP3^-$) were also significantly increased and GITR expression was increased on precursor Treg than WT mice (Fischer et al. 2017). GITR is a member of the tumor necrosis factor receptor superfamily (TNFRSF) and plays an important role in Treg development. It was shown that TNFRF agonists enhance the activation of IL-2R and STAT5 which promotes Treg precursor cells develop into Treg cells (Mahmud et al. 2014). Knockout of GITR impaired Treg development, inhibition of GITR, OX40, and TNFR2 in vivo resulted in a severe reduction in Treg development (Mahmud et al. 2014), suggesting a role for GITR in promoting Treg development. Thus, knockout of A20 in T cells leads to the elevated GITR expression and Treg development in the thymus (Fischer et al. 2017).

10.3.3.6 CYLD Inhibits Treg Development

The deubiquitinating enzyme CYLD does not only regulate thymocyte development toward $CD4^+$ and $CD8^+$ SP T cells, but also has an important role in Treg development. Knockout of exons 2 and 3 of the *Cyld* gene significantly increased peripheral Treg cells in mice, whereas thymic Treg were unaffected (Zhao et al. 2011). The knockout of CYLD (exons 2 and 3 deletion) was found to significant increase the in vitro TGF-β induced Treg differentiation from naive CD4 T cells (Zhao et al. 2011). Molecular mechanism studies revealed that CYLD binds SMAD7 and mediated deubiquitination lysine 360 and 374 of SMAD7, thereby inhibiting activation of kinases TAK1 and p38 and transcription factor AP-1 (Zhao et al. 2011). Knockout of SMAD7 or inhibition of p38 kinase activity inhibited Treg differentiation.

Using the mice lacking the exons 7 and 8 of *Cyld*, the investigators found another mechanism by which CYLD inhibited thymic Treg development. The exons 7 and 8 of *Cyld* encode the domains mediating the interaction of CYLD with NEMO or TRAF2 (Hovelmeyer et al. 2007), and that T cells isolated from mice knocking out exons 7 and 8 of *Cyld* (*Cyldex7/8*) have persistently activated NF-κB, and produce more proinflammatory factors such as IL-17A, IL-6, and IFN-γ than WT cells (Reissig et al. 2012). The thymic and peripheral Treg cells significant increase in *Cyldex7/8* mice, suggesting that CYLD inhibits Treg development. In addition, these *Cyldex7/8* Treg cells had reduced CD25 and CTLA-4 expression, and impaired suppressive function. Together, CYLD inhibits thymic Treg development on the one hand, and promotes Treg function on the other hand, thus affecting T cell activation (Reissig et al. 2012).

10.3.4 The Ubiquitin Regulates TEC Function and Self-tolerance

Positive and negative selection are highly regulated processes that rely on the well-organized thymic microenvironment composed of thymic stromal cells including thymic epithelial cells (TECs), endothelial cells, mesenchymal cells, and myeloid cells such as dendritic DCs and macrophages (Ladi et al. 2006; Han and Zúñiga-Pflücker 2021).

TECs are critical in the establishment of TCR diversity and self-tolerance, providing thymocytes with signals for migration, proliferation, differentiation, and survival (Anderson and Takahama 2012; Abramson and Anderson 2017). TECs play important roles in both positive and negative selection of αβT cells, Treg, and the production of other non-traditional T cells (Cosway et al. 2017; Kadouri et al. 2020). Based on the localization in the thymus, feature molecules and function, TECs are divided into cTECs and mTECs (Anderson and Takahama 2012; Takahama et al. 2017). cTECs are present in the thymic cortical region and constitutes a special thymic cortical microenvironment that supports the early development of T cells at the DN and DP stages and positive selection (Klein et al.

2014; Anderson and Takahama 2012; Wang et al. 2019). mTECs are located in the medullary region and play an important role in negative selection of T cells and self-tolerance (Derbinski et al. 2001; Danan-Gotthold et al. 2016). By randomly expressing peripheral tissue-specific antigen (TRA), mTECs assists T cells in negative selection, delete the auto-reactive T cells, and establish central tolerance, thereby prevent autoreactive T cells entering the periphery from attacking self-antigens and causing autoimmune disease (Mathis and Benoist 2004; Marx et al. 2021).

AIRE is a key transcription factor for TRA expression in mTECs, and mutation or deletion of AIRE leads to reduced TRA expression in mice, resulting in disturbed central tolerance establishment, and production of autoimmune T cell and autoimmune diseases (Anderson et al. 2002; Anderson et al. 2005; Perry et al. 2014; Oftedal et al. 2015). Loss-of-function mutations of *AIRE* also lead to the development of multiorgan autoimmune diseases in humans, such as auto immune endocrine disorder candidiasis ectodermal dystrophy (APECED) or autoimmune polyglandular syndrome type 1 (APS-1) (Nagamine et al. 1997; Zhu et al. 2017). AIRE has two plant homologous structural domains (PHDs) (Kumar et al. 2001) and the PHD has a similar structure and sequence to RING domain (Pascual et al. 2000). It has been shown that PHD1 of AIRE have E3 ubiquitin ligase activity, which might play a role in the establishment of central tolerance (Uchida et al. 2004). It is debatable whether AIRE is an E3 ubiquitin ligase, as other study indicated that the PHD of AIRE does not have E3 ubiquitin ligase activity (Bottomley et al. 2005).

10.3.4.1 TRAF6 Promotes mTEC Development

TRAF6 acts as an E3 ubiquitin ligase mediating the activation of classical NF-κB pathway downstream of TLR and TCR, and the knockout of TRAF6 led to thymic atrophy in mice. Further studies revealed that the distribution of mTEC in the thymus was disrupted and mature mTEC was reduced in TRAF6 KO mice, indicating that TRAF6 promotes mTEC maturation and organogenesis (Naito et al. 1999). Meanwhile, transplantation of TRAF6-KO embryos into thymus-deficient nude mice resulted in inflammation in the lungs, liver, pancreas, kidney, and splenomegaly, and the presence of high level of autoantibodies against the pancreas and lungs in the serum (Akiyama et al. 2005), suggesting that TRAF6 KO leads to the development of autoimmune diseases. The NF-κB signaling pathway plays a key role in the development of mTEC. NF-κB pathway is divided into classical and nonclassical pathway. Numerous studies have demonstrated the role of nonclassical NF-κB in mTEC maturation and function. Knockout of molecules such as RANK, CD40, LIGHT, and LTβR (which primarily activate nonclassical NF-κB signaling) in mice causes impaired mTEC development and function, resulting in various degrees of autoimmune disease symptoms (Abramson and Anderson 2017; Kadouri et al. 2020; Naito et al. 1999). In addition, knockout of RelB, the nonclassical NF-κB signaling transcription factor, resulted in reduced functional mTEC and decreased AIRE expression, and the KO mice showed the infiltration of inflammatory cell in multiple organs (Heino et al. 2000; Weih et al. 1995; Burkly et al. 1995). These suggest an important role for nonclassical NF-κB signaling in mTEC development.

Fetal thymic organ culture (FTOC) experiments using TRAF6 knockout embryos indicated that RANKL and CD40L stimulation failed to induce $UEA^{+}Aire^{+}$ mTEC production in TRAF6 KO cells, while the expression of AIRE and TRA were blocked (Akiyama et al. 2008). Further studies revealed that TRAF6 regulates RelB expression, in the TRAF6 KO thymic medullary cells, the expression level of RelB was significantly reduced compared to control cells (Akiyama et al. 2005). RANKL and CD40L stimulation of TRAF6 KO embryonic thymus also failed to induce RelB expression, suggesting that TRAF6 regulates mTEC development by affecting RelB expression (Akiyama et al. 2005; Akiyama et al. 2008). Further studies dominated that knockout TRAF6 in the TEC did not affect cTEC development, but led to a substantial reduction in mature functional mTEC and immature mTEC (Bonito et al. 2013). In addition, mice with TEC-specific deletion of TRAF6 showed reduced body weight, inflammatory responses in the liver, lung, and kidney, and presence of autoantibodies against the liver, lung, kidney, small intestine, and colon (Bonito et al. 2013). The transplantation of T cells isolated from the TRAF6 KO liver into immunodeficient mice resulted in the development of AIH symptoms (Bonito et al. 2013). These results suggest that TRAF6 directly regulates mTEC development and autoimmune disease development. However, the detailed molecular mechanism of how TRAF6 as an E3 ubiquitin ligase regulates RelB expression and mTEC development has not been well studied.

10.3.4.2 CYLD Promotes mTEC Development

The function of the deubiquitinating enzyme CYLD in the positive selection and Treg development has been discussed above, and studies have shown that CYLD also regulates mTEC development and negative selection (Reissig et al. 2015). In the thymus of *Cyldex7/8*, the development from immature TEC to mature mTEC is blocked, and cell number of mature mTEC is significantly reduced. Given the role of CYLD in T cell development, the T-cell specific CYLD knockout mice had the normal mTEC development, suggesting that CYLD directly regulates mTEC development rather than via T cells. The TCR of HY TCR transgenic mice (HYtg) specifically recognizes the peptide smcy of Y chromosome, so the thymocytes expressing HY-responsive TCR were deleted from the male HYtg thymus, and DP and SP cells are almost absent; meanwhile, the T-cell development in female HYtg mice only generate $CD8^+$ T cells (Kisielow et al. 1988). To investigate the effect of knockout CYLD (exons 7 and 8) on negative selection of thymocytes, HY TCR transgenic mice (HYtg) were bred with *Cyldex7/8* mice to generate HYtg *Cyldex7/8* mice. DP and $CD4^+$ SP and $CD8^+$ SP cells were detected in the thymus of HYtg *Cyldex7/8* male mice, indicating that knockout of *Cyld* (exons 7 and 8) caused the impaired thymic negative selection and the production of self-reactive T cells. It is suggested that CYLD plays a role in the self-tolerance and negative selection by affecting mTEC development and maturation (Reissig et al. 2015).

10.3.4.3 LUBAC Promotes TEC Survival

LUBAC also have multiple functions in T-cell development and TEC. Specific knockout of HOIL-1 or HOIP in TEC affects TEC development and survival, and T cell development (Jain et al. 2021). TEC specifically Rbck1 ($Hoil\text{-}1^{\Delta Foxn1}$) or Rnf31 ($Hoip^{\Delta Foxn1}$) knockout mice suffer from severe thymic atrophy in adult mice (8 weeks old), with significant reduced cell number of DN, DP, and SP cell (Jain et al. 2021). Within DN cells, that DN3 development was impaired in KO thymus (Jain et al. 2021). Additionally, TECs in adult $Hoil\text{-}1^{\Delta Foxn1}$ and $Hoip^{\Delta Foxn1}$ mice were significantly reduced, the immunofluorescence staining showed the defected histological structure in both the K8+ cortical and $K5^+$ or $UEA\text{-}1^+$ medullary regions in the thymus. More importantly, the AIRE+ mTECs were almost diminished with the concomitant increased $ERTR7^+$ fibroblasts (Jain et al. 2021), mTEC including $AIRE^+$ mTEC in embryonic and newborn Hoil-1 ΔFoxn1 mice were also reduced (Jain et al. 2021). RNA sequencing of WT and HOIL-1 KO TECs indicated that the expression of programmed cell death associated genes was altered in HOIL-1 KO TECs, such as the expression of *Mlkl* and *Casp8* was upregulated in HOIL-1 KO cTEC and $mTEC^{hi}$ ($MHCII^{hi}CD80^{hi}$) compared to control cells (Jain et al. 2021). When *Mlkl* and *Casp8* were knocked-out in $Hoil\text{-}1^{\Delta Foxn1}$ mice, the number of total thymus cells, DN, DP, and SP cells returned to normal levels, but the numbers of cTEC and mTEC only were partially recovered, suggesting that HOIL-1 may affect caspase-8/MLKL-mediated apoptosis by regulating caspase-8/MLKL (Jain et al. 2021). The effect of knockdown of *Mlkl* and *Casp8* genes in $Hoil\text{-}1^{\Delta Foxn1}$ mice on thymocyte development is not necessarily due to TEC, given that *Mlkl* and

Casp8 genes are knocked down systemically (Jain et al. 2021). More study is needed to investigate the function of LUBAC on AIRE⁺ mTEC and central tolerance.

10.4 Conclusion and Perspectives

T cells play an important role in adaptive immunity. Hematopoietic progenitor cells from bone marrow enter the thymus and undergo T-cell commitment, TCR rearrangement, β-selection, positive selection, negative selection, and finally become mature T cells with diverse TCR egressing to the periphery, constituting a diverse TCR pool that enables the host to recognize the various antigens (Boehm 2008; Manley et al. 2011; Savage et al. 2020; Zuniga-Pflucker 2004; Ma et al. 2013). In the past decades, the roles of ubiquitination in T cell activation and TCR signaling have been extensively studied, but the functions of ubiquitination in T-cell development are limited. Ubiquitination regulates T-cell development, especially in the positive and negative selection. A variety of E3 ubiquitin ligases regulate key molecules downstream of TCR signaling, ensuring that optimal TCR signaling for T-cell development. Meanwhile, studies have reported the critical roles of E3 ubiquitin ligases and deubiquitinating enzymes in Treg development, by regulating key signaling pathways such as IL-2 or TGF-β pathways. In addition, negative selection of thymocytes is required to eliminate auto-reactive T cell and to establish central tolerance, a process in which mTEC plays an essential role. mTEC development and function are controlled by E3 ubiquitin ligases and deubiquitinating enzymes through regulation of NF-κB signaling pathways. Impaired negative selection leads to the release of auto-reactive T cells into the periphery, which may cause autoimmune diseases. The investigation of how ubiquitin affect T-cell development and central tolerance is providing the key information to understand T cell development as well as the treatment of autoimmune diseases.

References

Abramson J, Anderson G (2017) Thymic epithelial cells. Annu Rev Immunol 35:85–118. https://doi.org/10.1146/annurev-immunol-051116-052320

Adoro S et al (2017) Post-translational control of T cell development by the ESCRT protein CHMP5. Nat Immunol 18:780–790. https://doi.org/10.1038/ni.3764

Aifantis I, Buer J, von Boehmer H, Azogui O (1997) Essential role of the pre-T cell receptor in allelic exclusion of the T cell receptor beta locus. Immunity 7:601–607. https://doi.org/10.1016/s1074-7613(00)80381-7

Aifantis I, Gounari F, Scorrano L, Borowski C, von Boehmer H (2001) Constitutive pre-TCR signaling promotes differentiation through Ca2+ mobilization and activation of NF-kappaB and NFAT. Nat Immunol 2:403–409. https://doi.org/10.1038/87704

Aifantis I, Mandal M, Sawai K, Ferrando A, Vilimas T (2006) Regulation of T-cell progenitor survival and cell-cycle entry by the pre-T-cell receptor. Immunol Rev 209:159–169. https://doi.org/10.1111/j.0105-2896.2006.00343.x

Akiyama T et al (2005) Dependence of self-tolerance on TRAF6-directed development of thymic stroma. Science 308:248–251. https://doi.org/10.1126/science.1105677

Akiyama T et al (2008) The tumor necrosis factor family receptors RANK and CD40 cooperatively establish the thymic medullary microenvironment and self-tolerance. Immunity 29:423–437. https://doi.org/10.1016/j.immuni.2008.06.015

Anderson G, Takahama Y (2012) Thymic epithelial cells: working class heroes for T cell development and repertoire selection. Trends Immunol 33:256–263. https://doi.org/10.1016/j.it.2012.03.005

Anderson MS et al (2002) Projection of an immunological self shadow within the thymus by the aire protein. Science 298:1395–1401. https://doi.org/10.1126/science.1075958

Anderson MS et al (2005) The cellular mechanism of Aire control of T cell tolerance. Immunity 23:227–239. https://doi.org/10.1016/j.immuni.2005.07.005

Aschenbrenner K et al (2007) Selection of Foxp3+ regulatory T cells specific for self antigen expressed and presented by Aire+ medullary thymic epithelial cells. Nat Immunol 8:351–358. https://doi.org/10.1038/ni1444

Awong G, Zúñiga-Pflücker JC (2014) Module in biomedical sciences. Elsevier, Amsterdam

Azad N et al (2010) Role of oxidative/nitrosative stress-mediated Bcl-2 regulation in apoptosis and malignant transformation. Ann N Y Acad Sci 1203:1–6. https://doi.org/10.1111/j.1749-6632.2010.05608.x

Bachmaier K et al (2000) Negative regulation of lymphocyte activation and autoimmunity by the molecular adaptor Cbl-b. Nature 403:211–216. https://doi.org/10.1038/35003228

Baravalle G et al (2011) Ubiquitination of CD86 is a key mechanism in regulating antigen presentation by dendritic cells. J Immunol 187:2966–2973. https://doi.org/10.4049/jimmunol.1101643
Beal AM, Ramos-Hernandez N, Riling CR, Nowelsky EA, Oliver PM (2011) TGF-beta induces the expression of the adaptor Ndfip1 to silence IL-4 production during iTreg cell differentiation. Nat Immunol 13:77–85. https://doi.org/10.1038/ni.2154
Berger MA et al (1997) Subunit composition of pre-T cell receptor complexes expressed by primary thymocytes: CD3 delta is physically associated but not functionally required. J Exp Med 186:1461–1467. https://doi.org/10.1084/jem.186.9.1461
Berndsen CE, Wolberger C (2014) New insights into ubiquitin E3 ligase mechanism. Nat Struct Mol Biol 21:301–307. https://doi.org/10.1038/nsmb.2780
Blom B et al (1999) Disruption of alpha beta but not of gamma delta T cell development by overexpression of the helix-loop-helix protein Id3 in committed T cell progenitors. EMBO J 18:2793–2802. https://doi.org/10.1093/emboj/18.10.2793
Boehm T (2008) Thymus development and function. Curr Opin Immunol 20:178–184. https://doi.org/10.1016/j.coi.2008.03.001
Boisson B et al (2015) Human HOIP and LUBAC deficiency underlies autoinflammation, immunodeficiency, amylopectinosis, and lymphangiectasia. J Exp Med 212:939–951. https://doi.org/10.1084/jem.20141130
Bonito AJ et al (2013) Medullary thymic epithelial cell depletion leads to autoimmune hepatitis. J Clin Invest 123:3510–3524. https://doi.org/10.1172/JCI65414
Bottomley MJ et al (2005) NMR structure of the first PHD finger of autoimmune regulator protein (AIRE1). Insights into autoimmune polyendocrinopathy-candidiasis-ectodermal dystrophy (APECED) disease. J Biol Chem 280:11505–11512. https://doi.org/10.1074/jbc.M413959200
Brandle D, Muller S, Muller C, Hengartner H, Pircher H (1994) Regulation of RAG-1 and CD69 expression in the thymus during positive and negative selection. Eur J Immunol 24:145–151. https://doi.org/10.1002/eji.1830240122
Brummelkamp TR, Nijman SM, Dirac AM, Bernards R (2003) Loss of the cylindromatosis tumour suppressor inhibits apoptosis by activating NF-kappaB. Nature 424:797–801. https://doi.org/10.1038/nature01811
Burchill MA, Yang J, Vogtenhuber C, Blazar BR, Farrar MA (2007) IL-2 receptor beta-dependent STAT5 activation is required for the development of Foxp3+ regulatory T cells. J Immunol 178:280–290. https://doi.org/10.4049/jimmunol.178.1.280
Burchill MA et al (2008) Linked T cell receptor and cytokine signaling govern the development of the regulatory T cell repertoire. Immunity 28:112–121. https://doi.org/10.1016/j.immuni.2007.11.022
Burkly L et al (1995) Expression of relB is required for the development of thymic medulla and dendritic cells. Nature 373:531–536. https://doi.org/10.1038/373531a0
Cabarrocas J et al (2006) Foxp3+ CD25+ regulatory T cells specific for a neo-self-antigen develop at the double-positive thymic stage. Proc Natl Acad Sci USA 103:8453–8458. https://doi.org/10.1073/pnas.0603086103
Caramalho I, Nunes-Cabaco H, Foxall RB, Sousa AE (2015) Regulatory T-cell development in the human thymus. Front Immunol 6:395. https://doi.org/10.3389/fimmu.2015.00395
Carrasco YR, Navarro MN, de Yebenes VG, Ramiro AR, Toribio ML (2002) Regulation of surface expression of the human pre-T cell receptor complex. Semin Immunol 14:325–334. https://doi.org/10.1016/s1044-5323(02)00065-9
Carrasco YR, Navarro MN, Toribio ML (2003) A role for the cytoplasmic tail of the pre-T cell receptor (TCR) alpha chain in promoting constitutive internalization and degradation of the pre-TCR. J Biol Chem 278:14507–14513. https://doi.org/10.1074/jbc.M204944200
Casero D et al (2015) Long non-coding RNA profiling of human lymphoid progenitor cells reveals transcriptional divergence of B cell and T cell lineages. Nat Immunol 16:1282–1291. https://doi.org/10.1038/ni.3299
Caton AJ et al (2014) Strength of TCR signal from self-peptide modulates autoreactive thymocyte deletion and Foxp3(+) Treg-cell formation. Eur J Immunol 44:785–793. https://doi.org/10.1002/eji.201343767
Ceredig R, Rolink T (2002) A positive look at double-negative thymocytes. Nat Rev Immunol 2:888–897. https://doi.org/10.1038/nri937
Chan S, Correia-Neves M, Benoist C, Mathis D (1998) CD4/CD8 lineage commitment: matching fate with competence. Immunol Rev 165:195–207. https://doi.org/10.1111/j.1600-065x.1998.tb01240.x
Chang JH et al (2014) TRAF3 regulates the effector function of regulatory T cells and humoral immune responses. J Exp Med 211:137–151. https://doi.org/10.1084/jem.20131019
Chiang YJ et al (2000) Cbl-b regulates the CD28 dependence of T-cell activation. Nature 403:216–220. https://doi.org/10.1038/35003235
Coornaert B et al (2008) T cell antigen receptor stimulation induces MALT1 paracaspase-mediated cleavage of the NF-kappaB inhibitor A20. Nat Immunol 9:263–271. https://doi.org/10.1038/ni1561
Cosway EJ et al (2017) Redefining thymus medulla specialization for central tolerance. J Exp Med 214:3183–3195. https://doi.org/10.1084/jem.20171000
Danan-Gotthold M, Guyon C, Giraud M, Levanon EY, Abramson J (2016) Extensive RNA editing and splicing increase immune self-representation diversity in medullary thymic epithelial cells. Genome Biol 17:219. https://doi.org/10.1186/s13059-016-1079-9
Dardalhon V et al (2008) IL-4 inhibits TGF-beta-induced Foxp3+ T cells and, together with TGF-beta, generates

IL-9+ IL-10+ Foxp3(-) effector T cells. Nat Immunol 9:1347–1355. https://doi.org/10.1038/ni.1677
De Gassart A et al (2008) MHC class II stabilization at the surface of human dendritic cells is the result of maturation-dependent MARCH I down-regulation. Proc Natl Acad Sci USA 105:3491–3496. https://doi.org/10.1073/pnas.0708874105
Derbinski J, Schulte A, Kyewski B, Klein L (2001) Promiscuous gene expression in medullary thymic epithelial cells mirrors the peripheral self. Nat Immunol 2:1032–1039. https://doi.org/10.1038/ni723
Deshaies RJ, Joazeiro CA (2009) RING domain E3 ubiquitin ligases. Annu Rev Biochem 78:399–434. https://doi.org/10.1146/annurev.biochem.78.101807.093809
Dik WA et al (2005) New insights on human T cell development by quantitative T cell receptor gene rearrangement studies and gene expression profiling. J Exp Med 201:1715–1723. https://doi.org/10.1084/jem.20042524
Dufner A et al (2015) The ubiquitin-specific protease USP8 is critical for the development and homeostasis of T cells. Nat Immunol 16:950–960. https://doi.org/10.1038/ni.3230
Duwel M et al (2009) A20 negatively regulates T cell receptor signaling to NF-kappaB by cleaving Malt1 ubiquitin chains. J Immunol 182:7718–7728. https://doi.org/10.4049/jimmunol.0803313
Dye BT, Schulman BA (2007) Structural mechanisms underlying posttranslational modification by ubiquitin-like proteins. Annu Rev Biophys Biomol Struct 36:131–150. https://doi.org/10.1146/annurev.biophys.36.040306.132820
Emmerich CH, Schmukle AC, Walczak H (2011) The emerging role of linear ubiquitination in cell signaling. Sci Signal 4:re5. https://doi.org/10.1126/scisignal.2002187
Falk I, Biro J, Kohler H, Eichmann K (1996) Proliferation kinetics associated with T cell receptor-beta chain selection of fetal murine thymocytes. J Exp Med 184:2327–2339. https://doi.org/10.1084/jem.184.6.2327
Fang D et al (2002) Dysregulation of T lymphocyte function in itchy mice: a role for Itch in TH2 differentiation. Nat Immunol 3:281–287. https://doi.org/10.1038/ni763
Fehling HJ, von Boehmer H (1997) Early alpha beta T cell development in the thymus of normal and genetically altered mice. Curr Opin Immunol 9:263–275. https://doi.org/10.1016/s0952-7915(97)80146-x
Fero ML et al (1996) A syndrome of multiorgan hyperplasia with features of gigantism, tumorigenesis, and female sterility in p27(Kip1)-deficient mice. Cell 85:733–744. https://doi.org/10.1016/s0092-8674(00)81239-8
Fischer JC et al (2017) A20 restrains thymic regulatory T cell development. J Immunol 199:2356–2365. https://doi.org/10.4049/jimmunol.1602102
Fontenot JD, Gavin MA, Rudensky AY (2003) Foxp3 programs the development and function of CD4+CD25+ regulatory T cells. Nat Immunol 4:330–336. https://doi.org/10.1038/ni904
Fontenot JD, Rasmussen JP, Gavin MA, Rudensky AY (2005) A function for interleukin 2 in Foxp3-expressing regulatory T cells. Nat Immunol 6:1142–1151. https://doi.org/10.1038/ni1263
Fu G et al (2013) Themis sets the signal threshold for positive and negative selection in T-cell development. Nature 504:441–445. https://doi.org/10.1038/nature12718
Fu G et al (2014) Fine-tuning T cell receptor signaling to control T cell development. Trends Immunol 35:311–318. https://doi.org/10.1016/j.it.2014.05.003
Fujimoto Y et al (2002) CD83 expression influences CD4+ T cell development in the thymus. Cell 108:755–767. https://doi.org/10.1016/s0092-8674(02)00673-6
Galy A, Verma S, Bárcena A, Spits H (1993) Precursors of CD3+CD4+CD8+ cells in the human thymus are defined by expression of CD34. Delineation of early events in human thymic development. J Exp Med 178:391–401. https://doi.org/10.1084/jem.178.2.391
Gascoigne NR, Rybakin V, Acuto O, Brzostek J (2016) TCR signal strength and T cell development. Annu Rev Cell Dev Biol 32:327–348. https://doi.org/10.1146/annurev-cellbio-111315-125324
Gerlach B et al (2011) Linear ubiquitination prevents inflammation and regulates immune signalling. Nature 471:591–596. https://doi.org/10.1038/nature09816
Germain RN (2002) T-cell development and the CD4-CD8 lineage decision. Nat Rev Immunol 2:309–322. https://doi.org/10.1038/nri798
Godfrey DI, Kennedy J, Suda T, Zlotnik A (1993) A developmental pathway involving four phenotypically and functionally distinct subsets of CD3-CD4-CD8-triple-negative adult mouse thymocytes defined by CD44 and CD25 expression. J Immunol 150:4244–4252
Godfrey DI, Kennedy J, Mombaerts P, Tonegawa S, Zlotnik A (1994) Onset of TCR-beta gene rearrangement and role of TCR-beta expression during CD3-CD4-CD8-thymocyte differentiation. J Immunol 152:4783–4792
Greenblatt MB et al (2015) CHMP5 controls bone turnover rates by dampening NF-kappaB activity in osteoclasts. J Exp Med 212:1283–1301. https://doi.org/10.1084/jem.20150407
Groettrup M et al (1993) A novel disulfide-linked heterodimer on pre-T cells consists of the T cell receptor beta chain and a 33 kd glycoprotein. Cell 75:283–294. https://doi.org/10.1016/0092-8674(93)80070-u
Guo F, Iclozan C, Suh WK, Anasetti C, Yu XZ (2008) CD28 controls differentiation of regulatory T cells from naive CD4 T cells. J Immunol 181:2285–2291. https://doi.org/10.4049/jimmunol.181.4.2285
Hacker H, Tseng PH, Karin M (2011) Expanding TRAF function: TRAF3 as a tri-faced immune regulator. Nat Rev Immunol 11:457–468. https://doi.org/10.1038/nri2998
Haddad R et al (2006) Dynamics of thymus-colonizing cells during human development.

Immunity 24:217–230. https://doi.org/10.1016/j.immuni.2006.01.008
Han J, Zúñiga-Pflücker JC (2021) A 2020 view of thymus stromal cells in T cell development. J Immunol 206:249–256. https://doi.org/10.4049/jimmunol.2000889
Hao B et al (2005) Structural basis of the Cks1-dependent recognition of p27(Kip1) by the SCF(Skp2) ubiquitin ligase. Mol Cell 20:9–19. https://doi.org/10.1016/j.molcel.2005.09.003
Hao QL et al (2008) Human intrathymic lineage commitment is marked by differential CD7 expression: identification of CD7-lymphc-myeloid thymic progenitors. Blood 111:1318–1326. https://doi.org/10.1182/blood-2007-08-106294
Harada Y et al (2010) Transcription factors Foxo3a and Foxol couple the E3 ligase Cbl-b to the induction of Foxp3 expression in induced regulatory T cells. J Exp Med 207:1381–1391. https://doi.org/10.1084/jem.20100004
Heino M et al (2000) RNA and protein expression of the murine autoimmune regulator gene (Aire) in normal, RelB-deficient and in NOD mouse. Eur J Immunol 30:1884–1893. https://doi.org/10.1002/1521-4141(200007)30:7<1884::AID-IMMU1884>3.0.CO;2-P
Hershko A, Ciechanover A (1998) The ubiquitin system. Annu Rev Biochem 67:425–479. https://doi.org/10.1146/annurev.biochem.67.1.425
Hoffman ES et al (1996) Productive T-cell receptor beta-chain gene rearrangement: coincident regulation of cell cycle and clonality during development in vivo. Genes Dev 10:948–962. https://doi.org/10.1101/gad.10.8.948
Hosokawa HA-O, Rothenberg EA-OX (2021) How transcription factors drive choice of the T cell fate. Nat Rev Immunol 21:162–176
Hovelmeyer N et al (2007) Regulation of B cell homeostasis and activation by the tumor suppressor gene CYLD. J Exp Med 204:2615–2627. https://doi.org/10.1084/jem.20070318
Hsieh CS, Lee HM, Lio CW (2012) Selection of regulatory T cells in the thymus. Nat Rev Immunol 12:157–167. https://doi.org/10.1038/nri3155
Hu H, Sun SC (2016) Ubiquitin signaling in immune responses. Cell Res 26:457–483. https://doi.org/10.1038/cr.2016.40
Huang F, Gu H (2008) Negative regulation of lymphocyte development and function by the Cbl family of proteins. Immunol Rev 224:229–238. https://doi.org/10.1111/j.1600-065X.2008.00655.x
Huang F et al (2006) Establishment of the major compatibility complex-dependent development of CD4+ and CD8+ T cells by the Cbl family proteins. Immunity 25:571–581. https://doi.org/10.1016/j.immuni.2006.08.021
Husnjak K, Dikic I (2012) Ubiquitin-binding proteins: decoders of ubiquitin-mediated cellular functions. Annu Rev Biochem 81:291–322. https://doi.org/10.1146/annurev-biochem-051810-094654
Itoh M et al (1999) Thymus and autoimmunity: production of CD25+CD4+ naturally anergic and suppressive T cells as a key function of the thymus in maintaining immunologic self-tolerance. J Immunol 162:5317–5326
Jain R et al (2021) Dual roles for LUBAC signaling in thymic epithelial cell development and survival. Cell Death Differ 28:2946–2956. https://doi.org/10.1038/s41418-021-00850-8
Joachims ML, Chain JL, Hooker SW, Knott-Craig CJ, Thompson LF (2006) Human alpha beta and gamma delta thymocyte development: TCR gene rearrangements, intracellular TCR beta expression, and gamma delta developmental potential–differences between men and mice. J Immunol 176:1543–1552. https://doi.org/10.4049/jimmunol.176.3.1543
Johnston JA, Bacon CM, Riedy MC, O'Shea JJ (1996) Signaling by IL-2 and related cytokines: JAKs, STATs, and relationship to immunodeficiency. J Leukoc Biol 60:441–452. https://doi.org/10.1002/jlb.60.4.441
Jones JM, Simkus C (2009) The roles of the RAG1 and RAG2 "non-core" regions in V(D)J recombination and lymphocyte development. Arch Immunol Ther Exp 57:105–116. https://doi.org/10.1007/s00005-009-0011-3
Jordan MS et al (2001) Thymic selection of CD4+CD25+ regulatory T cells induced by an agonist self-peptide. Nat Immunol 2:301–306. https://doi.org/10.1038/86302
Kadouri N, Nevo S, Goldfarb Y, Abramson J (2020) Thymic epithelial cell heterogeneity: TEC by TEC. Nat Rev Immunol 20:239–253. https://doi.org/10.1038/s41577-019-0238-0
Kerdiles YM et al (2009) Foxol links homing and survival of naive T cells by regulating L-selectin, CCR7 and interleukin 7 receptor. Nat Immunol 10:176–184. https://doi.org/10.1038/ni.1689
Kisielow P, Bluthmann H, Staerz UD, Steinmetz M, von Boehmer H (1988) Tolerance in T-cell-receptor transgenic mice involves deletion of nonmature CD4+8+ thymocytes. Nature 333:742–746. https://doi.org/10.1038/333742a0
Kiyokawa H et al (1996) Enhanced growth of mice lacking the cyclin-dependent kinase inhibitor function of p27(Kip1). Cell 85:721–732. https://doi.org/10.1016/s0092-8674(00)81238-6
Klein L, Kyewski B, Allen PM, Hogquist KA (2014) Positive and negative selection of the T cell repertoire: what thymocytes see (and don't see). Nat Rev Immunol 14:377–391. https://doi.org/10.1038/nri3667
Kossatz U et al (2004) Skp2-dependent degradation of p27kip1 is essential for cell cycle progression. Genes Dev 18:2602–2607. https://doi.org/10.1101/gad.321004
Kovalenko A et al (2003) The tumour suppressor CYLD negatively regulates NF-kappaB signalling by deubiquitination. Nature 424:801–805. https://doi.org/10.1038/nature01802
Kreslavsky T et al (2012) beta-Selection-induced proliferation is required for alphabeta T cell differentia-

tion. Immunity 37:840–853. https://doi.org/10.1016/j.immuni.2012.08.020

Krueger A, Ziętara N, Łyszkiewicz M (2017) T cell development by the numbers. Trends Immunol 38:128–139. https://doi.org/10.1016/j.it.2016.10.007

Ktorza S et al (1996) Complete TCR-delta rearrangements and partial (D-J) recombination of the TCR-beta locus in CD34+7+ precursors from human cord blood. J Immunol 156:4120–4127

Kumar PG et al (2001) The autoimmune regulator (AIRE) is a DNA-binding protein. J Biol Chem 276:41357–41364. https://doi.org/10.1074/jbc.M104898200

Kumar BV, Connors TJ, Farber DL (2018) Human T cell development, localization, and function throughout life. Immunity 48(2):202–213

Kurobe H et al (2006) CCR7-dependent cortex-to-medulla migration of positively selected thymocytes is essential for establishing central tolerance. Immunity 24:165–177. https://doi.org/10.1016/j.immuni.2005.12.011

Kyewski B, Derbinski J (2004) Self-representation in the thymus: an extended view. Nat Rev Immunol 4:688–698. https://doi.org/10.1038/nri1436

Ladi E, Yin X, Chtanova T, Robey EA (2006) Thymic microenvironments for T cell differentiation and selection. Nat Immunol 7:338–343. https://doi.org/10.1038/ni1323

Larkin J et al (2008) CD4+CD25+ regulatory T cell repertoire formation shaped by differential presentation of peptides from a self-antigen. J Immunol 180:2149–2157. https://doi.org/10.4049/jimmunol.180.4.2149

Laufer TM, DeKoning J, Markowitz JS, Lo D, Glimcher LH (1996) Unopposed positive selection and autoreactivity in mice expressing class II MHC only on thymic cortex. Nature 383:81–85. https://doi.org/10.1038/383081a0

Le J et al (2020) Single-cell RNA-Seq mapping of human thymopoiesis reveals lineage specification trajectories and a commitment spectrum in T cell development. Immunity 52:1105–1118. https://doi.org/10.1016/j.immuni.2020.05.010

Lee HM, Bautista JL, Hsieh CS (2011) Thymic and peripheral differentiation of regulatory T cells. Adv Immunol 112:25–71. https://doi.org/10.1016/B978-0-12-387827-4.00002-4

Legrand N et al (2006) Transient accumulation of human mature thymocytes and regulatory T cells with CD28 superagonist in "human immune system" Rag2(-/-) gammac(-/-) mice. Blood 108:238–245. https://doi.org/10.1182/blood-2006-01-0190

Levelt CN, Wang B, Ehrfeld A, Terhorst C, Eichmann K (1995) Regulation of T cell receptor (TCR)-beta locus allelic exclusion and initiation of TCR-alpha locus rearrangement in immature thymocytes by signaling through the CD3 complex. Eur J Immunol 25:1257–1261. https://doi.org/10.1002/eji.1830250519

Linette GP et al (1994) Bcl-2 is upregulated at the CD4+ CD8+ stage during positive selection and promotes thymocyte differentiation at several control points. Immunity 1:197–205. https://doi.org/10.1016/1074-7613(94)90098-1

Lio CW, Hsieh CS (2008) A two-step process for thymic regulatory T cell development. Immunity 28:100–111. https://doi.org/10.1016/j.immuni.2007.11.021

Liu YC (2007) The E3 ubiquitin ligase Itch in T cell activation, differentiation, and tolerance. Semin Immunol 19:197–205. https://doi.org/10.1016/j.smim.2007.02.003

Liu H et al (2016) Ubiquitin ligase MARCH 8 cooperates with CD83 to control surface MHC II expression in thymic epithelium and CD4 T cell selection. J Exp Med 213:1695–1703. https://doi.org/10.1084/jem.20160312

Liu H et al (2020) Ubiquitination of MHC Class II is required for development of regulatory but not conventional CD4(+) T cells. J Immunol 205:1207–1216. https://doi.org/10.4049/jimmunol.1901328

Liu C, Lan Y, Liu B, Zhang H, Hu H (2021) T cell development: old tales retold by single-cell RNA sequencing. Trends Immunol 42:165–175. https://doi.org/10.1016/j.it.2020.12.004

Lu M et al (2005) The earliest thymic progenitors in adults are restricted to T, NK, and dendritic cell lineage and have a potential to form more diverse TCRbeta chains than fetal progenitors. J Immunol 175:5848–5856. https://doi.org/10.4049/jimmunol.175.9.5848

Ma D, Wei Y, Liu F (2013) Regulatory mechanisms of thymus and T cell development. Dev Comp Immunol 39:91–102. https://doi.org/10.1016/j.dci.2011.12.013

Mahmud SA et al (2014) Costimulation via the tumor-necrosis factor receptor superfamily couples TCR signal strength to the thymic differentiation of regulatory T cells. Nat Immunol 15:473–481. https://doi.org/10.1038/ni.2849

Malyukova A et al (2007) The tumor suppressor gene hCDC4 is frequently mutated in human T-cell acute lymphoblastic leukemia with functional consequences for Notch signaling. Cancer Res 67:5611–5616. https://doi.org/10.1158/0008-5472.CAN-06-4381

Manley NR, Richie ER, Blackburn CC, Condie BG, Sage J (2011) Structure and function of the thymic microenvironment. Front Biosci 16:2461–2477. https://doi.org/10.2741/3866

Marx A et al (2021) Thymus and autoimmunity. Semin Immunopathol 43:45–64. https://doi.org/10.1007/s00281-021-00842-3

Masuda K et al (2005) Thymic anlage is colonized by progenitors restricted to T, NK, and dendritic cell lineages. J Immunol 174:2525–2532. https://doi.org/10.4049/jimmunol.174.5.2525

Mathis D, Benoist C (2004) Back to central tolerance. Immunity 20:509–516. https://doi.org/10.1016/s1074-7613(04)00111-6

Matsuki Y et al (2007) Novel regulation of MHC class II function in B cells. EMBO J 26:846–854. https://doi.org/10.1038/sj.emboj.7601556

Michie AM, Zúñiga-Pflücker JC (2002) Regulation of thymocyte differentiation: pre-TCR signals and beta-selection. Semin Immunol 14:311–323. https://doi.org/10.1016/s1044-5323(02)00064-7
Minami Y, Taniguchi T (1995) IL-2 signaling: recruitment and activation of multiple protein tyrosine kinases by the components of the IL-2 receptor. Curr Opin Cell Biol 7:156–162. https://doi.org/10.1016/0955-0674(95)80023-9
Nagamine K et al (1997) Positional cloning of the APECED gene. Nat Genet 17:393–398. https://doi.org/10.1038/ng1297-393
Naito A et al (1999) Severe osteopetrosis, defective interleukin-1 signalling and lymph node organogenesis in TRAF6-deficient mice. Genes Cells 4:353–362. https://doi.org/10.1046/j.1365-2443.1999.00265.x
Nakayama K et al (1996) Mice lacking p27(Kip1) display increased body size, multiple organ hyperplasia, retinal dysplasia, and pituitary tumors. Cell 85:707–720. https://doi.org/10.1016/s0092-8674(00)81237-4
Nakayama K et al (2004) Skp2-mediated degradation of p27 regulates progression into mitosis. Dev Cell 6:661–672. https://doi.org/10.1016/s1534-5807(04)00131-5
Naramura M, Kole HK, Hu RJ, Gu H (1998) Altered thymic positive selection and intracellular signals in Cbl-deficient mice. Proc Natl Acad Sci USA 95:15547–15552. https://doi.org/10.1073/pnas.95.26.15547
Naramura M et al (2002) c-Cbl and Cbl-b regulate T cell responsiveness by promoting ligand-induced TCR down-modulation. Nat Immunol 3:1192–1199. https://doi.org/10.1038/ni855
Ni X et al (2019) TRAF6 directs FOXP3 localization and facilitates regulatory T-cell function through K63-linked ubiquitination. EMBO J 38:e99766. https://doi.org/10.15252/embj.201899766
Nishiyama M, Nita A, Yumimoto K, Nakayama KI (2015) FBXL12-mediated degradation of ALDH3 is essential for trophoblast differentiation during placental development. Stem Cells 33:3327–3340. https://doi.org/10.1002/stem.2088
Nitta T, Nitta S, Lei Y, Lipp M, Takahama Y (2009) CCR7-mediated migration of developing thymocytes to the medulla is essential for negative selection to tissue-restricted antigens. Proc Natl Acad Sci USA 106:17129–17133. https://doi.org/10.1073/pnas.0906956106
O'Neil J et al (2007) FBW7 mutations in leukemic cells mediate NOTCH pathway activation and resistance to gamma-secretase inhibitors. J Exp Med 204:1813–1824. https://doi.org/10.1084/jem.20070876
Oftedal BE et al (2015) Dominant mutations in the autoimmune regulator AIRE are associated with common organ-specific autoimmune diseases. Immunity 42:1185–1196. https://doi.org/10.1016/j.immuni.2015.04.021
Oh J et al (2013) MARCH1-mediated MHCII ubiquitination promotes dendritic cell selection of natural regulatory T cells. J Exp Med 210:1069–1077. https://doi.org/10.1084/jem.20122695
Oliver PM et al (2006) Ndfip1 protein promotes the function of itch ubiquitin ligase to prevent T cell activation and T helper 2 cell-mediated inflammation. Immunity 25:929–940. https://doi.org/10.1016/j.immuni.2006.10.012
Onizawa M et al (2015) The ubiquitin-modifying enzyme A20 restricts ubiquitination of the kinase RIPK3 and protects cells from necroptosis. Nat Immunol 16:618–627. https://doi.org/10.1038/ni.3172
Ouyang W, Beckett O, Flavell RA, Li MO (2009) An essential role of the Forkhead-box transcription factor Foxo1 in control of T cell homeostasis and tolerance. Immunity 30:358–371. https://doi.org/10.1016/j.immuni.2009.02.003
Pacholczyk R, Ignatowicz H, Kraj P, Ignatowicz L (2006) Origin and T cell receptor diversity of Foxp3+CD4+CD25+ T cells. Immunity 25:249–259. https://doi.org/10.1016/j.immuni.2006.05.016
Panigada M et al (2002) Constitutive endocytosis and degradation of the pre-T cell receptor. J Exp Med 195:1585–1597. https://doi.org/10.1084/jem.20020047
Pao KC et al (2018) Activity-based E3 ligase profiling uncovers an E3 ligase with esterification activity. Nature 556:381–385. https://doi.org/10.1038/s41586-018-0026-1
Park Y, Jin HS, Aki D, Lee J, Liu YC (2014) The ubiquitin system in immune regulation. Adv Immunol 124:17–66. https://doi.org/10.1016/B978-0-12-800147-9.00002-9
Pascual J, Martinez-Yamout M, Dyson HJ, Wright PE (2000) Structure of the PHD zinc finger from human Williams-Beuren syndrome transcription factor. J Mol Biol 304:723–729. https://doi.org/10.1006/jmbi.2000.4308
Peltzer N et al (2014) HOIP deficiency causes embryonic lethality by aberrant TNFR1-mediated endothelial cell death. Cell Rep 9:153–165. https://doi.org/10.1016/j.celrep.2014.08.066
Penit C, Lucas B, Vasseur F (1995) Cell expansion and growth arrest phases during the transition from precursor (CD4-8-) to immature (CD4+8+) thymocytes in normal and genetically modified mice. J Immunol 154:5103–5113
Perry JSA et al (2014) Distinct contributions of Aire and antigen-presenting-cell subsets to the generation of self-tolerance in the thymus. Immunity 41:414–426. https://doi.org/10.1016/j.immuni.2014.08.007
Petrie HT, Zúñiga-Pflücker JC (2007) Zoned out: functional mapping of stromal signaling microenvironments in the thymus. Annu Rev Immunol 25:649–679. https://doi.org/10.1146/annurev.immunol.23.021704.115715
Picca CC et al (2006) Role of TCR specificity in CD4+ CD25+ regulatory T-cell selection. Immunol Rev 212:74–85. https://doi.org/10.1111/j.0105-2896.2006.00416.x

Pickart CM (2001) Mechanisms underlying ubiquitination. Annu Rev Biochem 70:503–533. https://doi.org/10.1146/annurev.biochem.70.1.503
Plum J et al (2008) Human intrathymic development: a selective approach. Semin Immunopathol 30:411–423. https://doi.org/10.1007/s00281-008-0135-2
Poliani PL, Fontana E, Roifman CM, Notarangelo LD (2013) zeta Chain-associated protein of 70 kDa (ZAP70) deficiency in human subjects is associated with abnormalities of thymic stromal cells: Implications for T-cell tolerance. J Allergy Clin Immunol 131(597-600):591–592. https://doi.org/10.1016/j.jaci.2012.11.002
Proietto AI et al (2008) Dendritic cells in the thymus contribute to T-regulatory cell induction. Proc Natl Acad Sci USA 105:19869–19874. https://doi.org/10.1073/pnas.0810268105
Qiao G et al (2013) T cell activation threshold regulated by E3 ubiquitin ligase Cbl-b determines fate of inducible regulatory T cells. J Immunol 191:632–639. https://doi.org/10.4049/jimmunol.1202068
Radtke F, Wilson A, Mancini SJ, MacDonald HR (2004) Notch regulation of lymphocyte development and function. Nat Immunol 5:247–253. https://doi.org/10.1038/ni1045
Ramon HE et al (2011) The ubiquitin ligase adaptor Ndfip1 regulates T cell-mediated gastrointestinal inflammation and inflammatory bowel disease susceptibility. Mucosal Immunol 4:314–324. https://doi.org/10.1038/mi.2010.69
Reavie L et al (2010) Regulation of hematopoietic stem cell differentiation by a single ubiquitin ligase-substrate complex. Nat Immunol 11:207–215. https://doi.org/10.1038/ni.1839
Reiley WW et al (2006) Regulation of T cell development by the deubiquitinating enzyme CYLD. Nat Immunol 7:411–417. https://doi.org/10.1038/ni1315
Reissig S et al (2012) The tumor suppressor CYLD controls the function of murine regulatory T cells. J Immunol 189:4770–4776. https://doi.org/10.4049/jimmunol.1201993
Reissig S et al (2015) The deubiquitinating enzyme CYLD regulates the differentiation and maturation of thymic medullary epithelial cells. Immunol Cell Biol 93:558–566. https://doi.org/10.1038/icb.2014.122
Res P, Spits H (1999) Developmental stages in the human thymus. Semin Immunol 11:39–46
Richards DM et al (2015) Treg cell differentiation: from thymus to peripheral tissue. Prog Mol Biol Transl Sci 136:175–205. https://doi.org/10.1016/bs.pmbts.2015.07.014
Rodewald HR, Ogawa M, Haller C, Waskow C, DiSanto JP (1997) Pro-thymocyte expansion by c-kit and the common cytokine receptor gamma chain is essential for repertoire formation. Immunity 6:265–272. https://doi.org/10.1016/s1074-7613(00)80329-5
Rowell EA, Wells AD (2006) The role of cyclin-dependent kinases in T-cell development, proliferation, and function. Crit Rev Immunol 26:189–212. https://doi.org/10.1615/critrevimmunol.v26.i3.10
Rusten TE, Vaccari T, Stenmark H (2011) Shaping development with ESCRTs. Nat Cell Biol 14:38–45. https://doi.org/10.1038/ncb2381
Saito T, Watanabe N (1998) Positive and negative thymocyte selection. Crit Rev Immunol 18:359–370. https://doi.org/10.1615/critrevimmunol.v18.i4.40
Sarmento LM et al (2005) Notch1 modulates timing of G1-S progression by inducing SKP2 transcription and p27 Kip1 degradation. J Exp Med 202:157–168. https://doi.org/10.1084/jem.20050559
Savage PA, Klawon DEJ, Miller CH (2020) Regulatory T cell development. Annu Rev Immunol 38:421–453. https://doi.org/10.1146/annurev-immunol-100219-020937
Seo W, Taniuchi I (2016) Transcriptional regulation of early T-cell development in the thymus. Eur J Immunol 46:531–538. https://doi.org/10.1002/eji.201545821
Shen HQ et al (2003) T/NK bipotent progenitors in the thymus retain the potential to generate dendritic cells. J Immunol 171:3401–3406. https://doi.org/10.4049/jimmunol.171.7.3401
Shim JH et al (2006) CHMP5 is essential for late endosome function and down-regulation of receptor signaling during mouse embryogenesis. J Cell Biol 172:1045–1056. https://doi.org/10.1083/jcb.200509041
Shortman K, Egerton M, Spangrude GJ, Scollay R (1990) The generation and fate of thymocytes. Semin Immunol 2:3–12
Six EM et al (2007) A human postnatal lymphoid progenitor capable of circulating and seeding the thymus. J Exp Med 204:3085–3093. https://doi.org/10.1084/jem.20071003
Sluimer J, Distel B (2018) Regulating the human HECT E3 ligases. Cell Mol Life Sci 75:3121–3141. https://doi.org/10.1007/s00018-018-2848-2
Smeets MF, Wiest DL, Izon DJ (2014) Fli-1 regulates the DN2 to DN3 thymocyte transition and promotes γδ T-cell commitment by enhancing TCR signal strength. Eur J Immunol 44:2617–2624. https://doi.org/10.1002/eji.201444442
Staal FJ, Wiekmeijer AS, Brugman MH, Pike-Overzet K (2016) The functional relationship between hematopoietic stem cells and developing T lymphocytes. Ann N Y Acad Sci 1370:36–44. https://doi.org/10.1111/nyas.12995
Starr TK, Jameson SC, Hogquist KA (2003) Positive and negative selection of T cells. Annu Rev Immunol 21:139–176. https://doi.org/10.1146/annurev.immunol.21.120601.141107
Strasser A, Harris AW, von Boehmer H, Cory S (1994) Positive and negative selection of T cells in T-cell receptor transgenic mice expressing a bcl-2 transgene. Proc Natl Acad Sci USA 91:1376–1380. https://doi.org/10.1073/pnas.91.4.1376
Sun SC (2008) Deubiquitylation and regulation of the immune response. Nat Rev Immunol 8:501–511. https://doi.org/10.1038/nri2337
Taghon T et al (2009) Notch signaling is required for proliferation but not for differentiation at a well-

defined beta-selection checkpoint during human T-cell development. Blood 113:3254–3263. https://doi.org/10.1182/blood-2008-07-168906

Tai X, Cowan M, Feigenbaum L, Singer A (2005) CD28 costimulation of developing thymocytes induces Foxp3 expression and regulatory T cell differentiation independently of interleukin 2. Nat Immunol 6:152–162. https://doi.org/10.1038/ni1160

Takahama Y, Ohigashi I, Baik S, Anderson G (2017) Generation of diversity in thymic epithelial cells. Nat Rev Immunol 17:295–305. https://doi.org/10.1038/nri.2017.12

Tanigaki K, Honjo T (2007) Regulation of lymphocyte development by notch signaling. Nat Immunol 8:451–456. https://doi.org/10.1038/ni1453

Teh CE et al (2016) Linear ubiquitin chain assembly complex coordinates late thymic T-cell differentiation and regulatory T-cell homeostasis. Nat Commun 7:13353. https://doi.org/10.1038/ncomms13353

Tetzlaff MT et al (2004) Defective cardiovascular development and elevated cyclin E and Notch proteins in mice lacking the Fbw7 F-box protein. Proc Natl Acad Sci USA 101:3338–3345. https://doi.org/10.1073/pnas.0307875101

Thien CB, Langdon WY (2001) Cbl: many adaptations to regulate protein tyrosine kinases. Nat Rev Mol Cell Biol 2:294–307. https://doi.org/10.1038/35067100

Thompson BJ et al (2007) The SCFFBW7 ubiquitin ligase complex as a tumor suppressor in T cell leukemia. J Exp Med 204:1825–1835. https://doi.org/10.1084/jem.20070872

Thompson BJ et al (2008) Control of hematopoietic stem cell quiescence by the E3 ubiquitin ligase Fbw7. J Exp Med 205:1395–1408. https://doi.org/10.1084/jem.20080277

Trigueros C et al (1998) Identification of a late stage of small noncycling pTalpha- pre-T cells as immediate precursors of T cell receptor alpha/beta+ thymocytes. J Exp Med 188:1401–1412. https://doi.org/10.1084/jem.188.8.1401

Trompouki E et al (2003) CYLD is a deubiquitinating enzyme that negatively regulates NF-kappaB activation by TNFR family members. Nature 424:793–796. https://doi.org/10.1038/nature01803

Tsagaratou A, Trompouki E, Grammenoudi S, Kontoyiannis DL, Mosialos G (2010) Thymocyte-specific truncation of the deubiquitinating domain of CYLD impairs positive selection in a NF-kappaB essential modulator-dependent manner. J Immunol 185:2032–2043. https://doi.org/10.4049/jimmunol.0903919

Tsukiyama T et al (2001) Down-regulation of p27Kip1 expression is required for development and function of T cells. J Immunol 166:304–312. https://doi.org/10.4049/jimmunol.166.1.304

Tsunematsu R et al (2004) Mouse Fbw7/Sel-10/Cdc4 is required for notch degradation during vascular development. J Biol Chem 279:9417–9423. https://doi.org/10.1074/jbc.M312337200

Tsvetkov LM, Yeh KH, Lee SJ, Sun H, Zhang H (1999) p27(Kip1) ubiquitination and degradation is regulated by the SCF(Skp2) complex through phosphorylated Thr187 in p27. Curr Biol 9:661–664. https://doi.org/10.1016/s0960-9822(99)80290-5

Tze LE et al (2011) CD83 increases MHC II and CD86 on dendritic cells by opposing IL-10-driven MARCH1-mediated ubiquitination and degradation. J Exp Med 208:149–165. https://doi.org/10.1084/jem.20092203

Uchida D et al (2004) AIRE functions as an E3 ubiquitin ligase. J Exp Med 199:167–172. https://doi.org/10.1084/jem.20031291

Ueno T et al (2004) CCR7 signals are essential for cortex-medulla migration of developing thymocytes. J Exp Med 200:493–505. https://doi.org/10.1084/jem.20040643

Venuprasad K et al (2006) Convergence of itch-induced ubiquitination with MEKK1-JNK signaling in Th2 tolerance and airway inflammation. J Clin Invest 116:1117–1126. https://doi.org/10.1172/JCI26858

Venuprasad K et al (2008) The E3 ubiquitin ligase itch regulates expression of transcription factor Foxp3 and airway inflammation by enhancing the function of transcription factor TIEG1. Nat Immunol 9:245–253. https://doi.org/10.1038/ni1564

Vereecke L, Beyaert R, van Loo G (2009) The ubiquitin-editing enzyme A20 (TNFAIP3) is a central regulator of immunopathology. Trends Immunol 30:383–391. https://doi.org/10.1016/j.it.2009.05.007

Voll RE et al (2000) NF-kappa B activation by the pre-T cell receptor serves as a selective survival signal in T lymphocyte development. Immunity 13:677–689. https://doi.org/10.1016/s1074-7613(00)00067-4

von Boehmer H (2005) Unique features of the pre-T-cell receptor alpha-chain: not just a surrogate. Nat Rev Immunol 5:571–577. https://doi.org/10.1038/nri1636

von Rohrscheidt J et al (2016) Thymic CD4 T cell selection requires attenuation of March8-mediated MHCII turnover in cortical epithelial cells through CD83. J Exp Med 213:1685–1694. https://doi.org/10.1084/jem.20160316

Walseng E et al (2010) Dendritic cell activation prevents MHC class II ubiquitination and promotes MHC class II survival regardless of the activation stimulus. J Biol Chem 285:41749–41754. https://doi.org/10.1074/jbc.M110.157586

Wang HX et al (2019) Thymic epithelial cells contribute to thymopoiesis and T cell development. Front Immunol 10:3099. https://doi.org/10.3389/fimmu.2019.03099

Weerkamp F et al (2006) Human thymus contains multipotent progenitors with T/B lymphoid, myeloid, and erythroid lineage potential. Blood 107:3131–3137. https://doi.org/10.1182/blood-2005-08-3412

Weih F et al (1995) Multiorgan inflammation and hematopoietic abnormalities in mice with a targeted disruption of RelB, a member of the NF-kappa B/Rel family. Cell 80:331–340. https://doi.org/10.1016/0092-8674(95)90416-6

Weissler KA, Caton AJ (2014) The role of T-cell receptor recognition of peptide:MHC complexes in the formation and activity of Foxp3(+) regulatory T cells. Immunol Rev 259:11–22. https://doi.org/10.1111/imr.12177

Wohlfert EA, Gorelik L, Mittler R, Flavell RA, Clark RB (2006) Cutting edge: deficiency in the E3 ubiquitin ligase Cbl-b results in a multifunctional defect in T cell TGF-beta sensitivity in vitro and in vivo. J Immunol 176:1316–1320. https://doi.org/10.4049/jimmunol.176.3.1316

Xie P, Kraus ZJ, Stunz LL, Liu Y, Bishop GA (2011) TNF receptor-associated factor 3 is required for T cell-mediated immunity and TCR/CD28 signaling. J Immunol 186:143–155. https://doi.org/10.4049/jimmunol.1000290

Xu S et al (2007) Substrate recognition and ubiquitination of SCFSkp2/Cks1 ubiquitin-protein isopeptide ligase. J Biol Chem 282:15462–15470. https://doi.org/10.1074/jbc.M610758200

Yamasaki S et al (2006) Mechanistic basis of pre-T cell receptor-mediated autonomous signaling critical for thymocyte development. Nat Immunol 7:67–75. https://doi.org/10.1038/ni1290

Yamashita I, Nagata T, Tada T, Nakayama T (1993) CD69 cell surface expression identifies developing thymocytes which audition for T cell antigen receptor-mediated positive selection. Int Immunol 5:1139–1150. https://doi.org/10.1093/intimm/5.9.1139

Yang-Iott KS et al (2010) TCR beta feedback signals inhibit the coupling of recombinationally accessible V beta 14 segments with DJ beta complexes. J Immunol 184:1369–1378. https://doi.org/10.4049/jimmunol.0900723

Yao Z et al (2007) Nonredundant roles for Stat5a/b in directly regulating Foxp3. Blood 109:4368–4375. https://doi.org/10.1182/blood-2006-11-055756

Yassai M, Gorski J (2000) Thymocyte maturation: selection for in-frame TCR alpha-chain rearrangement is followed by selection for shorter TCR beta-chain complementarity-determining region 3. J Immunol 165:3706–3712. https://doi.org/10.4049/jimmunol.165.7.3706

Yi Z, Lin WW, Stunz LL, Bishop GA (2014) The adaptor TRAF3 restrains the lineage determination of thymic regulatory T cells by modulating signaling via the receptor for IL-2. Nat Immunol 15:866–874. https://doi.org/10.1038/ni.2944

Yui MA, Rothenberg EV (2014) Developmental gene networks: a triathlon on the course to T cell identity. Nat Rev Immunol 14:529–545. https://doi.org/10.1038/nri3702

Zhang J et al (2006) Impaired regulation of NF-kappaB and increased susceptibility to colitis-associated tumorigenesis in CYLD-deficient mice. J Clin Invest 116:3042–3049. https://doi.org/10.1172/JCI28746

Zhao Y et al (2011) The deubiquitinase CYLD targets Smad7 protein to regulate transforming growth factor beta (TGF-beta) signaling and the development of regulatory T cells. J Biol Chem 286:40520–40530. https://doi.org/10.1074/jbc.M111.292961

Zhao Y et al (2015) E3 ubiquitin ligase Cbl-b regulates thymic-derived CD4+CD25+ regulatory T Cell development by targeting Foxp3 for ubiquitination. J Immunol 194:1639–1645. https://doi.org/10.4049/jimmunol.1402434

Zhao B et al (2019) Notch and the pre-TCR coordinate thymocyte proliferation by induction of the SCF subunits Fbxl1 and Fbxl12. Nat Immunol 20:1381–1392. https://doi.org/10.1038/s41590-019-0469-z

Zhu W et al (2017) A new mutation site in the AIRE gene causes autoimmune polyendocrine syndrome type 1. Immunogenetics 69:643–651. https://doi.org/10.1007/s00251-017-0995-5

Zuniga-Pflucker JC (2004) T-cell development made simple. Nat Rev Immunol 4:67–72. https://doi.org/10.1038/nri1257

GPSR Compliance

The European Union's (EU) General Product Safety Regulation (GPSR) is a set of rules that requires consumer products to be safe and our obligations to ensure this.

If you have any concerns about our products, you can contact us on ProductSafety@springernature.com

In case Publisher is established outside the EU, the EU authorized representative is:

Springer Nature Customer Service Center GmbH
Europaplatz 3
69115 Heidelberg, Germany

The manufacturer's authorised representative in the EU is Springer Nature Customer Service Centre GmbH, Europaplatz 3, 69115 Heidelberg, Germany. If you have any concerns regarding our products, please contact ProductSafety@springernature.com

Printed and bound by CPI Group (UK) Ltd, Croydon, CR0 4YY
30/09/2026
02225307-0020